Research, Invention, and Innovation Congress (RI2C 2022)

Held Online

E3S Web of Conferences Volume 355 (2022)

Bangkok, Thailand
4-5 August 2022

Editors:

Malinee Sriariyanun
Yu-Shen Cheng
Debraj Bhattacharyya
Wawat Rodiahwati
Elizabeth Jayex Panakkal

ISBN: 978-1-7138-5912-3

Printed from e-media with permission by:

Curran Associates, Inc.
57 Morehouse Lane
Red Hook, NY 12571

CURRAN ASSOCIATES INC.
proceedings
.com

Some format issues inherent in the e-media version may also appear in this print version.

No changes have been made to the content of these proceedings. There may be changes to pagination and minor adjustments for aesthetics.

Printed with permission by Curran Associates, Inc. (2023)

For additional information, please contact EDP Sciences – Web of Conferences
at the address below.

EDP Sciences – Web of Conferences
17, Avenue du Hoggar
Parc d'Activité de Courtabœuf
BP 112
F-91944 Les Ulis Cedex A
France

Phone: +33 (0) 1 69 18 75 75
Fax: +33 (0) 1 69 28 84 91

contact-edps@webofconferences.org

Additional copies of this publication are available from:

Curran Associates, Inc.
57 Morehouse Lane
Red Hook, NY 12571 USA
Phone: 845-758-0400
Fax: 845-758-2633
Email: curran@proceedings.com
Web: www.proceedings.com

TABLE OF CONTENTS

Preface

The 3rd Research, Invention, and Innovation Congress (Ri2C2022) is hosted by King Mongkut's University of Technology Bangkok (KMUTNB). Prof. Dr. Suchart Siengchin, the Advisory Chair of Ri2C2022, invited scholar researchers, engineers and industries to attend this event on 4-5 August 2022. The Ri2C2022 organized three main themes; i) Materials and Mechanics (iMM), ii) Electricals and Electronics (iEE), and iii) Sustainable Sciences (iSS). The iSS research topic covered the fields of sustainable energies, sustainable environments and related fields. With the supports from presenters worldwide, scientific committee, reviewers and editorial board members, all papers presented in Ri2C2019 and Ri2C2021 were published in E3S Web of conference.

Among the COVID19 pandemic situation, the social distancing manner brings us apart, and it is difficult time for attendee worldwide travel to present and share their research in Bangkok, Thailand. We therefore organize the online event to be the stage for our participants. In Ri2C2022 event, a total of 26 papers with the focused topics of Environmental Science and Engineering and Energy were peer-reviewed by at least 2 reviewers and accepted for iSS session. In this event, the presented papers were submitted by our international authors with the ratio of international papers at 42.3%, and this ratio was increased from Ri2C2021 event (36%). The organizer committee will continue to host this event again in 2023 to welcome international scholars and industries to bridge the connection in research activities.

Malinee Sriariyanun, Ph.D

Managing Editor

iSS-Ri2C2022

https://doi.org/10.1051/e3sconf/202235500002

Opening Address

Welcome Message from Advisory Chair

On behalf of King Mongkut's University of Technology North Bangkok (KMUTNB). We are greatly honored and pleased to welcome you all to "Research, Invention, and Innovation Congress (RI2C2022)", during 4-5 August 2022 in a Virtual Platform. I would like to begin by thanking KMUTNB colleagues for providing us with such a beautiful and impressive conference which one of the important features of the KMUTNB is that we will be able to bring all researchers together in a public private partnership. R&D involvement is of utmost importance when it comes to network and information exchanges. I would like also to express my appreciation to all speakers such as Prof.Dr. Sakdirat Kaewunruen from University of Birmingham, UK, Prof.Dr.-Ing. Edmund Haberstroh from RWTH Aachen University, GERMANY, Prof.Dr. Surin Khomfoi form King Mongkut's Institute of Technology Ladkrabang, THAILAND, and the last one Dr. Sadayuki Sato, Director Engineering Division of Yaskawa Electrics (THAILAND), and also I would like to thanks all of International Advisory Board for the efforts and believing in cooperation this year conference.

KMUTNB was established from the constructive cooperation between the Thai Government and the Federal Republic of Germany, it was initially known as the beginning of "Thai-German Technical School," where the objective was focused on producing highly skill technicians and engineers for Thailand. To expand the capacity, we established the Faculty of College of Industrial Technology, which is mainly focused on the practice related professions for industrials similar to the university of applied science in Germany. Our success is reflected by the German teaching style that has embedded in most of practice engineers throughout the country.

As we know, KMUTNB welcomes many visiting academics from overseas every year and arranged numerous seminars and conferences per year. This was an extensive network of educational and research institutions at our university. Such academic Conference is major step forward in promoting and enhancing cooperation in Education and I believe that your scientific contributions in this conference are valuable source for other researchers and educators to study and eventually influences the educational method or policy in the future.

We are sure that this Conference will be a memorable, highly educational, and "not-to-be-missed" event. We hope that this Conference will help you better understand and obtain the Knowledge on the Engineering Science and Innovative Technology. I hereby acknowledge and appreciate the efforts and cooperation from everyone. Thank you so much for coming and for your attention. So please enjoy the conference and have a good day. Thank you and Welcome.Development between KMUTNB and you all.

Professor Dr.-Ing.habil. Suchart Siengchin

President, King Mongkut's University of Technology North Bangkok

Advisory Chair, Ri2C2022

List of Conference Chairs

Conference Chair: 2022 Research, Invention, and Innovation Congress (RI2C2022)

1. Advisory Chair

   Suchart Siengchin, King Mongkut's University of Technology North Bangkok (KMUTNB), Thailand

2. Conference Chair

   Somrerk Chandra-ambhorn, KMUTNB, Thailand

   Kampanart Theinnoi, KMUTNB, Thailand

3. General Secretary

   Wanlop Harnnarongchai, KMUTNB, Thailand

4. International Advisory Board

   Sakdirat Kaewunruen, University of Birmingham, UK

   Edmund Haberstroh, RWTH Aachen University, Germany

   Bashir Ahmad, King Abdulaziz University, Saudi Arabia

   Rajini Nagarajan, Kalasalingam Academy of Research and Education, India

   Jyotishkumar Parameswaranpillai, Alliance University, India

   Vincent Aroulmoji, Mahendra Educational Institutions, India

   Yasushi Umeda, The University of Tokyo, Japan

   Athanasios Tsolakis, University of Birmingham, UK

   Nutapong Somjit, University of Leeds, UK

5. RI2C2022 Academic Chair

   Somrerk Chandra-ambhorn, KMUTNB, Thailand

   Kampanart Theinnoi, KMUTNB, Thailand

   Titipong Lertwiriyaprapa, KMUTNB, Thailand

   Rungsima Yeetsorn, KMUTNB, Thailand

   Wanlop Kitisatorn, KMUTNB, Thailand

   Sunanta Sodsee, KMUTNB, Thailand

   Pornsak Srisungsitthisunti, KMUTNB, Thailand

   Nonchanutt Chudpooti, KMUTNB, Thailand

   Watchareewan Jitsakul, KMUTNB, Thailand

   Chaniga Chuensangjun, KMUTNB, Thailand

6. Innovative Sustainable Science (iSS) Editors

Aim and Scope of iss2022

2022 Research, Invention, and Innovation Congress (RI2C2022) (https://www.ri2c-2022.com/home.aspx)

Session: Innovative Sustainable Science (iSS)

Venue: Online Conference at King Mongkut's University of Technology North Bangkok, Thailand

Conference date 4-5 August 2022

Aim and Scope

Main topic coverages: Environmental and Energy Technology

The other topics related to Environmental and Energy Technology include, but not limit to

- Environment
- Energy and Earth Sciences
- Soil Sciences
- Hydrology
- Geography
- Energy Engineering (Production, Distribution and Storage)
- Renewable Energy
- Sustainable Development
- Natural Resource Management

Editorial policies

Evaluation process

All full papers will be subjected to double blind peer-reviews with at least 2 reviewers. The evaluation will be based on the relevancy to the conference scope, structure of the manuscript, logical and scientific merits, innovative, presentation, format and ethic of publications. The communication between the editorial office and reviewers is kept confidential. Reviewers should not disclose any information. After review of the manuscript by at least two independent reviewers, the decision will be made by the editor, which can be categorized in three conditions as follows; Accept without changes, Revisions Required, and Reject.The results of reviewing process will be informed via email from the conference organizer.

Paper submission guidelines

The full paper must be written in English language. Full papers should be original and unpublished research or review in topic areas covered by the iSS-Ri2C theme with the length 7-10 pages. The key compositions of the full research papers include 1) Abstract, 2) Introduction, 3) Materials and methods, 4) Results and discussion, 5) Conclusion, 6) Acknowledgement, 7) References.

Ethics in publishing

Iss2022's Publication Ethics is based on the guidelines and standards developed by the Committee on Publication Ethics (COPE : https://publicationethics.org/).

Plagiarism

Plagiarism and self-copying contents are absolutely prohibited in iss's proceeding. All submitted manuscript will be screened with plagiarism detection software. The screened manuscript by software and editor will be proceeded to reviewing process.

Opening session: Advisory Chair opening

General Chair opening

Plenary Speakers

STRI KMUTNB virtual congress
Passion for the Future
4-5 August, 2022
Research, Invention and Innovation Congress RI²C
Trends in Rubber and Plastics Manufacturing
Prof. Dr.-Ing. Edmund Haberstroh
RWTH Aachen University and DIK Hannover
RWTH AACHEN UNIVERSITY
DIK
Keynote Speaker III
Prof.Dr.-Ing. Edmund Haberstroh
RWTH Aachen University, Germany
Trends in Rubber and Plastics
Manufacturing Technology

Railway Research and Innovation for
Global Challenges, and Benchmarking
for Excellence
Dr Sakdirat Kaewunruen
School of Engineering, University of Birmingham
Keynote Speaker IV
Prof.Dr. Sakdirat Kaewunruen
University of Birmingham, UK
Railway research and innovation for
global challenges,
and the benchmarking for excellence

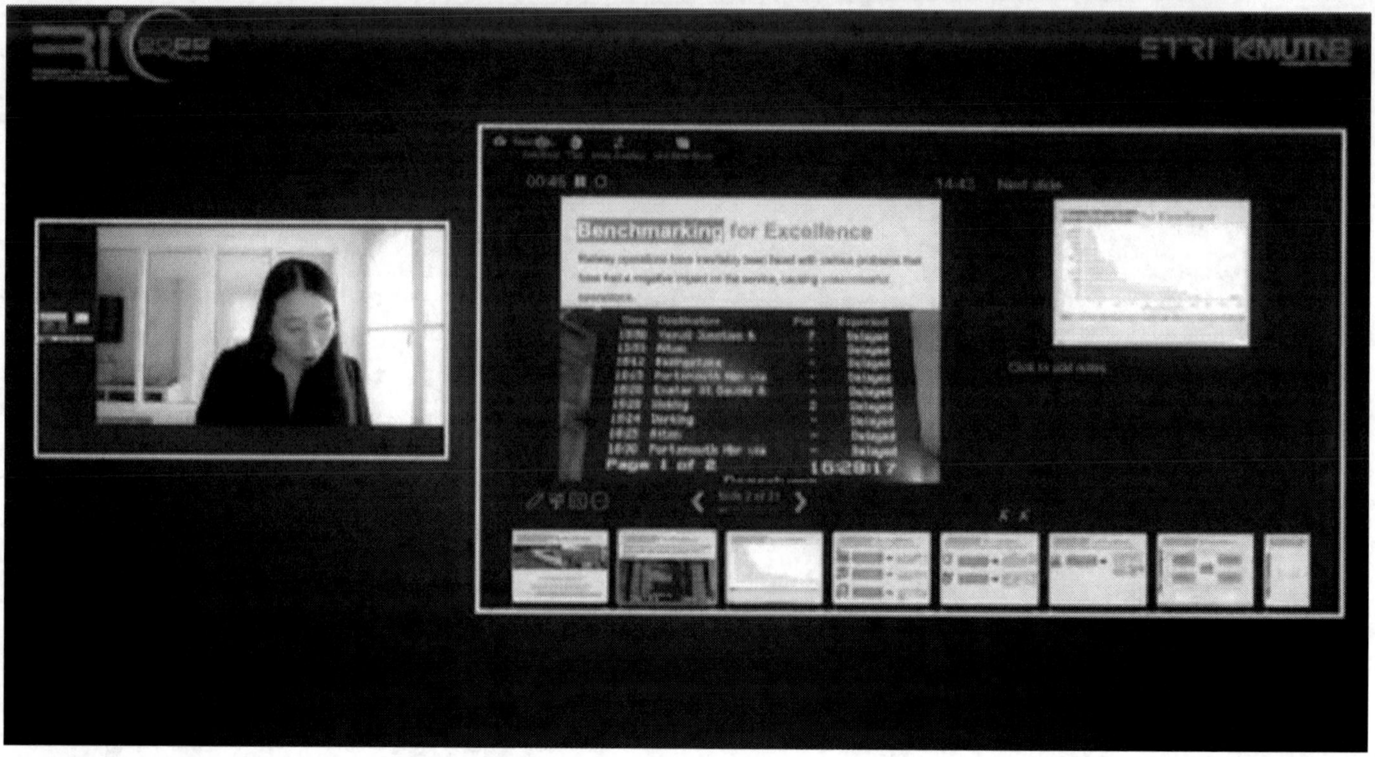
Benchmarking for Excellence
Page 1 of 2
00:45
14:43
Next slide

4 August 2022. ISS Section 1, Room 1. Visual platform. Time 13.15-17.00

Session Chairs: Malinee Sriariyanun

Co-chair: Elizabeth Jayex Panakkal

Secretary : Nichaphat Kitiborwornkul

Iss2022-01

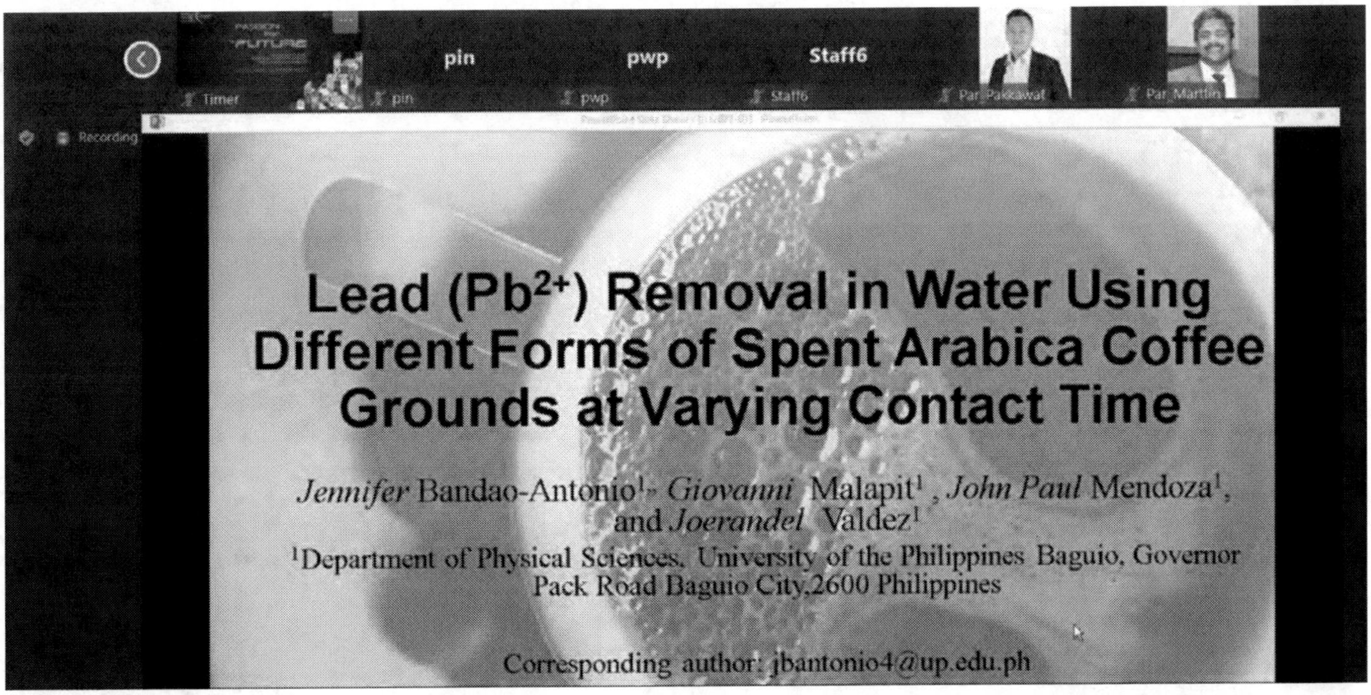

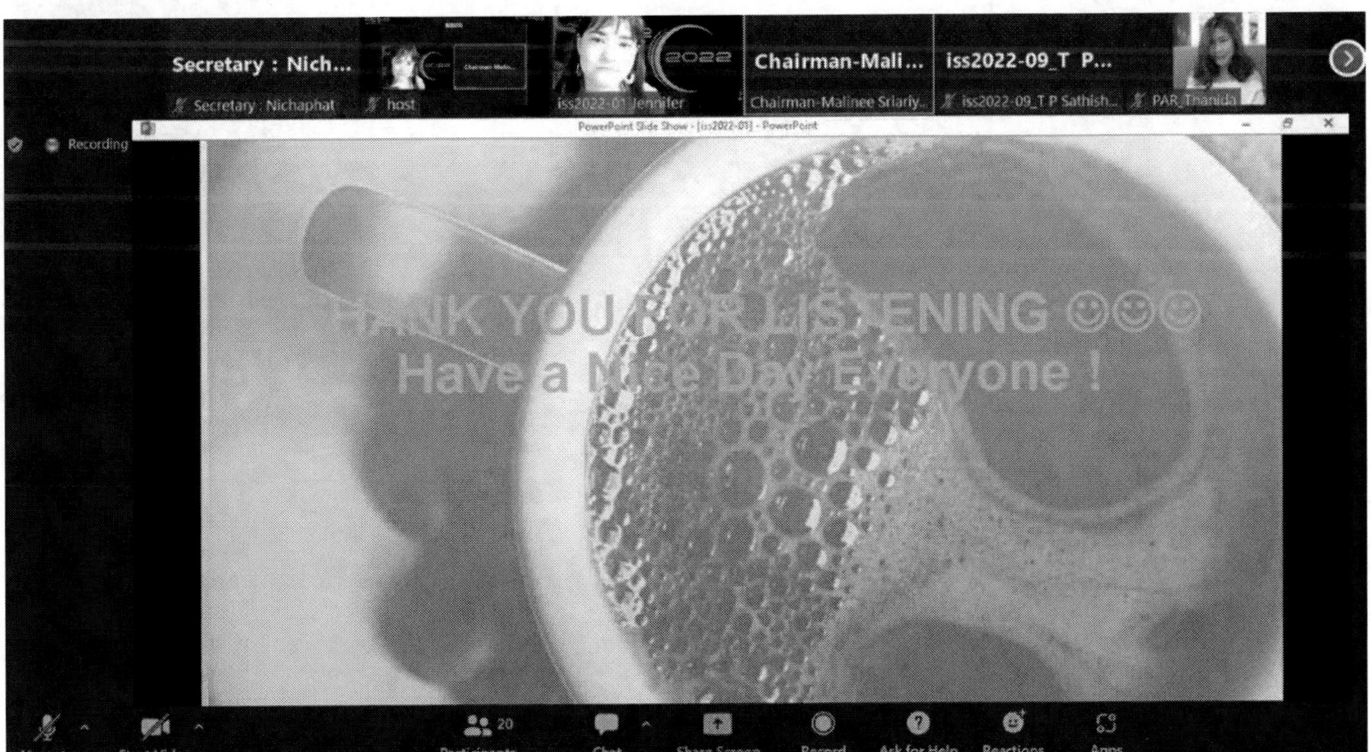

Iss2022-08

Secretary : Nich...
Chairman-Mali...
iss2022-09_T P...
Cultivated Banana
Cultivated bananas can be grown everywhere in Thailand all year round.
Unmute
Start Video
Participants
Chat
Share Screen
Record
Ask for Help
Reactions
Apps

03:10
pin
THANK YOU

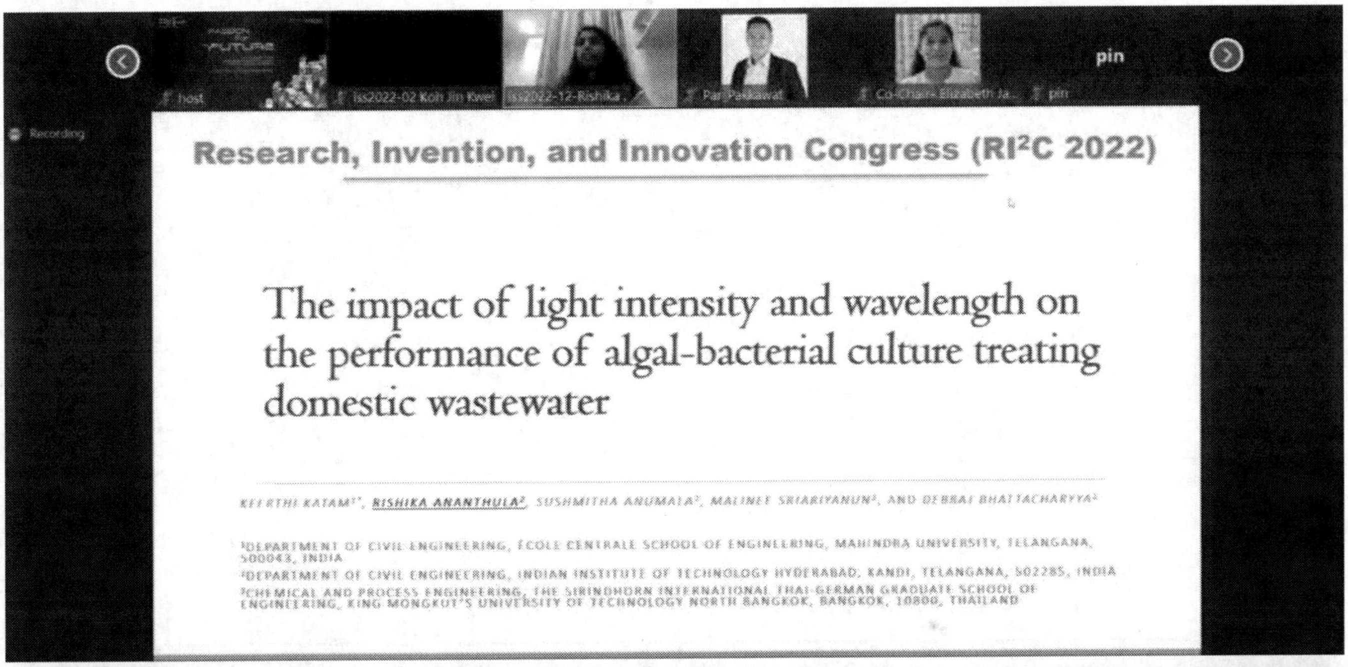
pin
Recording
Research, Invention, and Innovation Congress (RI²C 2022)
The impact of light intensity and wavelength on the performance of algal-bacterial culture treating domestic wastewater
KEERTHI KATAM1*, RISHIKA ANANTHULA2, SUSHMITHA ANUMALA2, MALINEE SRIARIYANUN3, AND DEBRAJ BHATTACHARYYA2
1DEPARTMENT OF CIVIL ENGINEERING, ECOLE CENTRALE SCHOOL OF ENGINEERING, MAHINDRA UNIVERSITY, TELANGANA, 500043, INDIA
2DEPARTMENT OF CIVIL ENGINEERING, INDIAN INSTITUTE OF TECHNOLOGY HYDERABAD, KANDI, TELANGANA, 502285, INDIA
3CHEMICAL AND PROCESS ENGINEERING, THE SIRINDHORN INTERNATIONAL THAI-GERMAN GRADUATE SCHOOL OF ENGINEERING, KING MONGKUT'S UNIVERSITY OF TECHNOLOGY NORTH BANGKOK, BANGKOK, 10800, THAILAND

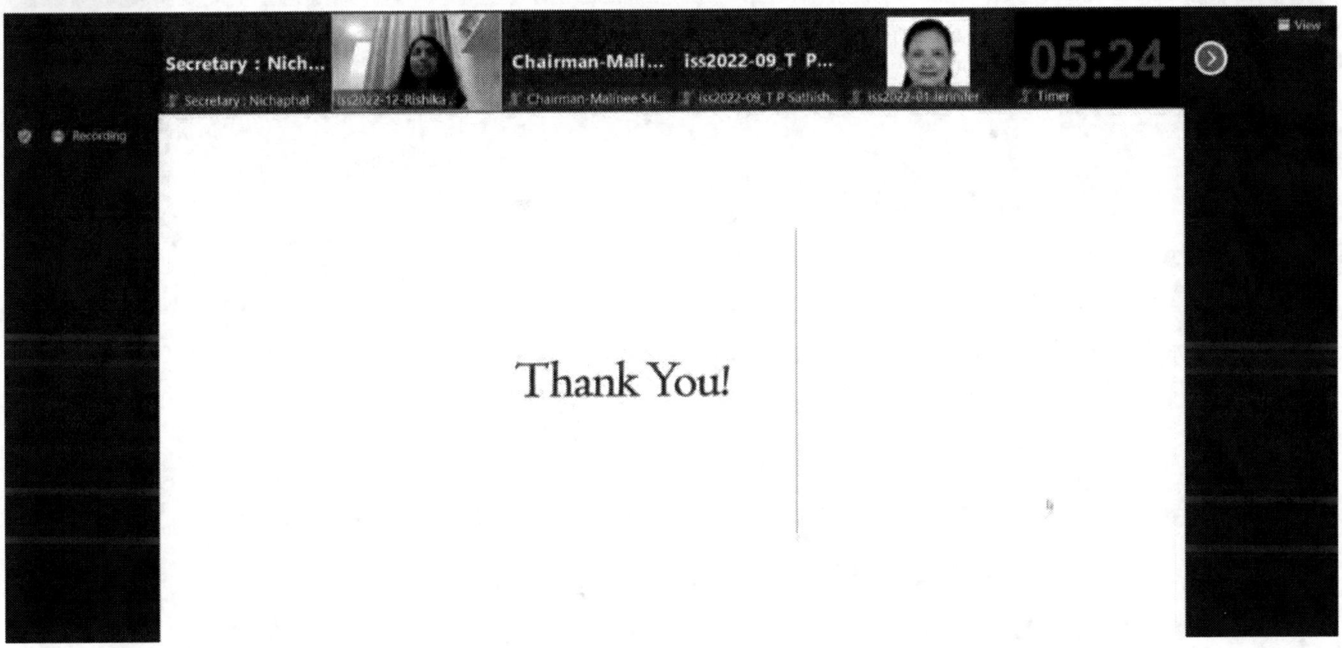
Secretary : Nich...
Chairman-Mali...
iss2022-09_T P...
05:24
View
Recording
Thank You!

Iss2022-19

Zoom Meeting
Secretary : Nich...
iss2022-12-Rish...
Chairman-Mali...
Recording
PASSION For THE FUTURE
4-5 AUGUST, 2022
RESEARCH, INVENTION AND INNOVATION CONGRESS
A VIRTUAL CONFERENCE
Research, Invention and Innovation Congress
Identification of bacterial species from healthy wood of *Aquilaria crassna* using colony PCR
Vanarat Phakeenuya, Bowonlak Phetsri, Thidarat Khunsuksaen, Weeradej Meeinkuirt, and Theerawut Phusantisampan
4:08 PM
8/4/2022

Secretary : Nich...
Chairman-Mali...
03:30
Recording
THANK YOU FOR
YOUR ATTENTION

Iss2022-02

iss2022-09_T P...
pin
Recording
Research Invention and Innovation Congress (RI2) 2022, Bangkok
A comparative study of pH and temperature on rheological behaviour between Polyacrylamide (PAM) and its modified PAM
Paper ID: iSS2022-02
Jin Kwei Koh1, Chin Wei Lai1,*, Mohd Rafie Johan1, Sanjay Mavinkere Rangappa2, Suchart Siengchin2
1Nanotechnology & Catalysis Research Centre (NANOCAT), Level 3, Block A, Institute for Advanced Studies, University of Malaya (UM), 50603, Kuala Lumpur, Malaysia.
2Natural Composite Research Group Lab, Department of Materials and Production Engineering, The Sirindhorn International Thai-German Graduate School of Engineering (TGGS), King Mongkut's University of Technology North Bangkok (KMUTNB), Bangkok, Thailand.

Iss2022-09

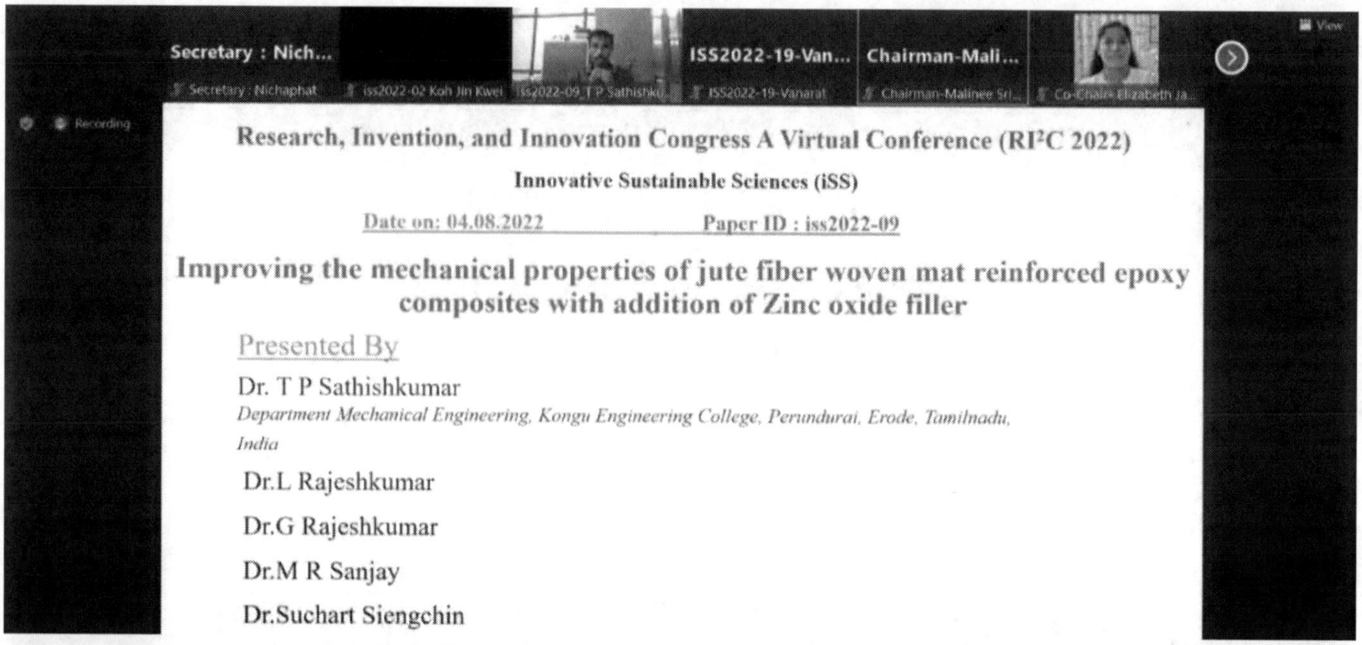
Secretary : Nich...
ISS2022-19-Van...
Chairman-Mali...
Secretary : Nichaphat
iss2022-02 Koh Jin Kwei
iss2022-09_T P Sathishku...
ISS2022-19-Vanarat
Chairman-Malinee Sri...
Co-Chair-Elizabeth Ja...
Recording
Research, Invention, and Innovation Congress A Virtual Conference (RI²C 2022)
Innovative Sustainable Sciences (iSS)
Date on: 04.08.2022
Paper ID : iss2022-09
Improving the mechanical properties of jute fiber woven mat reinforced epoxy composites with addition of Zinc oxide filler
Presented By
Dr. T P Sathishkumar
Department Mechanical Engineering, Kongu Engineering College, Perundurai, Erode, Tamilnadu, India
Dr.L Rajeshkumar
Dr.G Rajeshkumar
Dr.M R Sanjay
Dr.Suchart Siengchin

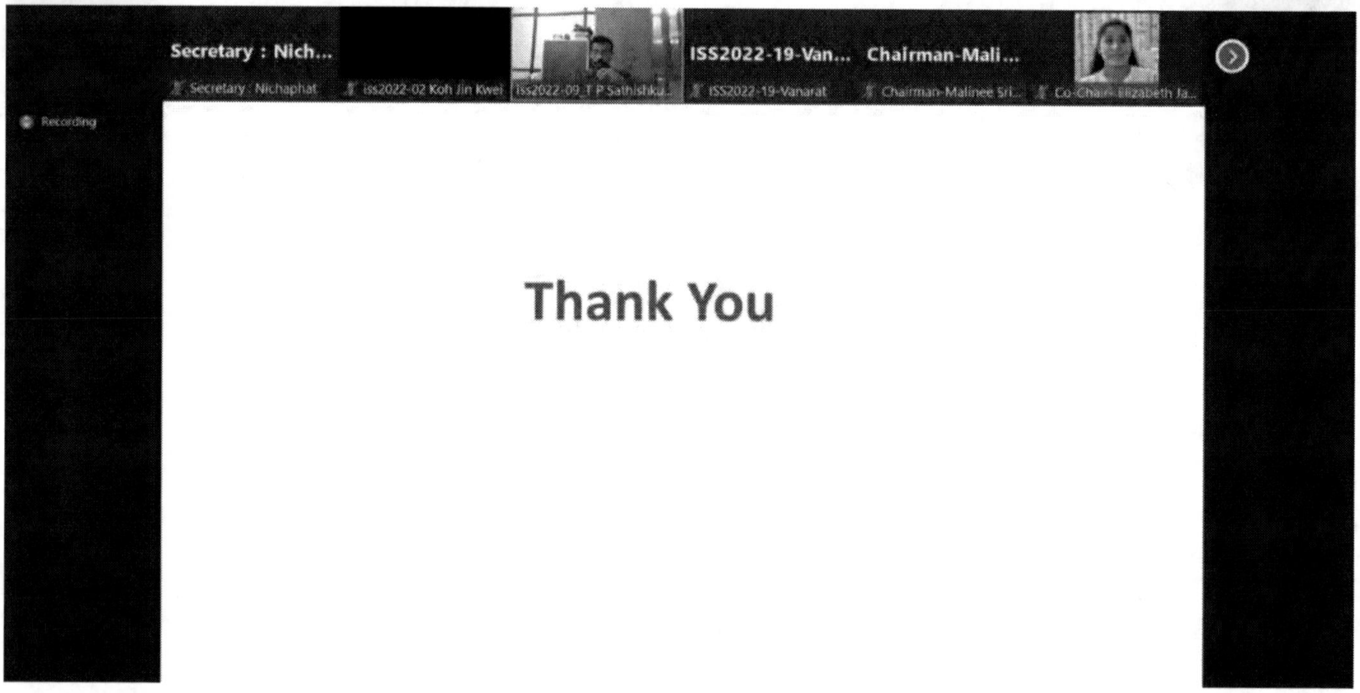
Secretary : Nich...
ISS2022-19-Van...
Chairman-Mali...
Secretary : Nichaphat
iss2022-02 Koh Jin Kwei
iss2022-09_T P Sathishku...
ISS2022-19-Vanarat
Chairman-Malinee Sri...
Co-Chair-Elizabeth Ja...
Recording
Thank You

Iss2022-10

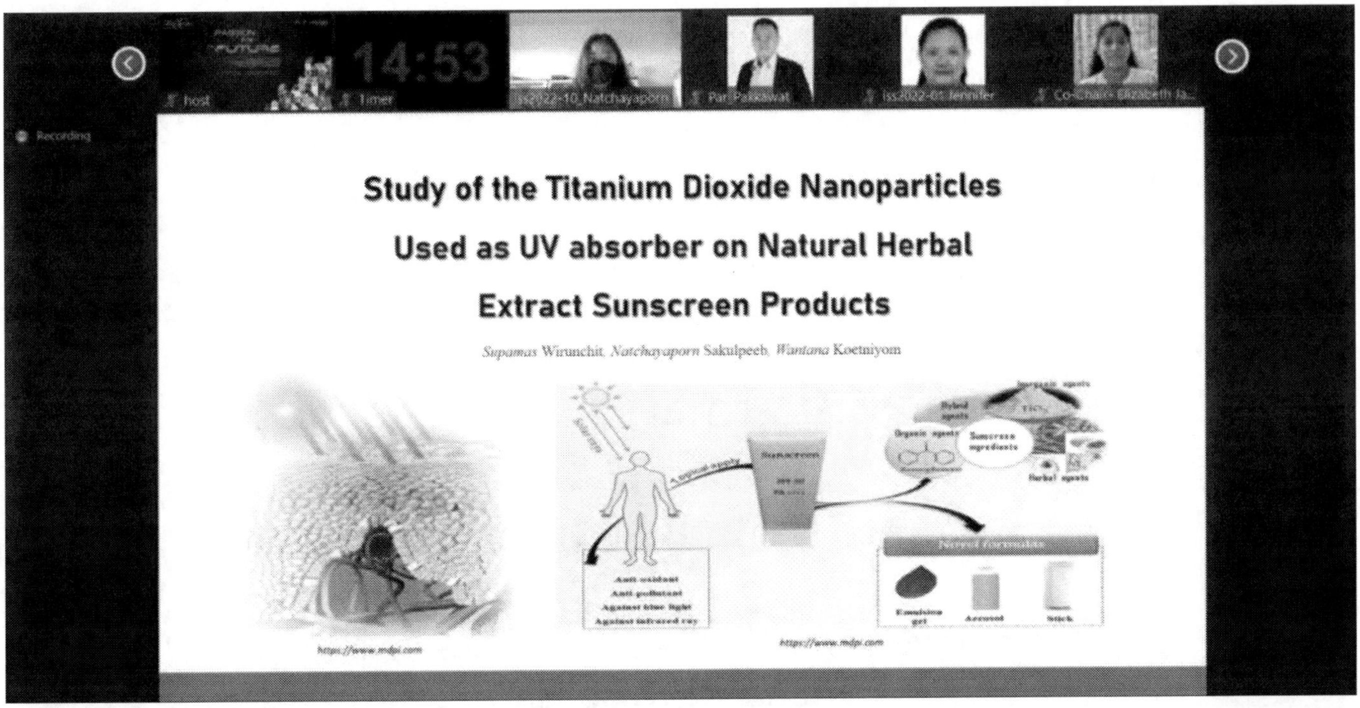
host
Timer
14:53
Iss2022-10_Natchayaporn
Pari Pakkawat
Iss2022-01 Jennifer
Recording
Study of the Titanium Dioxide Nanoparticles
Used as UV absorber on Natural Herbal
Extract Sunscreen Products
Supamas Wirunchit, Natchayaporn Sakulpeeb, Wantana Koetniyom
Anti oxidant
Anti pollutant
Against blue light
Against infrared ray
Novel formulate
Emulsion gel
Aerosol
Stick
Herbal agents
https://www.mdpi.com
https://www.mdpi.com

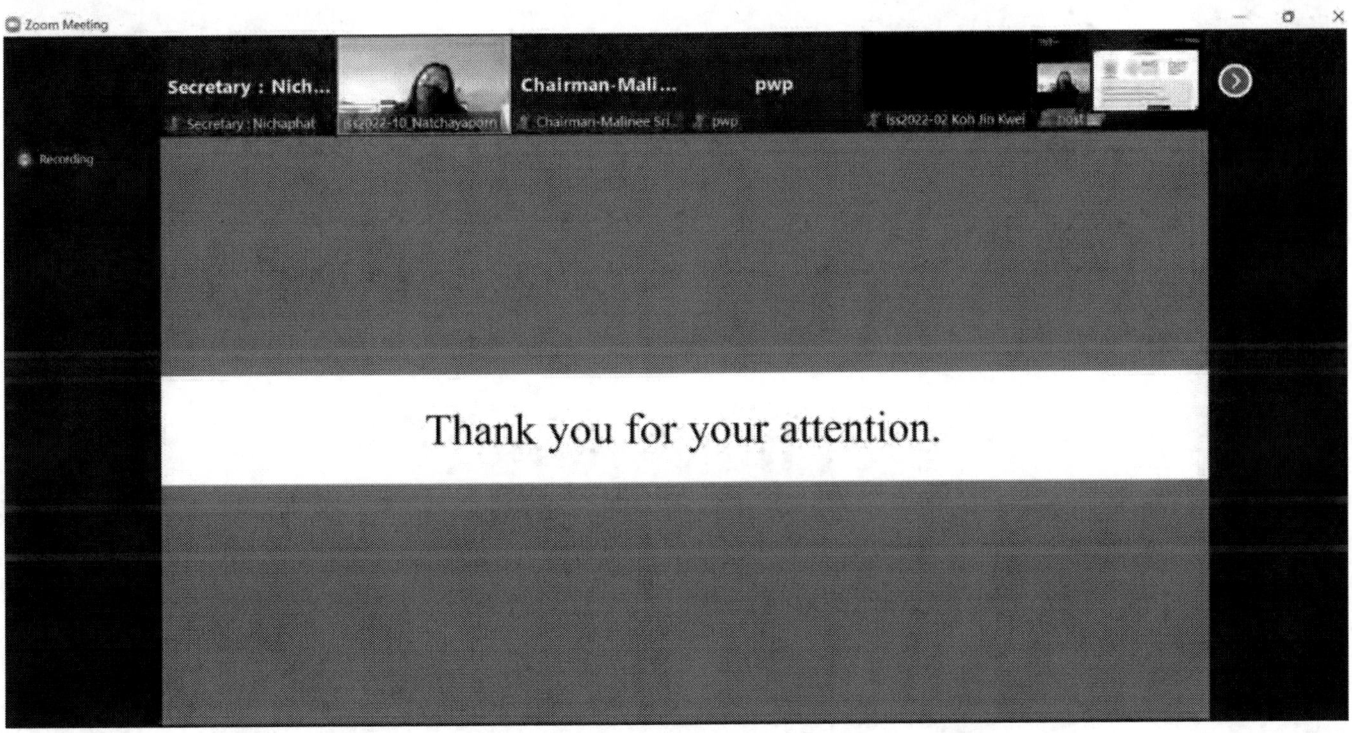
Zoom Meeting
Secretary : Nich...
Chairman-Mali...
pwp
Secretary : Nichaphat
Iss2022-10_Natchayaporn
Chairman-Malinee Sri...
pwp
Iss2022-02 Koh Jin Kwei
host
Recording
Thank you for your attention.

5 August 2022. ISS Section 2, Room 1. Visual platform. Time 09.00-14.45

Session Chairs: Malinee Sriariyanun

Co-chair: Prapakorn Tantayotai

Secretary : Nichaphat Kitiborwornkul

Iss2022-25

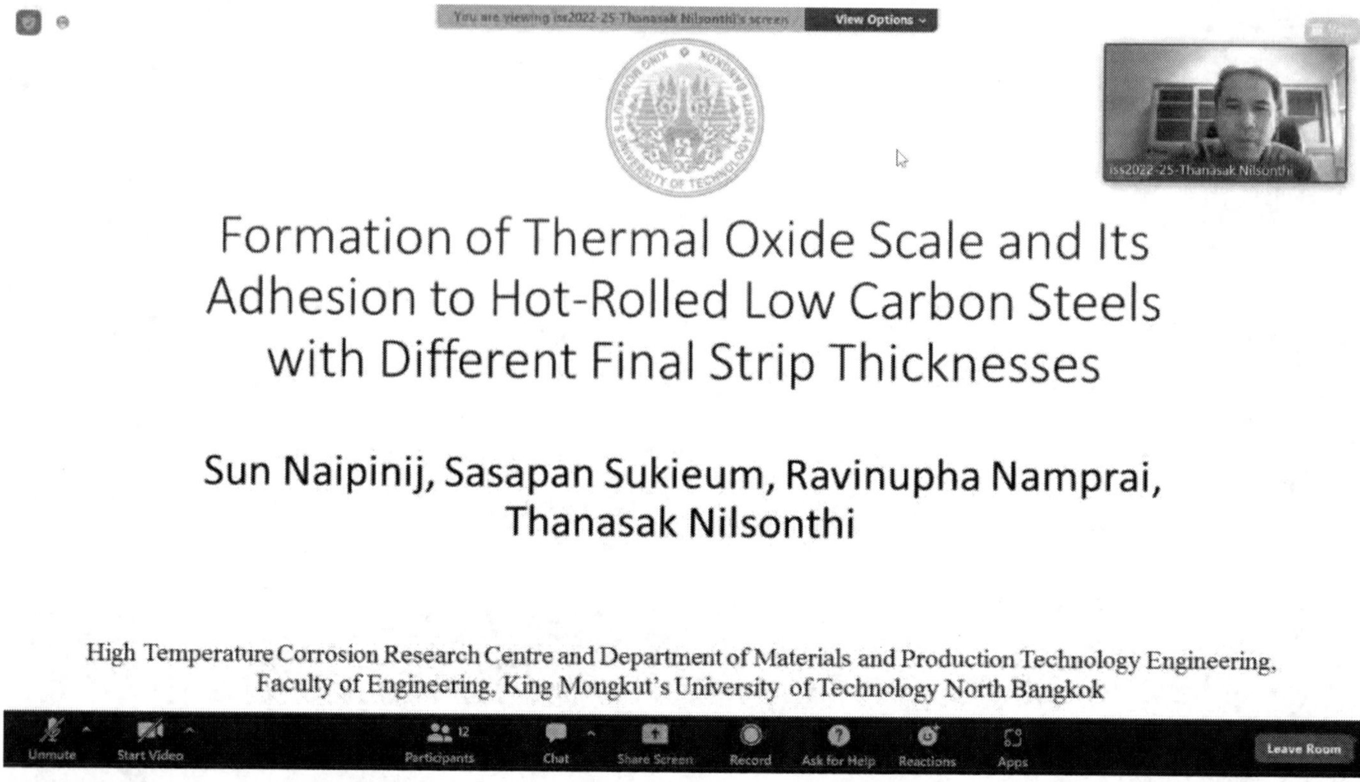

Iss2022-26

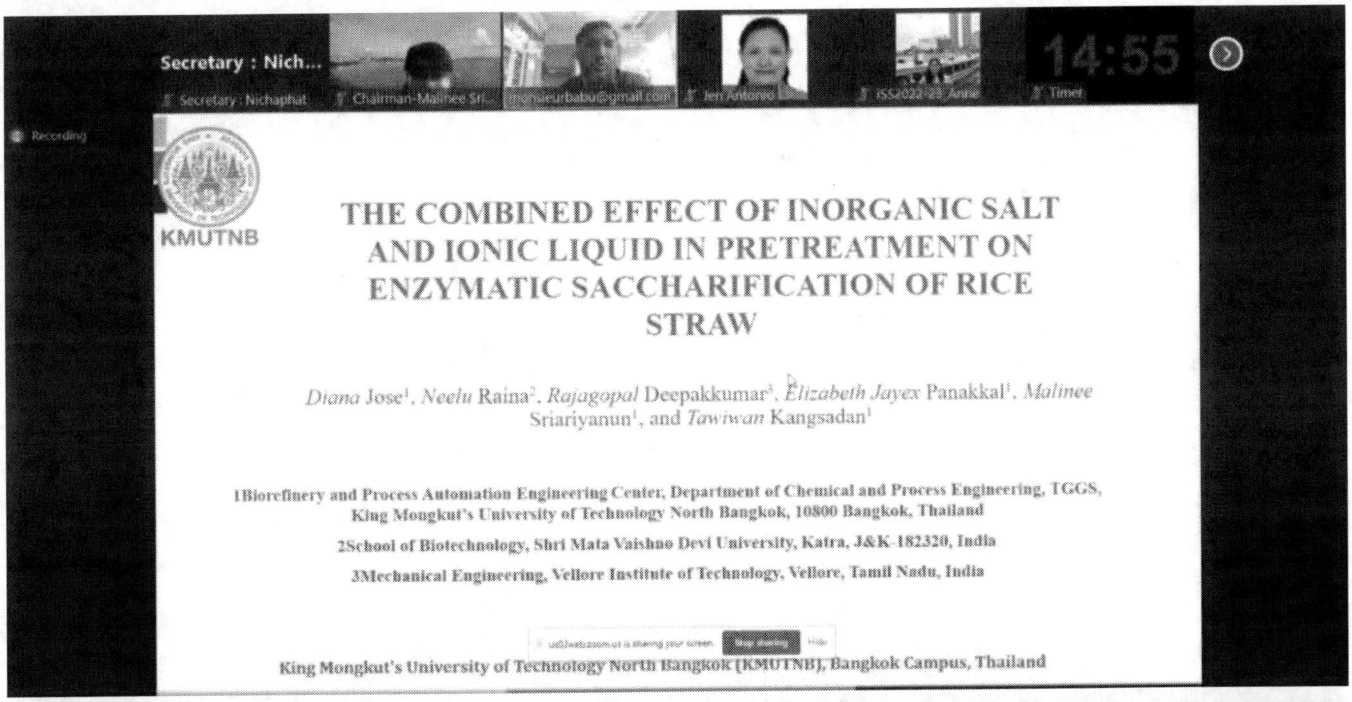
Secretary : Nich...
Secretary : Nichaphat
Chairman-Malinee Sri...
monsieurbabu@gmail.com
Jen Antonio
ISS2022-23_Anne
Timer
14:55
Recording
KMUTNB
THE COMBINED EFFECT OF INORGANIC SALT AND IONIC LIQUID IN PRETREATMENT ON ENZYMATIC SACCHARIFICATION OF RICE STRAW
Diana Jose1, Neelu Raina2, Rajagopal Deepakkumar3, Elizabeth Jayex Panakkal1, Malinee Sriariyanun1, and Tawiwan Kangsadan1
1Biorefinery and Process Automation Engineering Center, Department of Chemical and Process Engineering, TGGS, King Mongkut's University of Technology North Bangkok, 10800 Bangkok, Thailand
2School of Biotechnology, Shri Mata Vaishno Devi University, Katra, J&K-182320, India
3Mechanical Engineering, Vellore Institute of Technology, Vellore, Tamil Nadu, India
Stop sharing
Hide
King Mongkut's University of Technology North Bangkok (KMUTNB), Bangkok Campus, Thailand

Secretary : Nich...
Secretary : Nichaphat
Chairman-Malinee Sri...
monsieurbabu@gmail.com
ISS2022-23_Anne
Kriangsak Ketp...
Kriangsak Ketpang
Timer
03:54
Recording
Thank You
Stop sharing
Hide
17

Iss2022-20

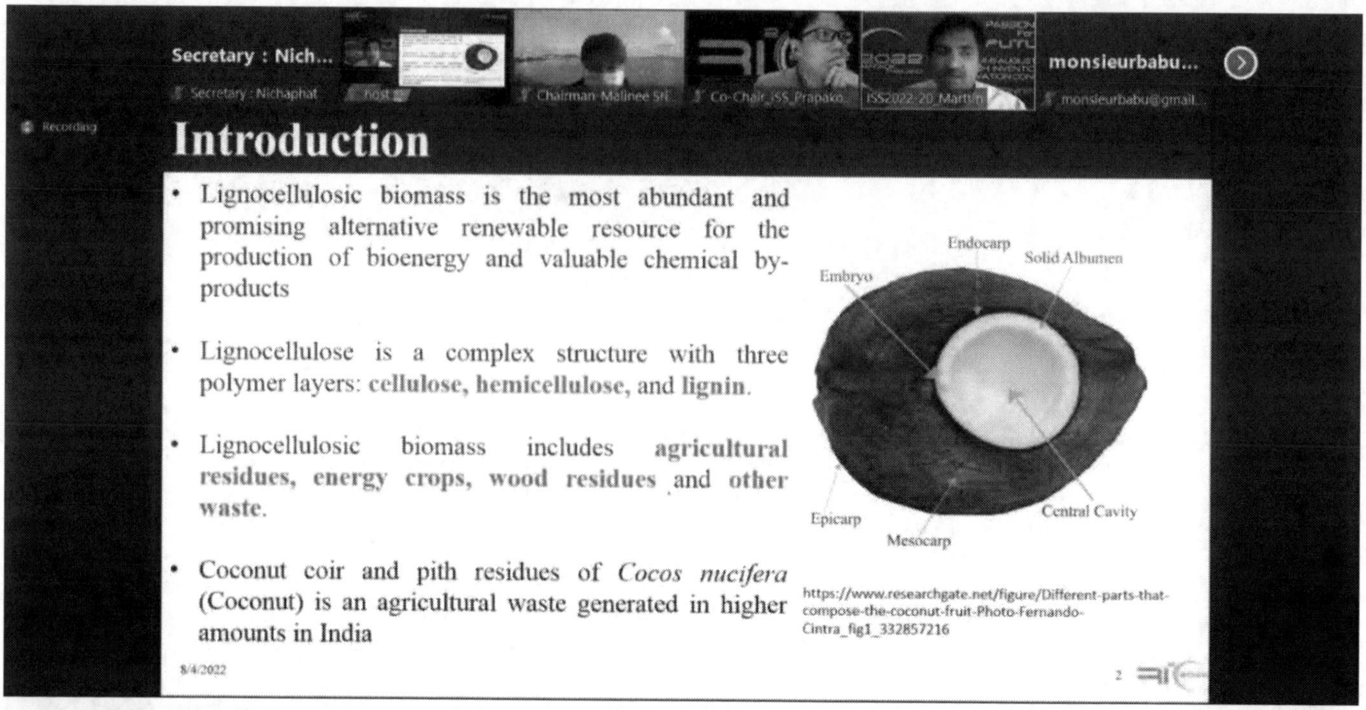
Secretary : Nich...
Secretary : Nichaphat
host
Chairman-Malinee Sri...
Co-Chair_ISS_Prapako...
ISS2022-20_Martin
monsieurbabu...
monsieurbabu@gmail...
Recording
Introduction
• Lignocellulosic biomass is the most abundant and promising alternative renewable resource for the production of bioenergy and valuable chemical by-products
• Lignocellulose is a complex structure with three polymer layers: cellulose, hemicellulose, and lignin.
• Lignocellulosic biomass includes agricultural residues, energy crops, wood residues and other waste.
• Coconut coir and pith residues of Cocos nucifera (Coconut) is an agricultural waste generated in higher amounts in India
Endocarp
Solid Albumen
Embryo
Epicarp
Mesocarp
Central Cavity
https://www.researchgate.net/figure/Different-parts-that-compose-the-coconut-fruit-Photo-Fernando-Cintra_fig1_332857216
8/4/2022
2

Secretary : Nich...
Secretary : Nichaphat
host
Chairman-Malinee Sriariy...
Co-Chair_ISS_Prapako...
ISS2022-20_Martin
monsieurbabu...
monsieurbabu@gmail...
View
Recording
Dilute Inorganic Acid Pretreatment of Mixed Residues of Cocos nucifera (Coconut) for recovery of reducing sugar: Optimization studies
THANK YOU
2022
BANGKOK, THAILAND
Research, Invention, and Innovation Congress
Unmute
Start Video
Participants
23
Chat
Share Screen
Record
Ask for Help
Reactions
Apps
Leave Room

Iss2022-22

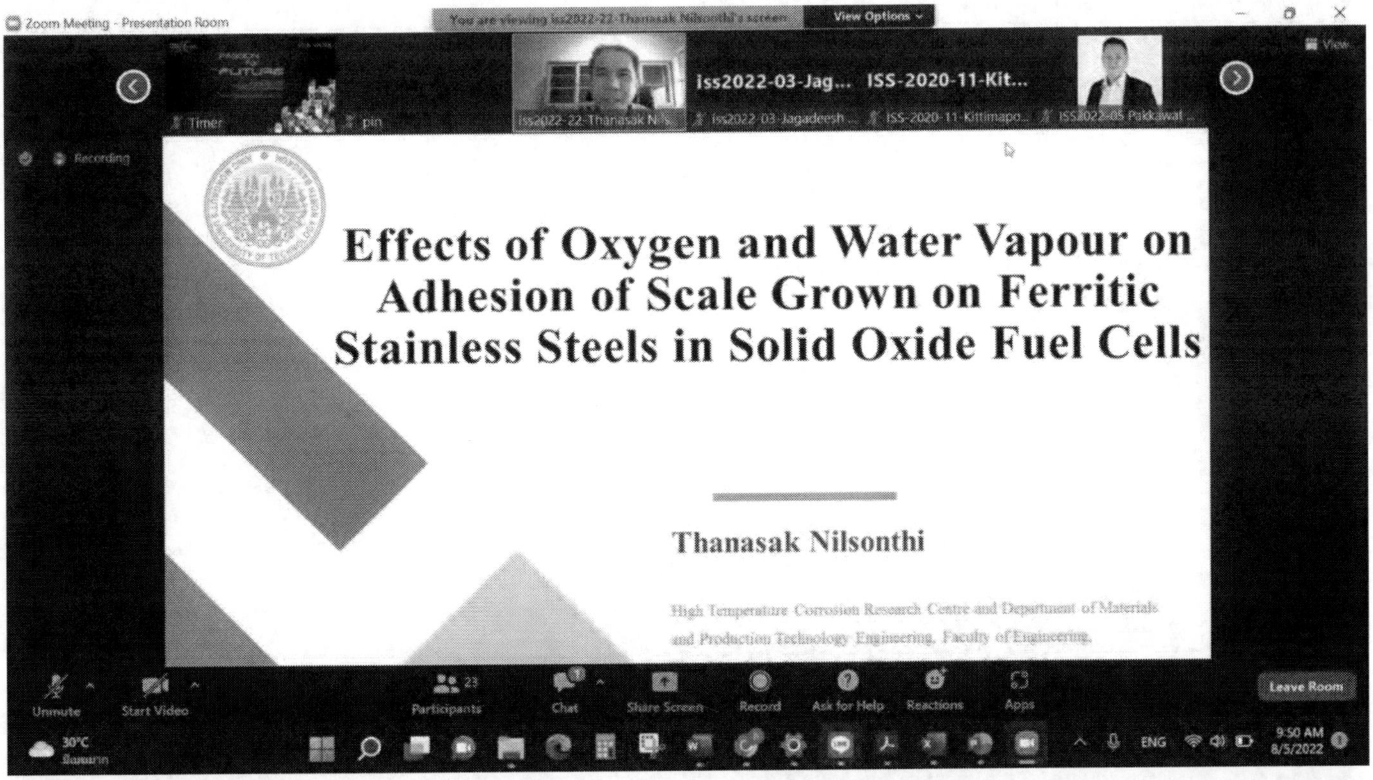
Zoom Meeting - Presentation Room
Effects of Oxygen and Water Vapour on Adhesion of Scale Grown on Ferritic Stainless Steels in Solid Oxide Fuel Cells
Thanasak Nilsonthi
High Temperature Corrosion Research Centre and Department of Materials and Production Technology Engineering, Faculty of Engineering,
Recording
Unmute
Start Video
Participants
Chat
Share Screen
Record
Ask for Help
Reactions
Apps
Leave Room
9:50 AM
8/5/2022

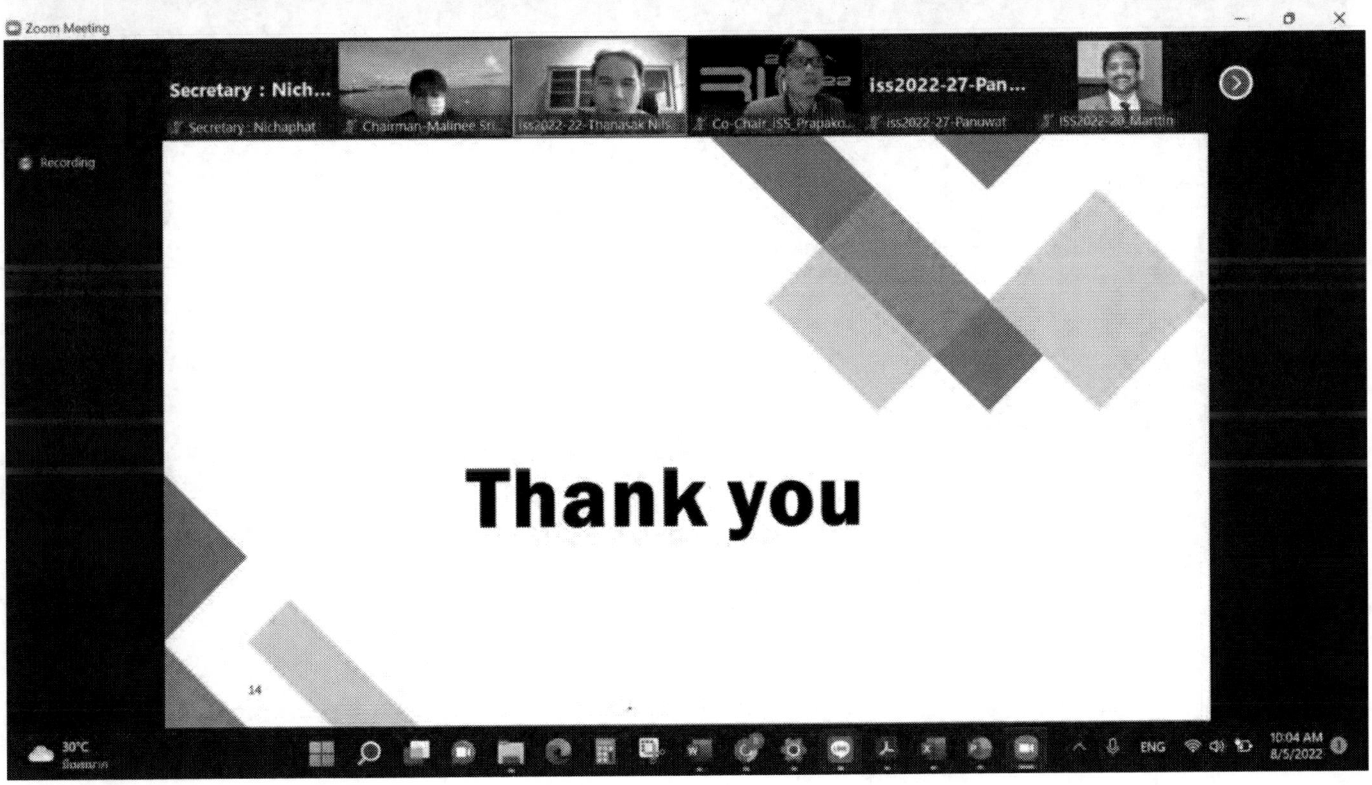
Zoom Meeting
Secretary : Nich...
iss2022-27-Pan...
Recording
Thank you
14
10:04 AM
8/5/2022

Iss2022-03

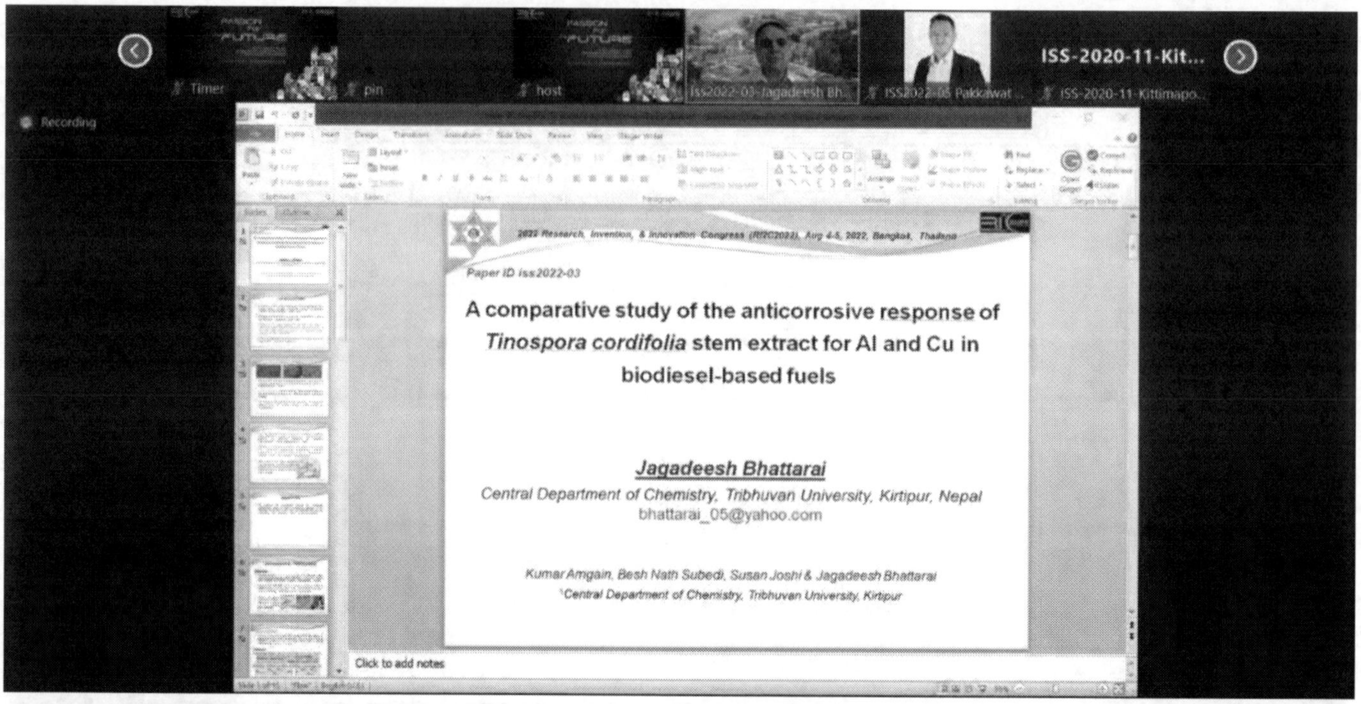
ISS-2020-11-Kit...
Recording
2022 Research, Invention, & Innovation Congress (RI2C2022), Aug 4-5, 2022, Bangkok, Thailand
Paper ID iss2022-03
A comparative study of the anticorrosive response of Tinospora cordifolia stem extract for Al and Cu in biodiesel-based fuels
Jagadeesh Bhattarai
Central Department of Chemistry, Tribhuvan University, Kirtipur, Nepal
bhattarai_05@yahoo.com
Kumar Amgain, Besh Nath Subedi, Susan Joshi & Jagadeesh Bhattarai
Central Department of Chemistry, Tribhuvan University, Kirtipur
Click to add notes

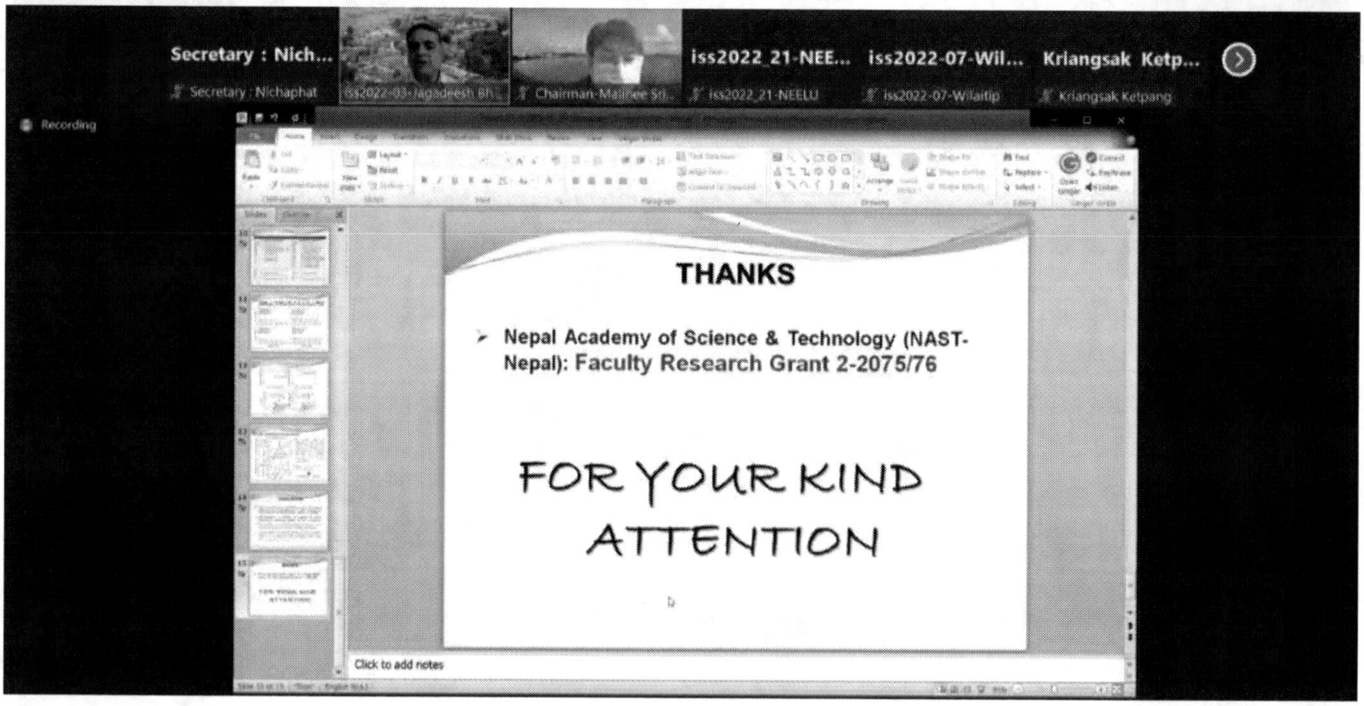
Secretary : Nich...
iss2022_21-NEE...
iss2022-07-Wil...
Kriangsak Ketp...
Secretary : Nichaphat
Recording
THANKS
Nepal Academy of Science & Technology (NAST-Nepal): Faculty Research Grant 2-2075/76
FOR YOUR KIND ATTENTION
Click to add notes

After Break

Iss2022-11

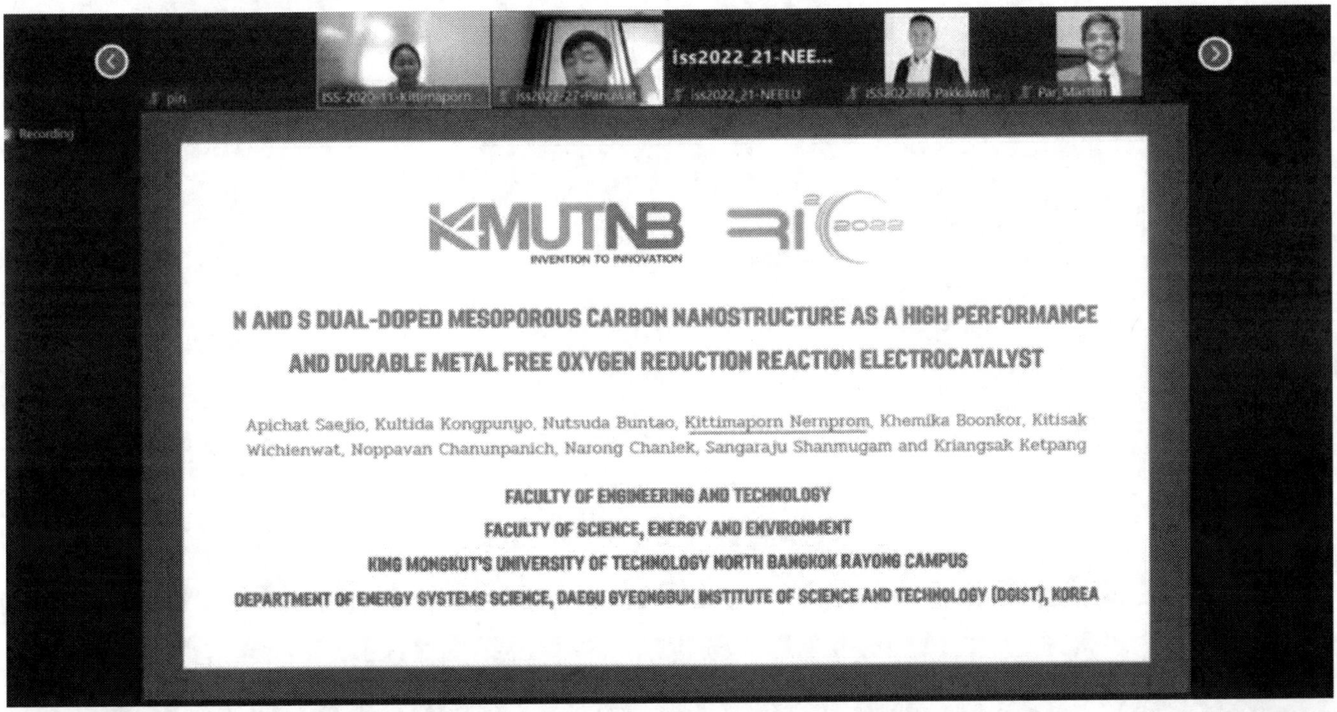

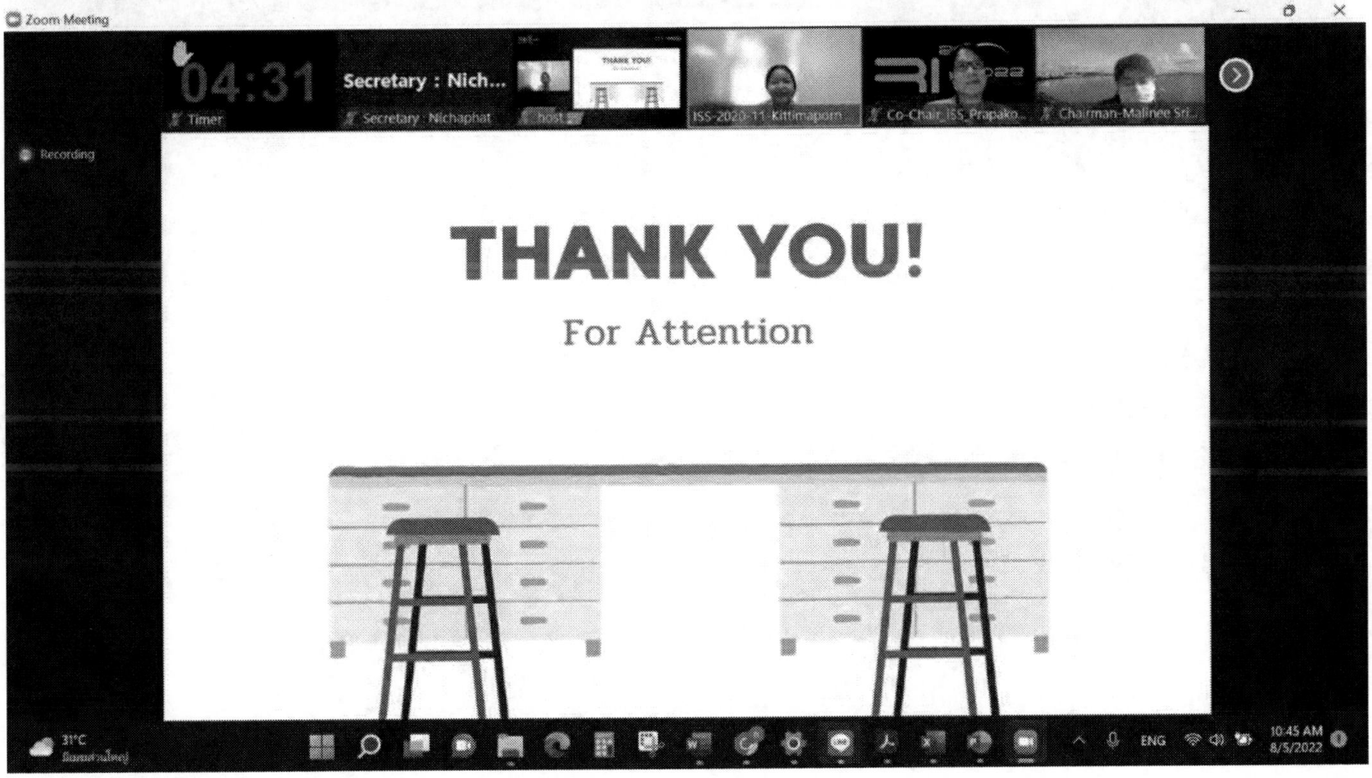

Iss2022-27

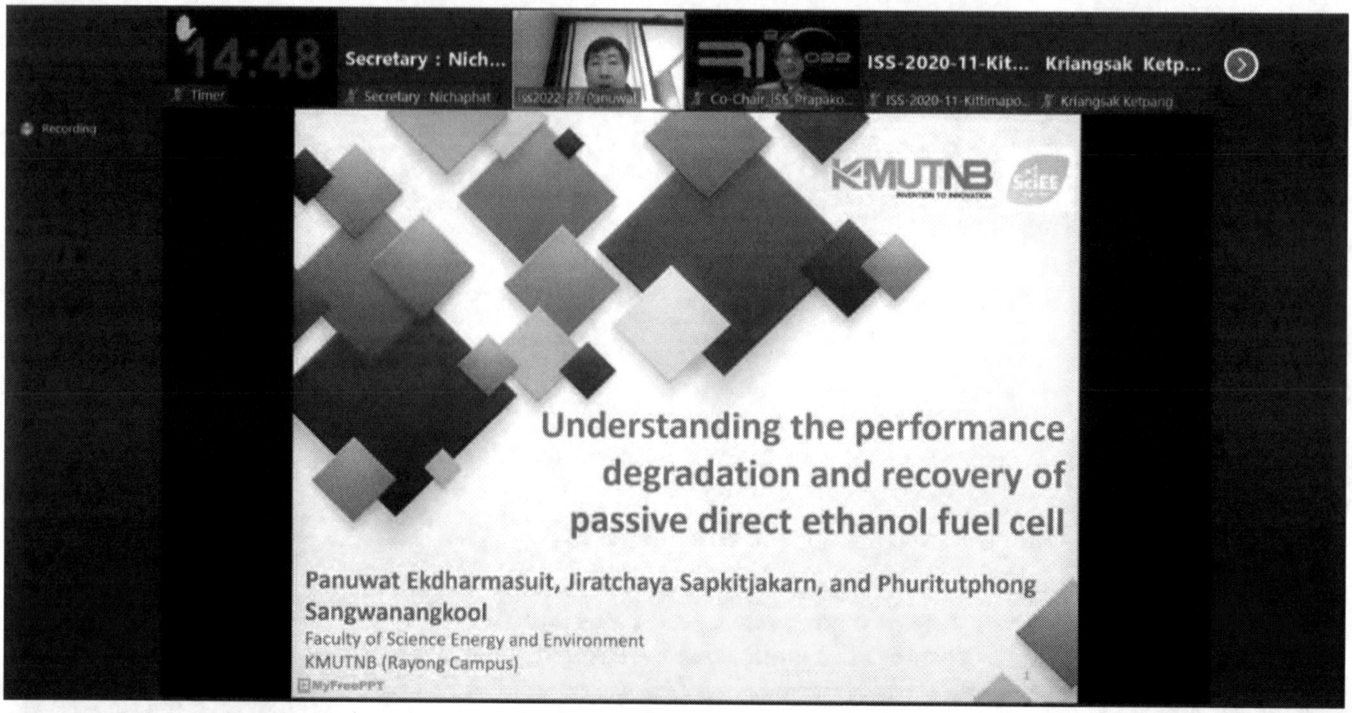
14:48
Timer
Secretary : Nich...
Secretary : Nichaphat
Iss2022-27-Panuwat
Co-Chair_ISS_Prapako...
ISS-2020-11-Kit...
ISS-2020-11-Kittimapo...
Kriangsak Ketp...
Kriangsak Ketpang
Recording
KMUTNB
INVENTION TO INNOVATION
SciEE
Understanding the performance degradation and recovery of passive direct ethanol fuel cell
Panuwat Ekdharmasuit, Jiratchaya Sapkitjakarn, and Phuritutphong Sangwanangkool
Faculty of Science Energy and Environment
KMUTNB (Rayong Campus)
MyFreePPT

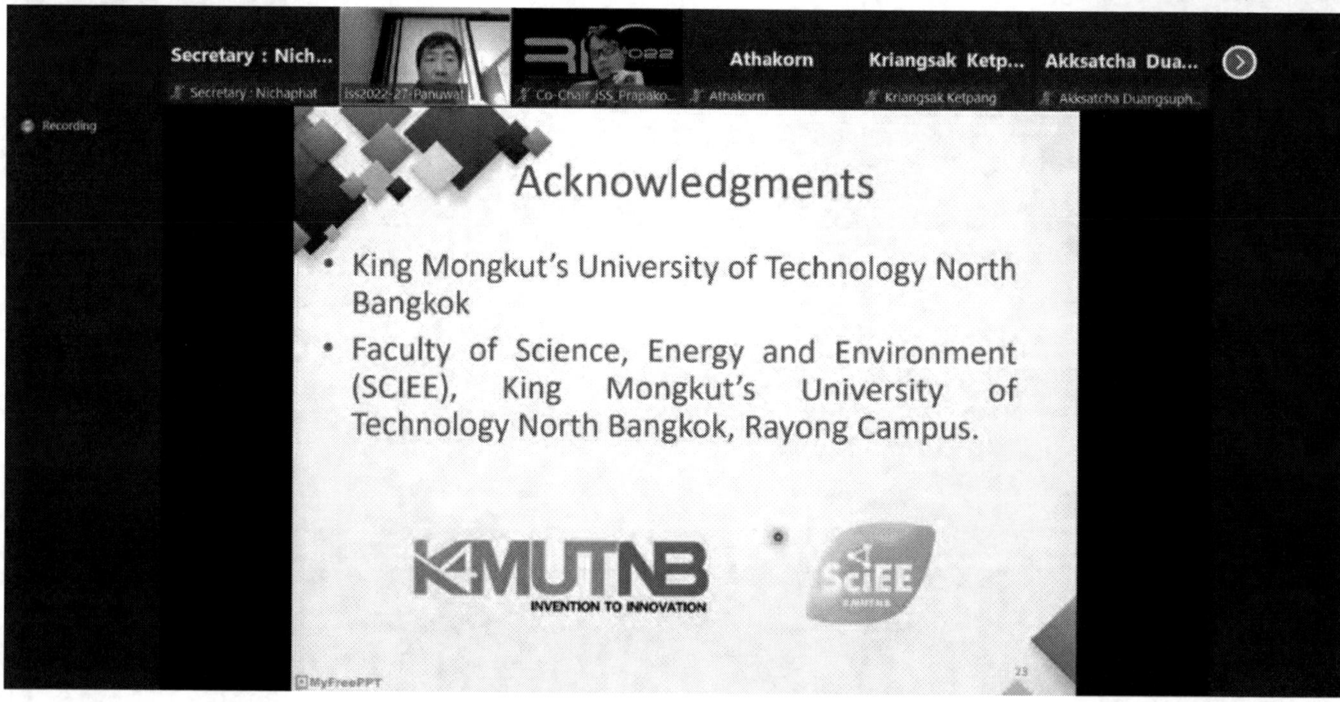
Secretary : Nich...
Secretary : Nichaphat
Iss2022-27-Panuwat
Co-Chair_ISS_Prapako...
Athakorn
Kriangsak Ketp...
Kriangsak Ketpang
Akksatcha Dua...
Akksatcha Duangsuph...
Recording
Acknowledgments
King Mongkut's University of Technology North Bangkok
Faculty of Science, Energy and Environment (SCIEE), King Mongkut's University of Technology North Bangkok, Rayong Campus.
KMUTNB
INVENTION TO INNOVATION
SciEE
23
MyFreePPT

Iss2022-04

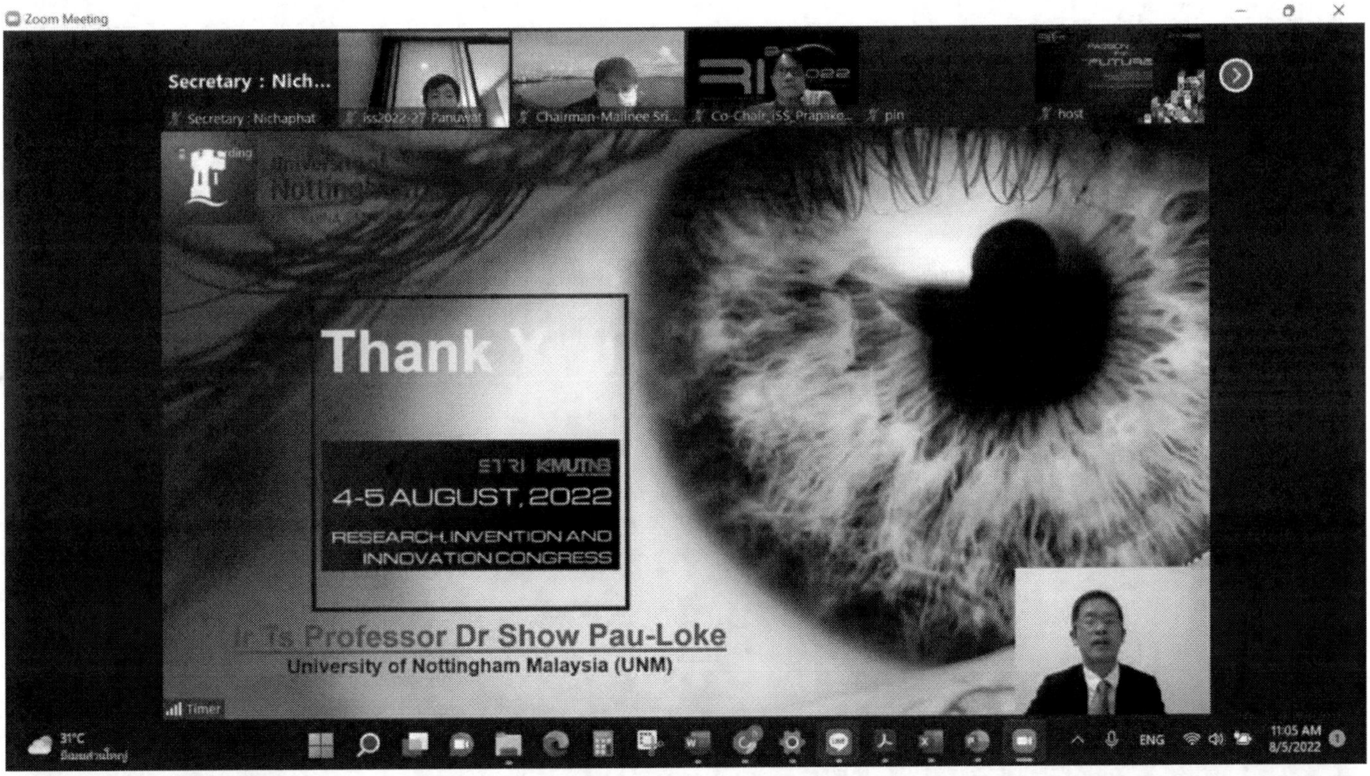
Zoom Meeting
Secretary : Nich...
Secretary : Nichaphat
iss2022-27-Panuwat
Chairman-Mallnee Sri...
Co-Chair_ISS_Prapako...
pin
host
Thank Y
4-5 AUGUST, 2022
RESEARCH, INVENTION AND INNOVATION CONGRESS
Ir Ts Professor Dr Show Pau-Loke
University of Nottingham Malaysia (UNM)
Timer
31°C
ENG
11:05 AM
8/5/2022

Zoom Meeting
Secretary : Nich...
iss2022-27-Pan...
Athakorn
Secretary : Nichaphat
Chairman-Malinee Sri...
Co-Chair_ISS_Prapako...
iss2022-27-Panuwat
Athakorn
pin
University of Nottingham
UK | CHINA | MALAYSIA
International Research Collaborators
Others:
Taiwan
YunTech
New Zealand
THE UNIVERSITY OF WAIKATO
United States
NREL
The University Of Sheffield.
MONASH University
China
Japan
Vietnam
JAIST
KOBE
Timer
31°C
ENG
11:16 AM
8/5/2022

Iss2022-05

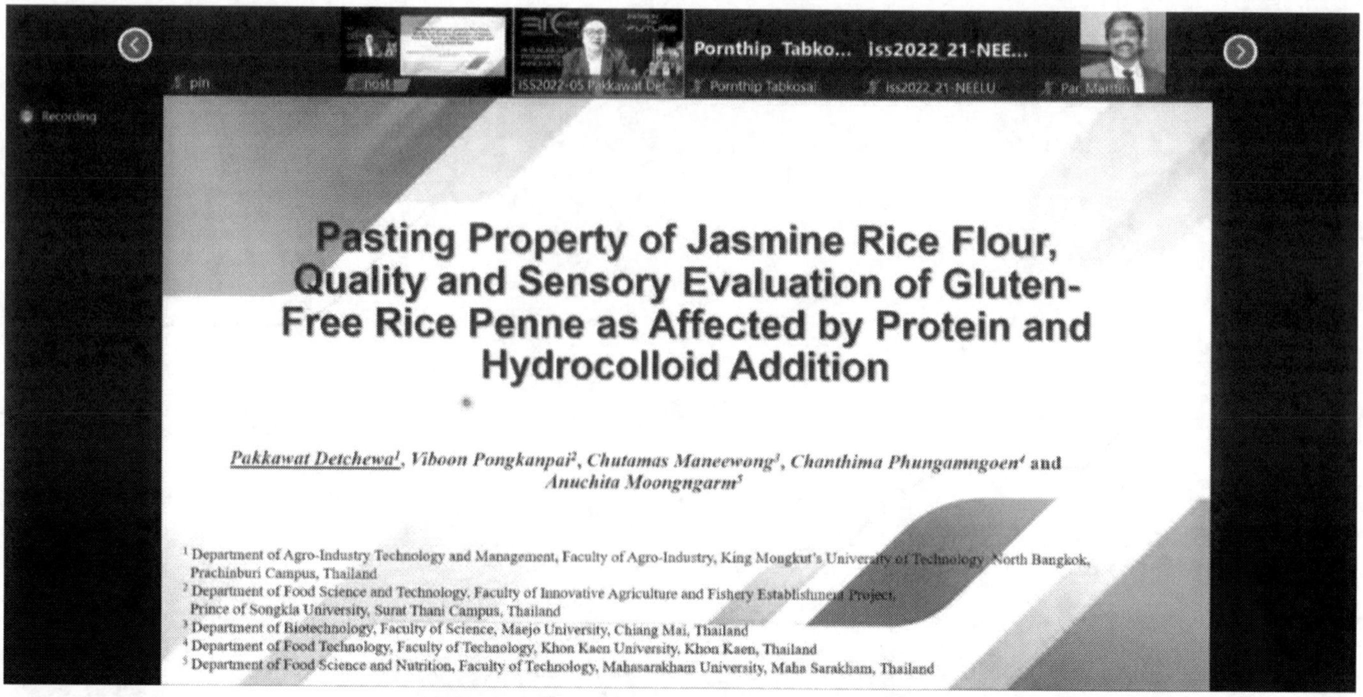
Pornthip Tabko...
Iss2022_21-NEE...
Recording
Pasting Property of Jasmine Rice Flour, Quality and Sensory Evaluation of Gluten-Free Rice Penne as Affected by Protein and Hydrocolloid Addition
Pakkawat Detchewa[1], Viboon Pongkanpai[2], Chutamas Maneewong[3], Chanthima Phungamngoen[4] and Anuchita Moongngarm[5]
[1] Department of Agro-Industry Technology and Management, Faculty of Agro-Industry, King Mongkut's University of Technology North Bangkok, Prachinburi Campus, Thailand
[2] Department of Food Science and Technology, Faculty of Innovative Agriculture and Fishery Establishment Project, Prince of Songkla University, Surat Thani Campus, Thailand
[3] Department of Biotechnology, Faculty of Science, Maejo University, Chiang Mai, Thailand
[4] Department of Food Technology, Faculty of Technology, Khon Kaen University, Khon Kaen, Thailand
[5] Department of Food Science and Nutrition, Faculty of Technology, Mahasarakham University, Maha Sarakham, Thailand

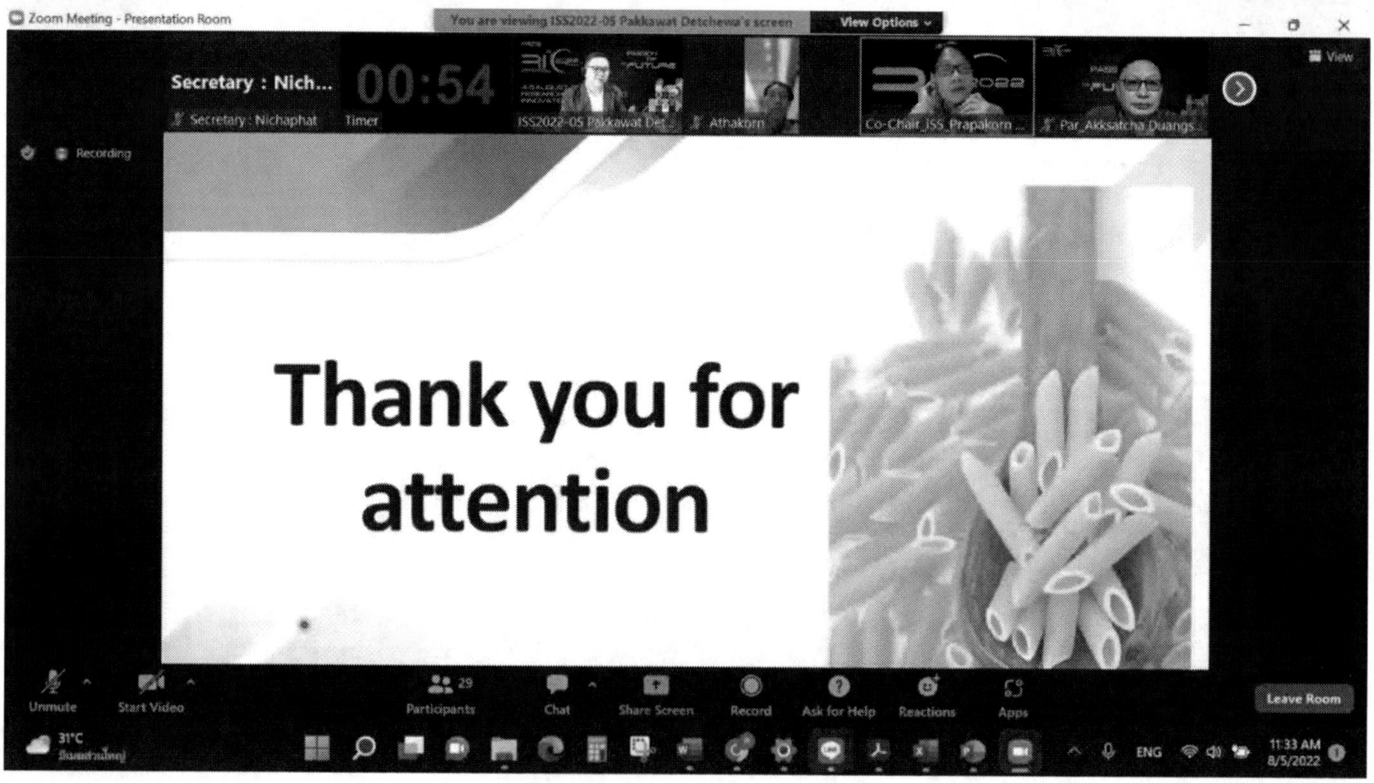
Zoom Meeting - Presentation Room
View Options
Secretary : Nich...
00:54
Secretary : Nichaphat
Timer
Athakorn
Recording
Thank you for attention
Unmute
Start Video
Participants
Chat
Share Screen
Record
Ask for Help
Reactions
Apps
Leave Room
31°C
ENG
11:33 AM
8/5/2022

Iss2022-07

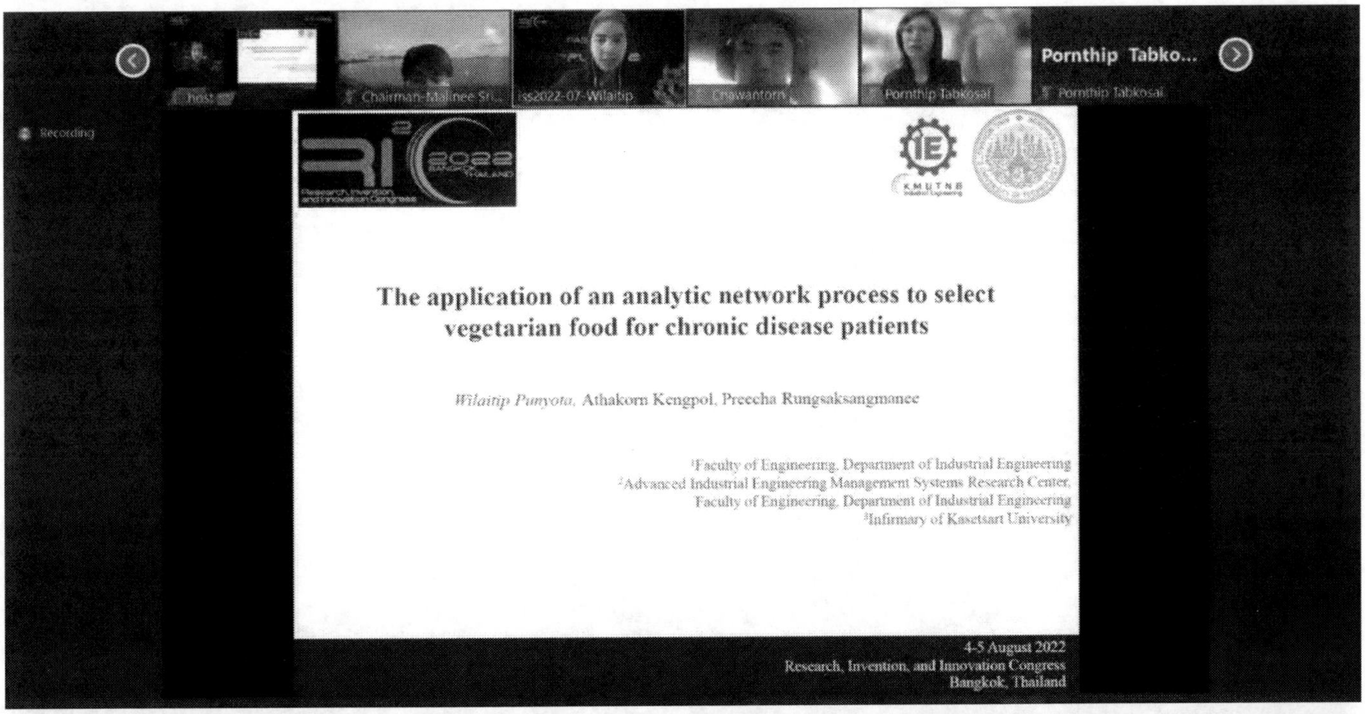

host
Chairman-Malinee Sri...
iss2022-07-Wilaitip
Chawantorn
Pornthip Tabkosai
Pornthip Tabko...
Recording
Research, Invention, and Innovation Congress
KMUTNB
The application of an analytic network process to select vegetarian food for chronic disease patients
Wilaitip Punyota, Athakorn Kengpol, Preecha Rungsaksangmanee
[1]Faculty of Engineering, Department of Industrial Engineering
[2]Advanced Industrial Engineering Management Systems Research Center, Faculty of Engineering, Department of Industrial Engineering
[3]Infirmary of Kasetsart University
4-5 August 2022
Research, Invention, and Innovation Congress
Bangkok, Thailand

05:43
Timer
Secretary : Nich...
Secretary : Nichaphat
Co-Chair ISS-Prapako...
iss2022-07-Wilaitip
pin
host
Recording
KMUTNB
Thank you for your attention
Krabi, Thailand

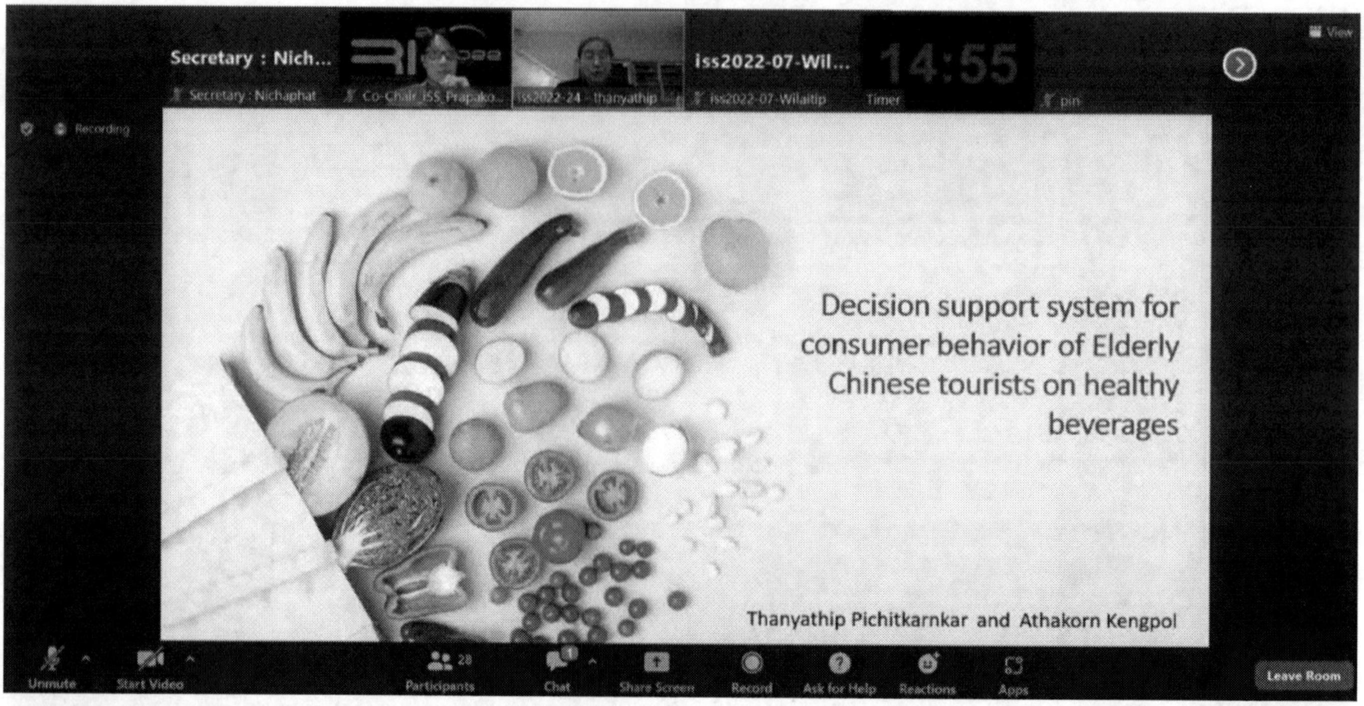
Secretary : Nich...
Secretary : Nichaphat
Co-Chair_ISS_Prapako...
iss2022-24 - thanyathip
iss2022-07-Wil...
iss2022-07-Wilaitip
14:55
Timer
pin
View
Recording
Decision support system for consumer behavior of Elderly Chinese tourists on healthy beverages
Thanyathip Pichitkarnkar and Athakorn Kengpol
Unmute
Start Video
28
Participants
Chat
Share Screen
Record
Ask for Help
Reactions
Apps
Leave Room

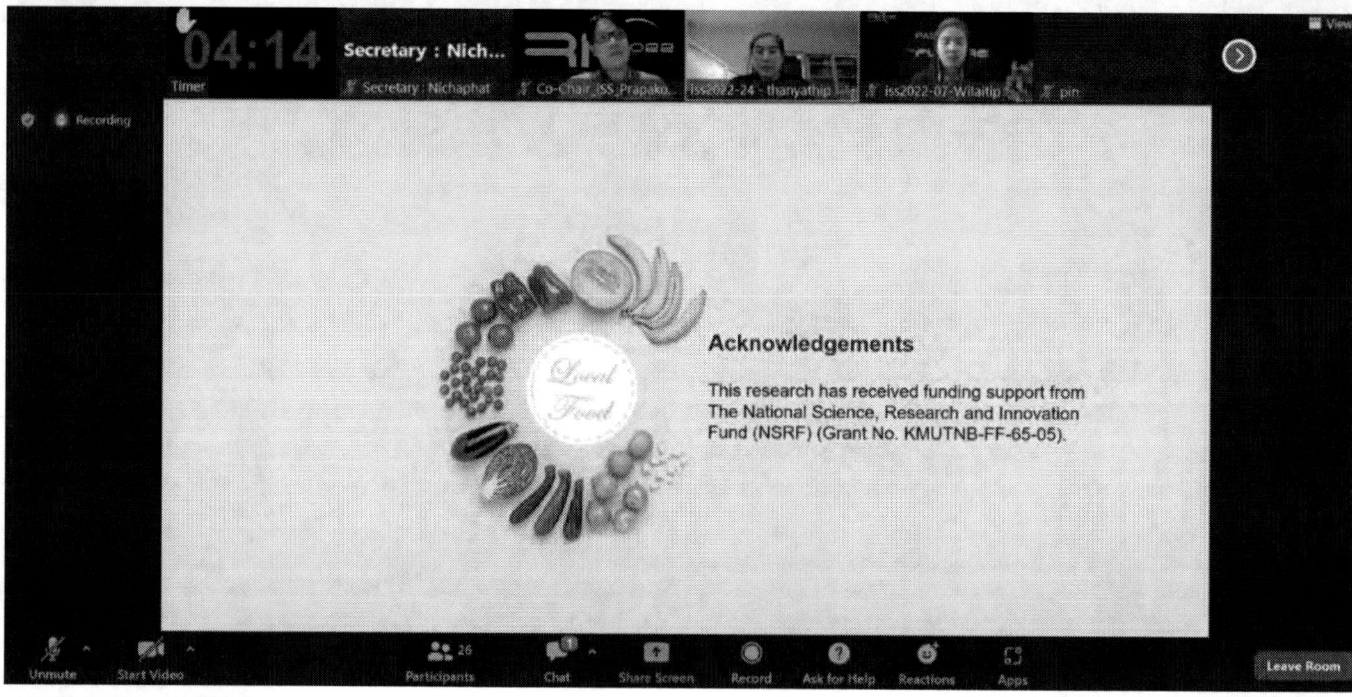
04:14
Timer
Secretary : Nich...
Secretary : Nichaphat
Co-Chair_ISS_Prapako...
iss2022-24 - thanyathip
iss2022-07-Wilaitip
pin
View
Recording
Local Food
Acknowledgements
This research has received funding support from The National Science, Research and Innovation Fund (NSRF) (Grant No. KMUTNB-FF-65-05).
Unmute
Start Video
26
Participants
Chat
Share Screen
Record
Ask for Help
Reactions
Apps
Leave Room

Iss2022-22

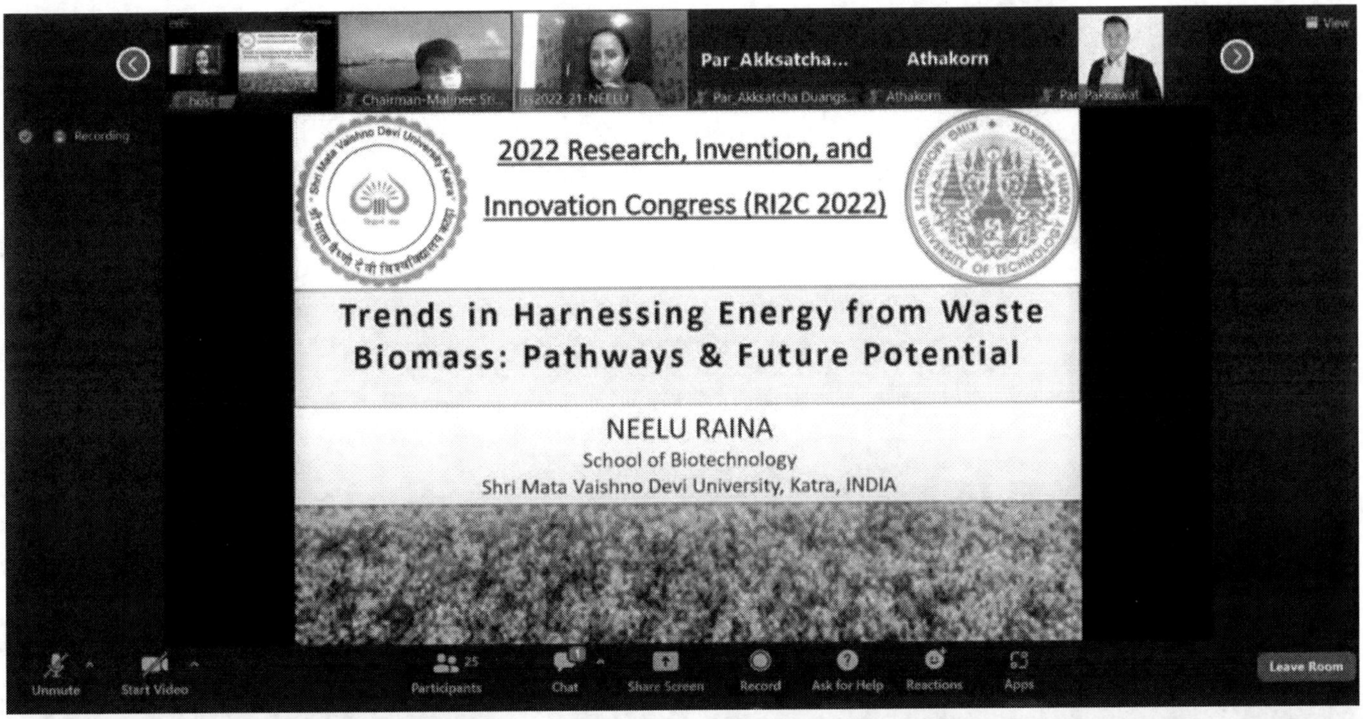
Par_Akksatcha...
Athakorn
Recording
2022 Research, Invention, and
Innovation Congress (RI2C 2022)
Trends in Harnessing Energy from Waste
Biomass: Pathways & Future Potential
NEELU RAINA
School of Biotechnology
Shri Mata Vaishno Devi University, Katra, INDIA
Unmute
Start Video
Participants
Chat
Share Screen
Record
Ask for Help
Reactions
Apps
Leave Room

05:33
Timer
Secretary : Nich...
iss2022-24 - tha...
iss2022-07-Wil...
Recording
Do!
ACKNOWLEDGEMENT

Iss2022-13

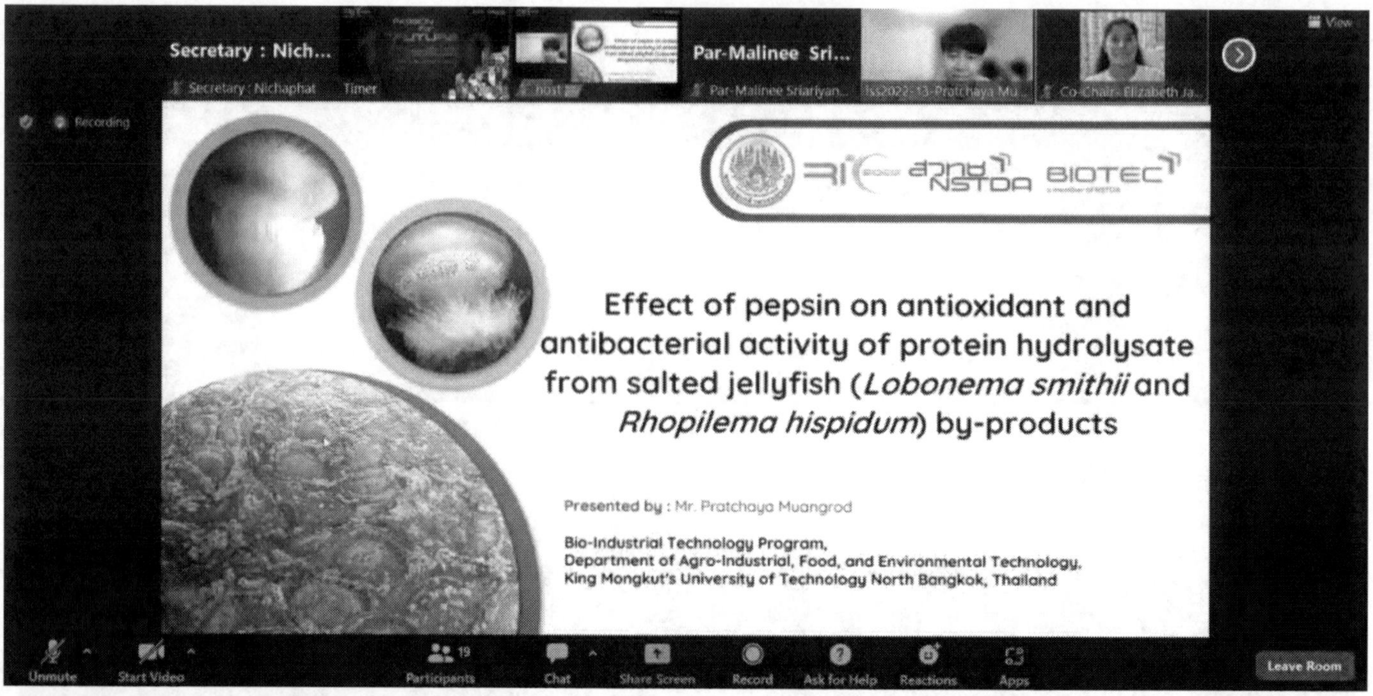
Secretary : Nich...
Par-Malinee Sri...
Recording
BIOTEC
Effect of pepsin on antioxidant and antibacterial activity of protein hydrolysate from salted jellyfish (*Lobonema smithii* and *Rhopilema hispidum*) by-products
Presented by : Mr. Pratchaya Muangrod
Bio-Industrial Technology Program,
Department of Agro-Industrial, Food, and Environmental Technology,
King Mongkut's University of Technology North Bangkok, Thailand
Unmute
Start Video
Participants
Chat
Share Screen
Record
Ask for Help
Reactions
Apps
Leave Room

03:18
Timer
Secretary : Nich...
Par-Malinee Sri...
Recording
Acknowledgments
BIOTEC
CSP
• Assoc. Prof. Benjawan Thumthanaruk, Ph.D.
Department of Agro-Industrial, Food and Environmental Technology, Faculty of Applied Science, KMUTNB
• Prof. Vilai Rungsardthong, Ph.D.
Department of Agro-Industrial, Food and Environmental Technology, Faculty of Applied Science, KMUTNB
• Sittiruk Roytrakul, Ph.D.
Proteomics Research Laboratory, National Center for Genetic Engineering and Biotechnology (BIOTEC)
• Assoc. Prof. Savitri Vatanyoopaisarn, Ph.D.
Department of Agro-Industrial, Food and Environmental Technology, Faculty of Applied Science, KMUTNB
• Benjamaporn Wonganu, Ph.D.
Department of Biotechnology, Faculty of Applied Science, KMUTNB
• King Mongkut's University of Technology North Bangkok and National Science and Technology Development Agency, Thailand (Contract no. B.020-2563)
• National Research Council of Thailand through the NRCT Senior Research Scholar Program (Contract no. 814-2020)
19

Iss2022-14

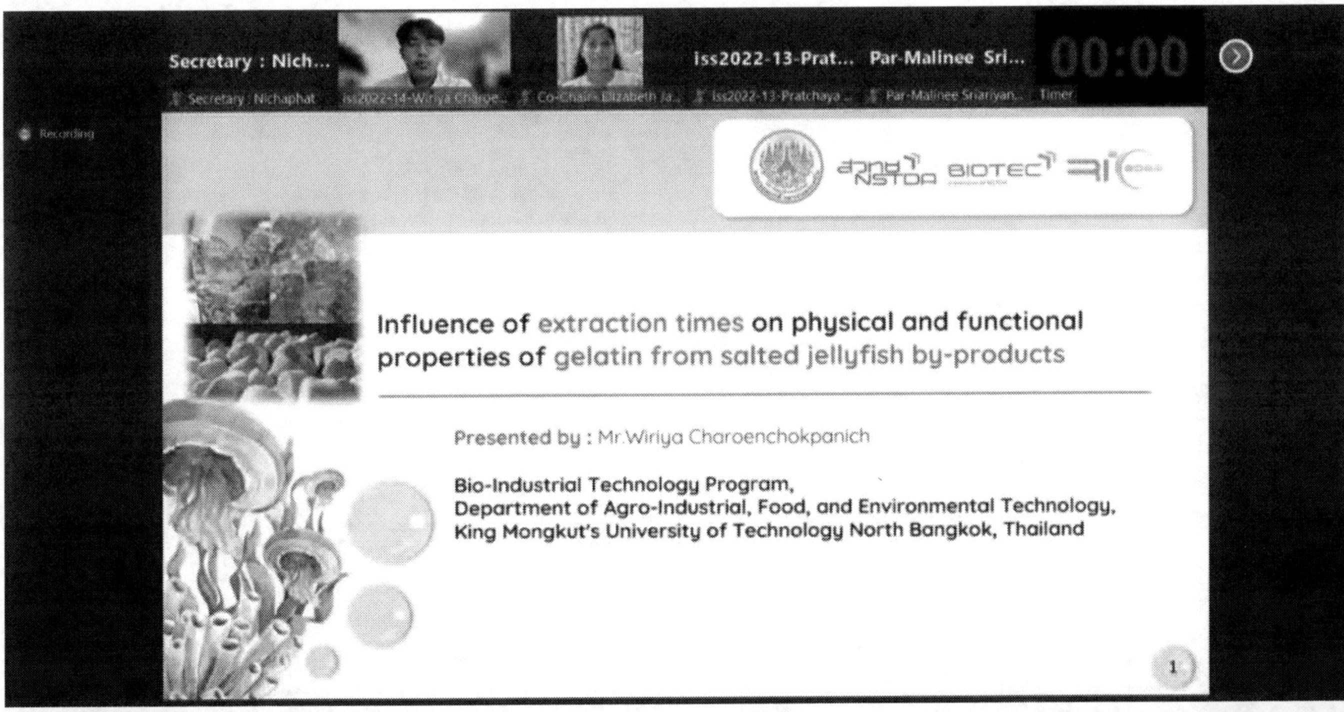
Secretary : Nich...
iss2022-13-Prat...
Par-Malinee Sri...
00:00
Recording
BIOTEC
Influence of extraction times on physical and functional properties of gelatin from salted jellyfish by-products
Presented by : Mr.Wiriya Charoenchokpanich
Bio-Industrial Technology Program,
Department of Agro-Industrial, Food, and Environmental Technology,
King Mongkut's University of Technology North Bangkok, Thailand
1

04:45
Timer
Secretary : Nich...
iss2022-13-Prat...
Recording
Acknowledgement
AFET
NSTDA
BIOTEC
• Assoc. Prof. Benjawan Thumthanaruk, Ph.D.
Department of Agro-Industrial, Food and Environmental Technology, Faculty of Applied Science, KMUTNB
• Sittiruk Roytrakul, Ph.D.
Laboratory/Research Unit: National Center for Genetic Engineering and Biotechnology, National Science and Technology Development Agency
• Prof. Vilai Rungsardthong, Ph.D.
Department of Agro-Industrial, Food and Environmental Technology, Faculty of Applied Science, KMUTNB
• Benjamaporn Wonganu, Ph.D.
Department of Biotechnology, Faculty of Applied Science, KMUTNB
• Assoc. Prof. Savitri Vatanyoopaisarn, Ph.D.
Department of Agro-Industrial, Food and Environmental Technology, Faculty of Applied Science, KMUTNB
• King Mongkut's University of Technology North Bangkok and National Science and Technology Development Agency, Thailand (Contract no. B.021-2563)
• National Research Council of Thailand through the NRCT Senior Research Scholar Program (Contract no. 814-2020)
20

Iss2022-15

Zoom Meeting
Secretary : Nich...
Iss2022-13-Prat...
Secretary : Nichaphat
chair babu
iss2022-15-Pinida Joradon
Iss2022-14-Wiriya Cha...
Co-Chair-Elizabeth Ja...
iss2022-13-Pratchaya ...
Recording
RI 2022
KMUTNB
NANOTEC
A comparative study of conventional and supercritical carbon dioxide extraction methods for the recovery of bioactive compound from Lion's Mane mushroom (Hericium erinaceus)
Presented by : Miss Pinida Joradon
Coauthor : Professor Dr. Vilai Rungsardthong
: Dr. Uracha Ruktanonchai
Department of Agro-Industrial, Food and Environmental Technology, Faculty of Applied Science, Food and Agro-industrial Research Center, King Mongkut's University of Technology North Bangkok, Thailand
32°C
ENG
1:32 PM
8/5/2022

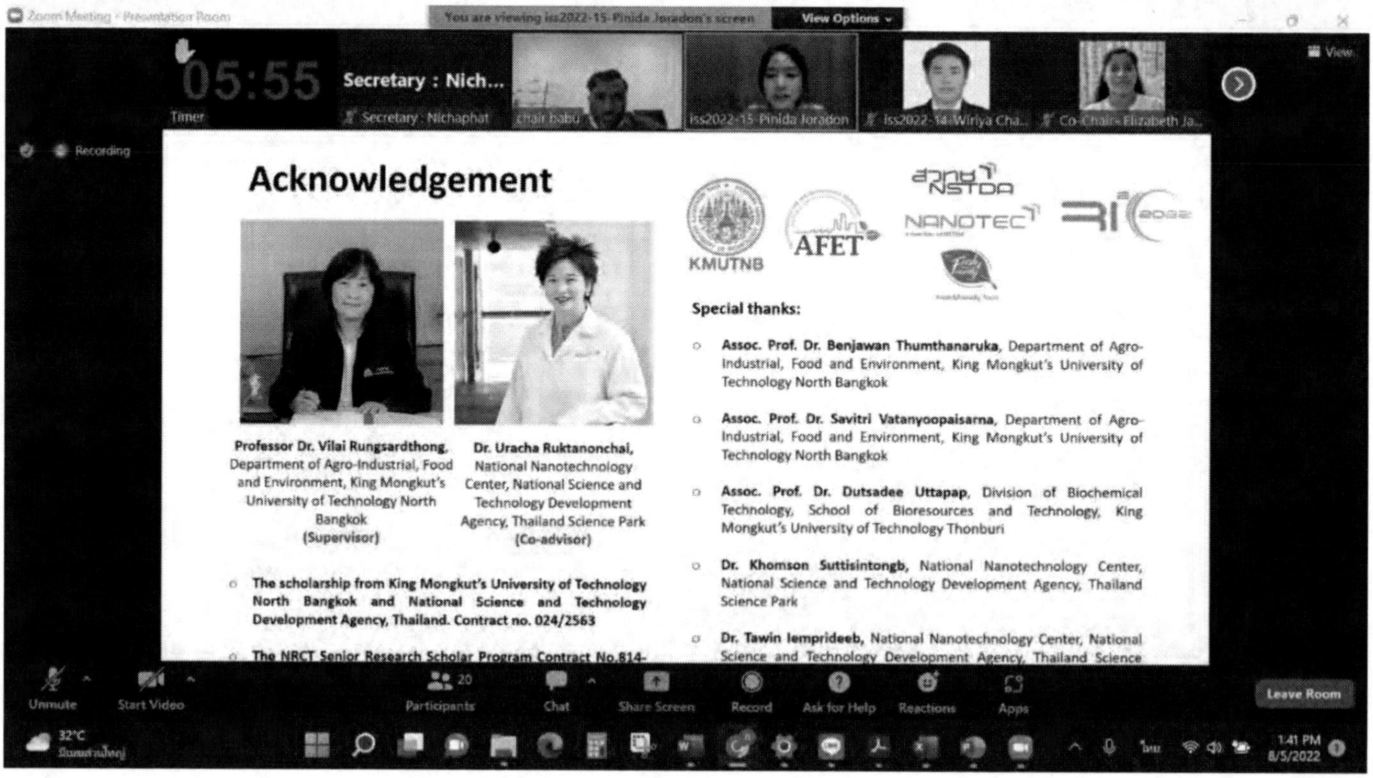
Zoom Meeting - Presentation Room
You are viewing iss2022-15-Pinida Joradon's screen
View Options
05:55
Timer
Secretary : Nich...
Secretary : Nichaphat
chair babu
iss2022-15-Pinida Joradon
Iss2022-14-Wiriya Cha...
Co-Chair-Elizabeth Ja...
Recording
Acknowledgement
KMUTNB
AFET
NSTDA
NANOTEC
RI 2022
Professor Dr. Vilai Rungsardthong, Department of Agro-Industrial, Food and Environment, King Mongkut's University of Technology North Bangkok (Supervisor)
Dr. Uracha Ruktanonchai, National Nanotechnology Center, National Science and Technology Development Agency, Thailand Science Park (Co-advisor)
The scholarship from King Mongkut's University of Technology North Bangkok and National Science and Technology Development Agency, Thailand. Contract no. 024/2563
The NRCT Senior Research Scholar Program Contract No.814-
Special thanks:
Assoc. Prof. Dr. Benjawan Thumthanaruka, Department of Agro-Industrial, Food and Environment, King Mongkut's University of Technology North Bangkok
Assoc. Prof. Dr. Savitri Vatanyoopaisarna, Department of Agro-Industrial, Food and Environment, King Mongkut's University of Technology North Bangkok
Assoc. Prof. Dr. Dutsadee Uttapap, Division of Biochemical Technology, School of Bioresources and Technology, King Mongkut's University of Technology Thonburi
Dr. Khomson Suttisintongb, National Nanotechnology Center, National Science and Technology Development Agency, Thailand Science Park
Dr. Tawin Iemprideeb, National Nanotechnology Center, National Science and Technology Development Agency, Thailand Science
Unmute
Start Video
Participants
Chat
Share Screen
Record
Ask for Help
Reactions
Apps
Leave Room
32°C
1:41 PM
8/5/2022

Iss2022-16

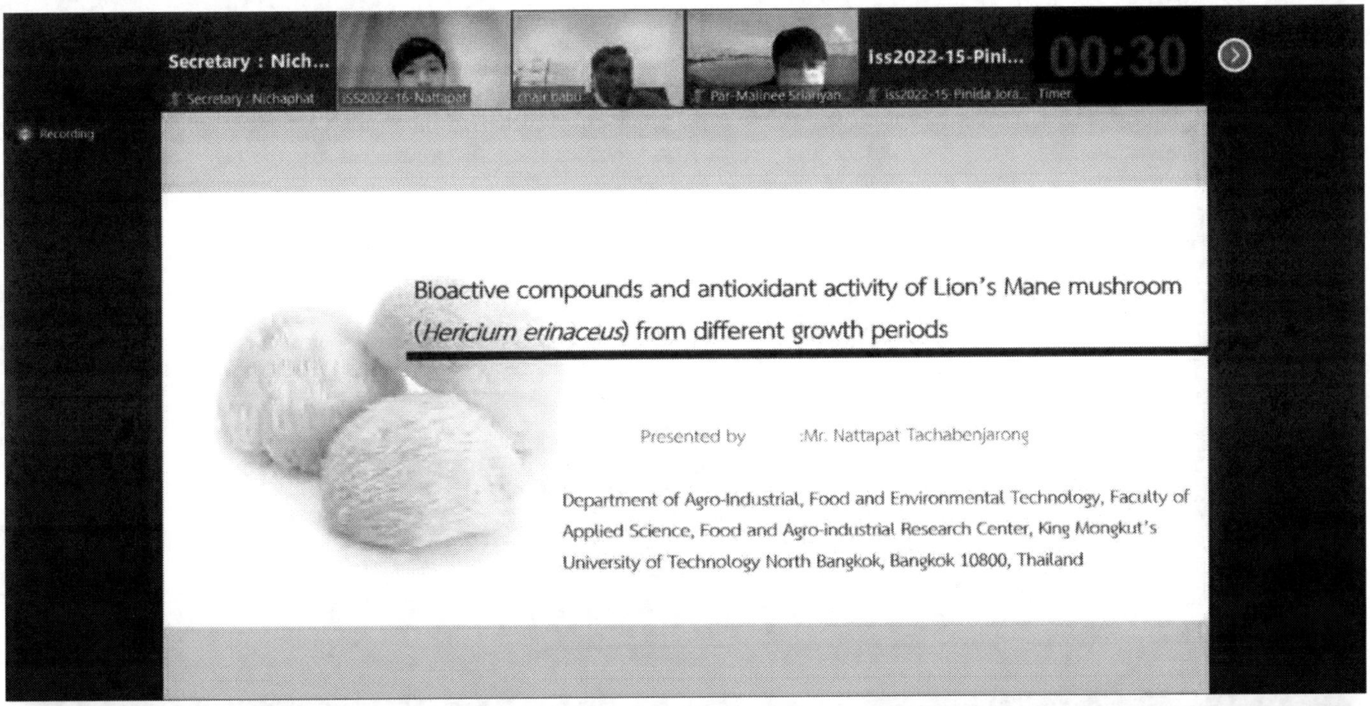
Secretary : Nich...
Secretary : Nichaphat
iss2022-16-Nattapat
chair babu
Par-Malinee Sriariyan...
iss2022-15-Pini...
iss2022-15-Pinida Jora...
00:30
Timer
Recording
Bioactive compounds and antioxidant activity of Lion's Mane mushroom (*Hericium erinaceus*) from different growth periods
Presented by :Mr. Nattapat Tachabenjarong
Department of Agro-Industrial, Food and Environmental Technology, Faculty of Applied Science, Food and Agro-industrial Research Center, King Mongkut's University of Technology North Bangkok, Bangkok 10800, Thailand

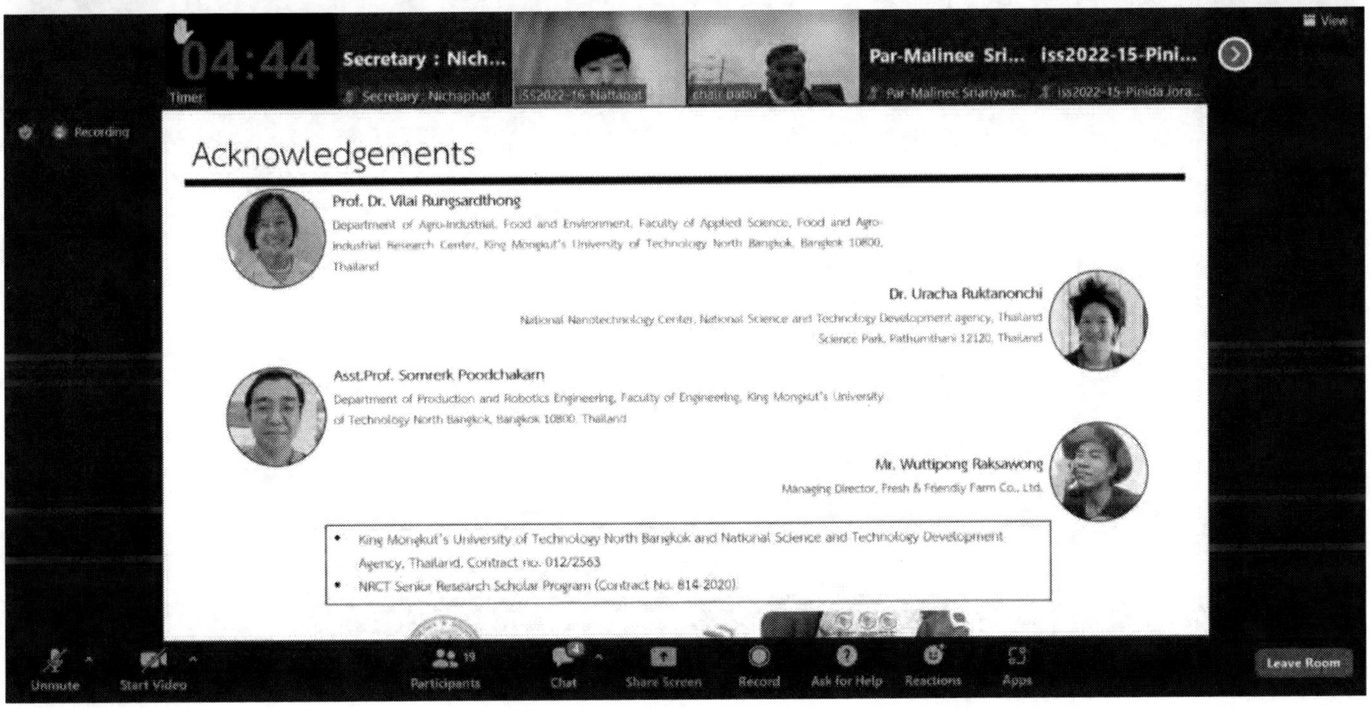
04:44
Timer
Secretary : Nich...
Secretary : Nichaphat
iss2022-16-Nattapat
chair babu
Par-Malinee Sri...
Par-Malinee Sriariyan...
iss2022-15-Pini...
iss2022-15-Pinida Jora...
View
Recording
Acknowledgements
Prof. Dr. Vilai Rungsardthong
Department of Agro-Industrial, Food and Environment, Faculty of Applied Science, Food and Agro-industrial Research Center, King Mongkut's University of Technology North Bangkok, Bangkok 10800, Thailand
Dr. Uracha Ruktanonchi
National Nanotechnology Center, National Science and Technology Development agency, Thailand Science Park, Pathumthani 12120, Thailand
Asst.Prof. Somrerk Poodchakarn
Department of Production and Robotics Engineering, Faculty of Engineering, King Mongkut's University of Technology North Bangkok, Bangkok 10800, Thailand
Mr. Wuttipong Raksawong
Managing Director, Fresh & Friendly Farm Co., Ltd.
King Mongkut's University of Technology North Bangkok and National Science and Technology Development Agency, Thailand, Contract no. 012/2563
NRCT Senior Research Scholar Program (Contract No. 814-2020)
Unmute
Start Video
Participants
Chat
Share Screen
Record
Ask for Help
Reactions
Apps
Leave Room

Iss2022-17

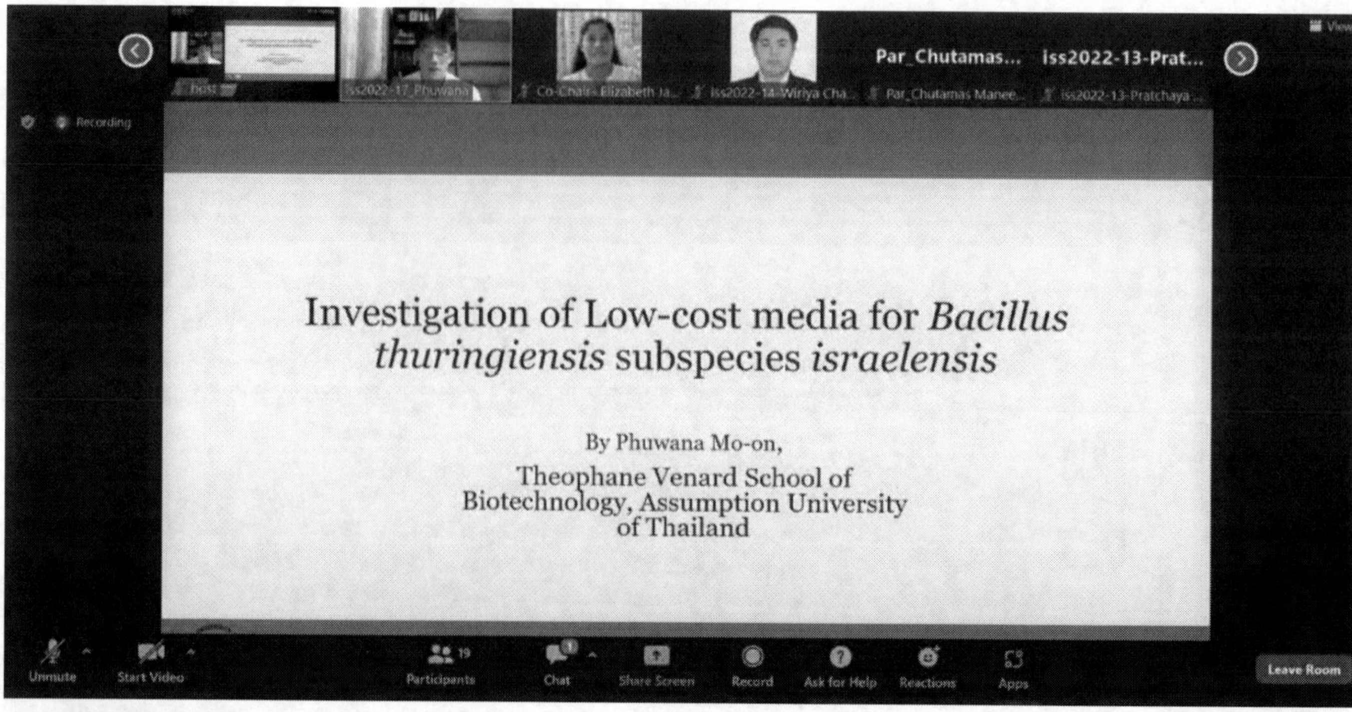
Par_Chutamas...
iss2022-13-Prat...
Recording
Investigation of Low-cost media for *Bacillus thuringiensis* subspecies *israelensis*
By Phuwana Mo-on,
Theophane Venard School of Biotechnology, Assumption University of Thailand
Unmute
Start Video
Participants
Chat
Share Screen
Record
Ask for Help
Reactions
Apps
Leave Room

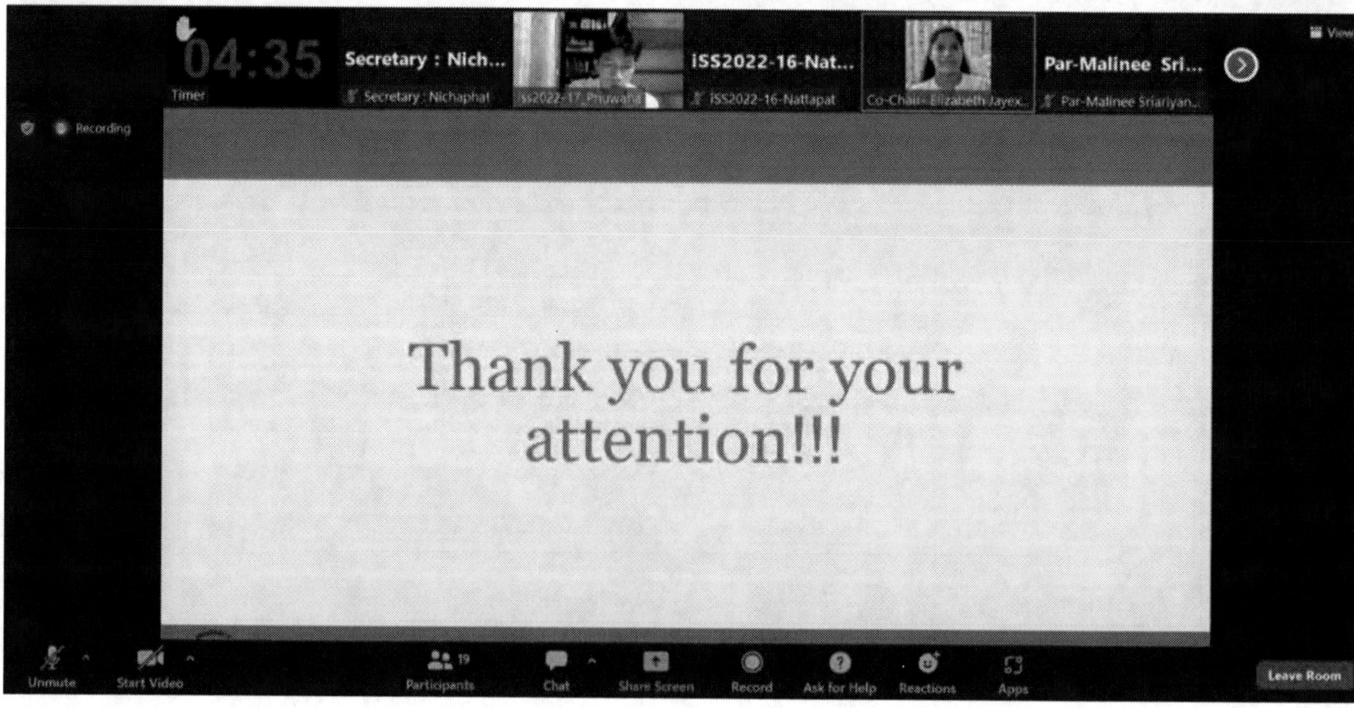
04:35
Timer
Secretary : Nich...
iSS2022-16-Nat...
Par-Malinee Sri...
Recording
Thank you for your attention!!!
Unmute
Start Video
Participants
Chat
Share Screen
Record
Ask for Help
Reactions
Apps
Leave Room

Iss2022-18

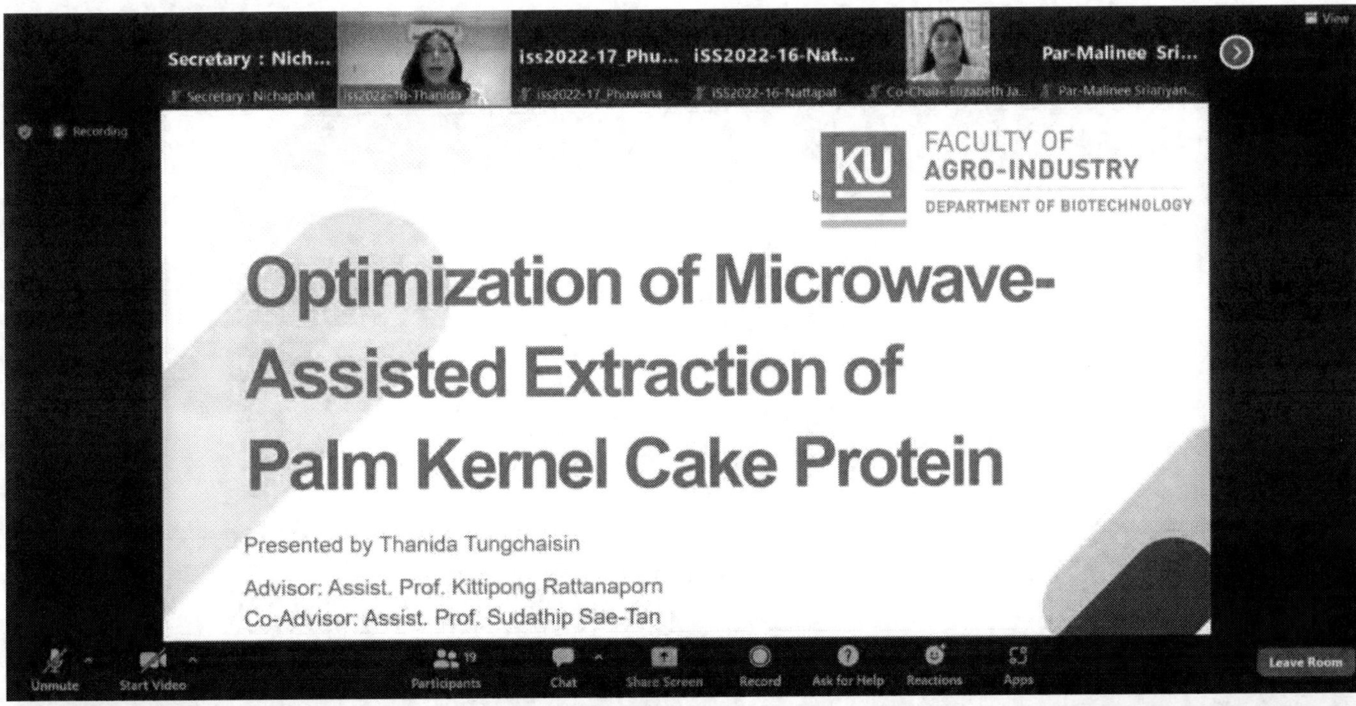
Secretary : Nich...
iss2022-17_Phu...
iSS2022-16-Nat...
Par-Malinee Sri...
Secretary : Nichaphat
iss2022-18-Thanida
iss2022-17_Phuwana
iSS2022-16-Nattapat
Co-Chair- Elizabeth Ja...
Par-Malinee Sriariyan...
Recording
KU
FACULTY OF
AGRO-INDUSTRY
DEPARTMENT OF BIOTECHNOLOGY
Optimization of Microwave-Assisted Extraction of Palm Kernel Cake Protein
Presented by Thanida Tungchaisin
Advisor: Assist. Prof. Kittipong Rattanaporn
Co-Advisor: Assist. Prof. Sudathip Sae-Tan
Unmute
Start Video
19
Participants
Chat
Share Screen
Record
Ask for Help
Reactions
Apps
Leave Room

Secretary : Nich...
iss2022-17_Phu...
iSS2022-16-Nat...
Par-Malinee Sri...
Secretary : Nichaphat
iss2022-18-Thanida
iss2022-17_Phuwana
iSS2022-16-Nattapat
Co-Chair- Elizabeth Ja...
Par-Malinee Sriariyan...
Recording
KU
FACULTY OF
AGRO-INDUSTRY
DEPARTMENT OF BIOTECHNOLOGY
THANK YOU
for your attention

Iss2022-23

Secretary : Nich...
14:54
iss2022-17_Phu...
Recording
Paddy straw: an economical substrate for oyster mushroom (*Pleurotus ostreatus*) cultivation
Anne Sahithi Somavarapu Thomas[1], Sona .A[2], Vinodhini Shanmugam[2], Marttin Paulraj Gundupalli[3, 4] Malinee Sriariyanun[3, 4*]
[1] School of Biosciences and Technology (SBST), Vellore Institute of Technology (VIT) Vellore, Tamilnadu, India
[2]Department of Biotechnology, D.K.M College for Women, Vellore, Tamilnadu, India
[3]Department of Chemical and Process Engineering, The Sirindhorn International Thai-German Graduate School of Engineering (TGGS), King Mongkut's University of Technology North Bangkok (KMUTNB), Bangkok, Thailand
[4]Biorefinery and Process Automation Engineering Center (BPAEC), King Mongkut's University of Technology North Bangkok (KMUTNB), Bangkok, Thailand
iSS 2022-23
TGGS
8/5/2022

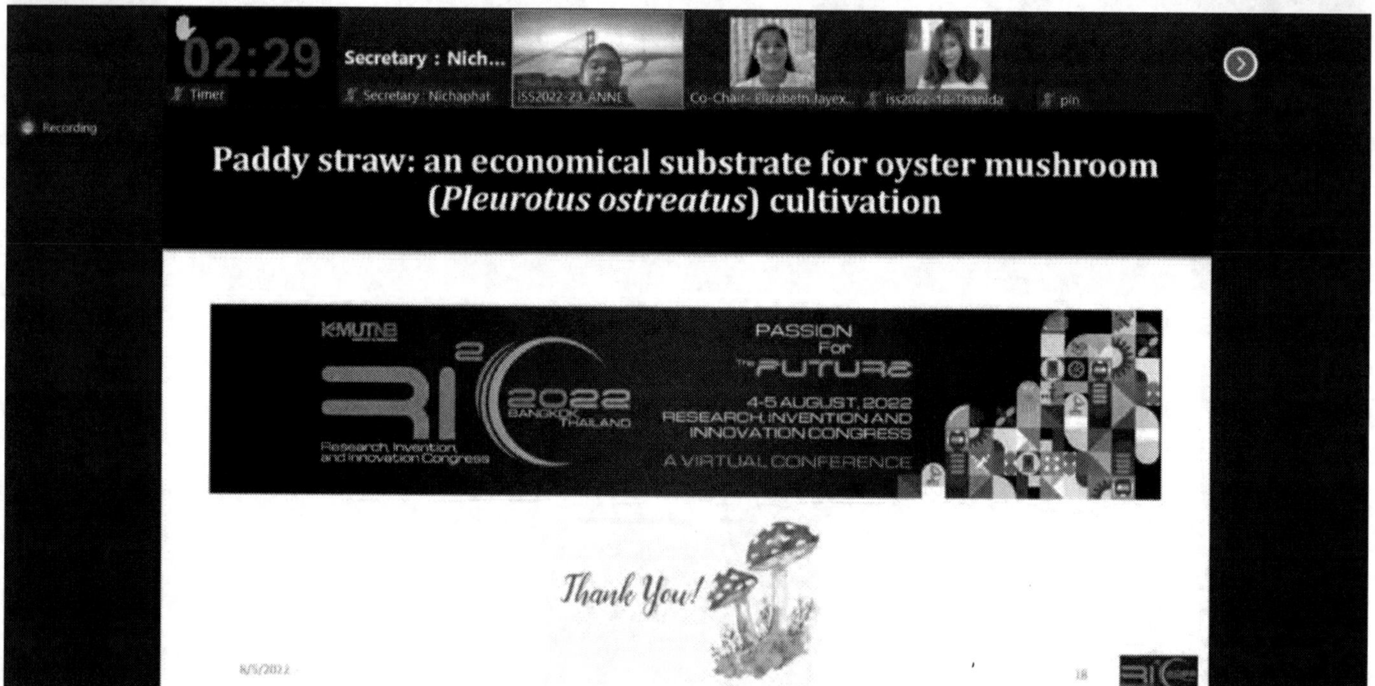
02:29
Timer
Secretary : Nich...
Recording
Paddy straw: an economical substrate for oyster mushroom (*Pleurotus ostreatus*) cultivation
KMUTNB
2022
BANGKOK THAILAND
Research, Invention, and Innovation Congress
PASSION For The FUTURE
4-5 AUGUST, 2022
RESEARCH, INVENTION AND INNOVATION CONGRESS
A VIRTUAL CONFERENCE
Thank You!
8/5/2022
18

Statement of Peer review

In submitting conference proceedings to *Web of Conferences*, the editors of the proceedings certify to the Publisher that

1. They adhere to its **Policy on Publishing Integrity** in order to safeguard good scientific practice in publishing.
2. All articles have been subjected to peer review administered by the proceedings editors.
3. Reviews have been conducted by expert referees, who have been requested to provide unbiased and constructive comments aimed, whenever possible, at improving the work.
4. Proceedings editors have taken all reasonable steps to ensure the quality of the materials they publish and their decision to accept or reject a paper for publication has been based only on the merits of the work and the relevance to the journal.

Title, date and place of the conference

Title: 2022 Research, Invention, and Innovation Congress (RI2C2022)
Date: 4-5 August 2022
Place: Online conference at King Mongkut's University of Technology North Bangkok

Proceedings editor(s):

Managing Editor: Malinee Sriariyanun, King Mongkut's University of Technology North Bangkok, Thailand

Editor members
1. Yushen Cheng, National Yunlin University of Science and Technology, Taiwan
2. Debraj Bhattacharyya, IIT Hyderabad, India
3. Wawat Rodiahwati, University of New England, Australia
4. Elizabeth Jayex Panakkal, King Mongkut's University of Technology North Bangkok, Thailand

Date and editor's signature 8 August 2022

Trends in harnessing energy from waste biomass: pathways & future potential

Neelu Raina[1,*], *Malinee* Sriariyanun[2], *Sukanya* Areeya[2], *Atin Kumar* Pathak[3], *Garima* Slathia[1], and *Himanshu* Saraswat[4]

[1]School of Biotechnology, Shri Mata Vaishno Devi University, Katra, J&K-182320, India
[2]Biorefinery and Process Automation Engineering Center, Department of Mechanical and Process Engineering, Sirindhorn International Thai-German Graduate School of Engineering, King Mongkut's University of Technology North Bangkok, Bangkok, Thailand
[3]School of Energy Management, Faculty of Engineering, Shri Mata Vaishno Devi University, Katra, J&K-182320, India
[4] Centre for Systems Biology and Bioinformatics, Panjab University, Chandigarh-160015, India

Abstract. Demand for fuel energy is continually on the rise. There is also a constant challenge involved to ensure that all our energy needs are fulfilled. Persistent overconsumption of conventional fossil fuels due to the rise in global population aided by economic expansion has resulted in reduction of fossil fuel reserves. This has fuelled the need to boost research efforts on renewable and sustainable bioenergy feedstocks. Since bioenergy utilizes organic matter; therefore, it is an economically viable and clean solution, which can minimize our reliance on non-renewable resources. The bioprocessing of lignocellulosic biomass to produce bio-based products under biorefinery setup is gaining global attention. The main challenge however remains to strike a balance between energy harvesting and economic viability with minimum environmental impacts. The development of zero-waste lignocellulosic biorefinery aligns completely with the idea of sustainable development without increasing carbon footprint. This concept is self-sustainable. It also advocates re-usage or recycling of waste; of which using lignocellulosic biomass waste is a major thrust. Improving the techno-economic efficiency of currently employed pretreatment methods and looking for combined pretreatment strategies will prove to be a stepping stone in the commercialization of zero-waste lignocellulosic biorefineries. This review investigates the most widespread pretreatment types, highlighs their advantages/disadvantages, and reviews the current status and technological advances in the bioconversion process of LCB into bioenergy in a biorefinery set-up.

Keyword. Biorefinery, Bioethanol, Lignocellulose, Pretreatment, Bioconversion

1 Introduction

The rising of global population is expected to reach 9.2 billion by the year 2050 [1]. Simultaneously, by the year 2030, the predicted enhancement in world energy demand is likely to reach around 53% [2]. Coal, oil and natural gas, the major forms of energy that we use today, are non-renewable and thus bound by limited availability. Gobbling up of non-renewable resources by industrial evolution has eventually brought fossil fuel reservoirs to the verge of exhaustion. Their consumption at this pace has severely degraded the environment as well. Therefore, the sharp rise in the demand for energy resources is far beyond the handling capacity of fossil-based resources. The extravagant use of fossil fuels as the sole source of energy has given rise to the dire need for gradual energy shift. This shift is driving us away from the traditional sources of energy towards alternative resources which are sustainable in the long run.

1.1 Is bioenergy the ultimate answer to our rising energy needs?

Biomass as a resource is present in large quantities. Its sustainable utilization offers a variety of benefits ranging from sequestration of carbon to generation of bioenergy. It has the capacity to serve diverse purposes as part of the integrated biorefinery system. In addition, it yields numerous benefits by strengthening the socio-economic structure of the society. Production of renewable and sustainable biofuels is thus gaining overwhelming interest globally. Biofuels are important from an economic

*Corresponding author: ancient99india@gmail.com

point of view as they can support a stable, secure and low-cost energy supply. Bioethanol is currently the front runner in the race among all the biofuels. It presents promising opportunities in terms of a green, carbon-neutral, environmentally safe, and clean alternative to fossil-derived fuels. The utilization of bioethanol blended gasoline in the transportation sector reduces the consumption of fossil fuels and the concomitant release of greenhouse gases (GHGs) in the environment. Its chemical-physical properties (evaporation, enthalpy and octane number) are similar to that of gasoline. In terms of chemical production, ethanol can also be used as a platform chemical itself. Its bioconversion *via* microbial fermentation can utilize a wide array of biomass types (simple sugars, starchy biomass, and lignocellulosic biomass) with the potential co-production of several industrially significant compounds. The presence of many macromolecular organic compounds in a biorefinery process acts as platform chemical/s for molecules of greater interest; besides decreasing the cost of associated biofuels.

The most promising molecular substrate for the production of bioethanol is lignocellulose. It is composed of cellulose, hemicellulose and lignin. Cellulose and hemicellulose are the reservoirs of sugar polymers, which are fermentable in nature. These sugars can be converted into bioethanol by the use of microorganisms. On the other hand, lignin is an aromatic polymer. The transition from fossil fuels towards biomass-based fuels has presented lignocellulosic biomass (LCB) as the favorable biomass feedstock. Since the annual worldwide production of LCB is estimated to be around 1×10^{10} metric tons (MT); LCB thus accounts for approximately half of the total biomass reserves present on planet earth [3]. LCB generated from various sectors (forestry, agriculture, agro-industrial and industrial) acts as precursors for the generation of cellulosic bioethanol (2nd generation) accompanied by a wide array of biochemicals. Also, bioethanol thus generated does not contribute to the price surge of food when compared to the conventional sugar/starch-based bioethanol (1st generation).

1.2 Lignocellulosic biomass waste as a source of bioenergy

Biorefinery approach presents an opportunity to integrate production of biofuels and industrially useful bio-based platform chemicals from LCB. It assimilates many conversion techniques over a single platform. The major focus of the lignocellulose biorefinery that tends to enhance its economic viability is the maximum utilization of the majority of LCB (*i.e.* cellulose, hemicellulose and lignin). It is accompanied by minimum or zero wastage from the system thus supporting the concept of circular bioeconomy (CB). However, the recalcitrant nature (presence of lignin inhibits LCB solubilization *via* enzymatic saccharification) of LCB is the main obstacle in its valorization to bioethanol and other value-added products. The bioconversion of LCB thus depends on many steps involved- a) pretreatment, b) hydrolysis, and c) fermentation. The present review entails the different steps involved in the valorization of LCB for bioethanol production along with various value-added products generated. Also, it brings forward future perspectives, recent technological advances (biotechnological) and significant challenges encountered in bioethanol production that need to be highlighted. A carbon-neutral biorefinery system using LCB as feedstock opens the door for a transition from linear production of biomass to a closed-loop system emphasizing zero-waste bioeconomy [4]. However, the scientific fraternity has explored several bottlenecks keeping in mind the "zero-waste" concept using LCB as substrate.

Waste forms an inherent and unavoidable constituent of any process that can be ascribed to various factors *i.e.* process inefficiencies, usability of resources and discarding of non-essential parts of the feedstock. The use of residual biomass available locally contributes to enhanced energy productivity and sustainable growth. Residual materials generated from forestry, agriculture and agro-industrial sectors amount to millions of tons of waste every year. These agro-industrial transformations need to be managed and have a disposal cost associated with them, therefore raising serious economic and environmental concerns. Also, their underutilization limits all possibilities of maximum utilization of the biomass. Therefore, the utilization of agricultural waste is an excellent choice for biofuel and biochemical production. Converting unwanted agricultural waste streams into valuable resources by their bioconversion requires improved technologies, which support practical applications for sustainable development. A total of 174.1 million tons of residues are generated per year in Thailand and the projected capacity for bioethanol production is estimated to be around 20,213.5 million liters per year. Nevertheless, only 31.2% of the fuel demand from the transportation sector can be met using this volume [5]. This calls for the need to improve bioenergy production. Looking for alternative energy sources can aid in the mitigation of GHGs in the environment.

The strategy to systematically utilize LB waste for bioethanol production strongly depends on its composition. Lignin, the aromatic polymer is a major impediment in the fermentation of cellulose and hemicellulose and the subsequent release of sugars and valuable chemicals. Therefore, it becomes necessary to loosen the lignin shield from the accompanying organic matrix. Recent studies have shown that the primary aim of the researchers during pretreatment is reducing cellulose crystallinity, cleavage of H-bonds and Van der Waals forces and solubilization of the hemicellulose. Many pretreatment methods have been used to separate the lignin polymer from the rest of the polysaccharide pool. Their primary focus is on the release of maximum concentration of fermentable sugars. This is accompanied by the production of platform chemicals which are of importance from an economic point of view. Pretreatment studies using acids, alkali, organosolv, deep eutectic solvents and combined pretreatment methods among others have been developed. Acids result in deconstruction of lignin, hemicellulose and cellulose solubilization, and destruction of β(1-4) glycosidic bonds. On the other hand, alkali solubilizes the recalcitrant layer of lignin and hemicellulose depending upon the chemical concentration, temperature and pressure conditions [6]. Ionic liquid pretreatment (imidazolium, choline and chloride acetate, pyridinium) depends on the selection of anions and cations in the reaction mixture. β-O-4 bonds are converted into ion dipole bonds thus dissolving the cellulose layer and resulting in the separation of lignin and hemicellulose components. Organosolv pretreatment dissolves β-O-aryl, 4-O-methylglucuronic acid, α-O-aryl in between lignin and hemicellulose. Recently the effect of more than two solvents has been explored during pretreatment studies usually operated at a high temperature of 200 °C [7]. Deep eutectic solvents are considered green, non-toxic solvents comprising of a hydrogen bond acceptor (HBA) and a hydrogen bond donor (HBD) at varied molar concentrations. During the pretreatment process, a new H-bond is formed between the hydroxyl group present in lignin and the carbohydrate complex. This results in hydrolysis of lignin-carbohydrate complex (LCC) thereby causing cleavage of ether bonds between lignin and hemicellulose. Combined pretreatment studies have been successful in terms of product recovery at multiple stages. For example, studies have shown that hot water and deep eutectic solvents have resulted in hemicellulose and lignin recovery from LCB [8].

Also, pretreatment methods have been designed to minimize the loss of polysaccharides and the formation of sugar derivatives from the substrate. This helps in obtaining near theoretical sugar yield from the biomass and decreases enzyme load to be used in the enzymatic hydrolysis step. However, until now pretreatment has proven to be challenging and a costly step with only a few reaching the commercial scale. An effective pretreatment method plays a prime role in increasing the digestibility of LCB. Pretreatment technology has proven crucial for decreasing the cost of production involved and making the overall process more sustainable. Table 1 highlights key features of some of the major pretreatment types studied at lab, pilot and full-level scales.

The subsequent step that follows after pretreatment is hydrolysis. Hydrolysis is required for the conversion of polymeric carbohydrates such as starch, cellulose and hemicellulose into reducing/fermentable sugars, since yeast cannot directly utilize carbohydrates as substrate. Enzymes can specifically hydrolyze both cellulose and hemicellulose polymers in order to produce soluble sugars at low temperatures (45-50 °C). Nowadays "*de-novo* enzyme designing models" and a series of "omics" based approaches (metatranscriptomics, metagenomics, metaproteomics and metabolomics) are serving as enzyme discovery tools that aid in designing highly useful enzyme cocktails. Recombinant DNA technology and site-directed mutagenesis have also helped in the advancement of the enzyme development sector [1]. Efficient bioconversion of carbohydrates present in the lignocellulosic residue into bioethanol can be attained by optimizing certain operational parameters of the hydrolysis process, such as biomass loading, additives in the hydrolysate medium, enzyme loading, shaking speed and the incubation time.

After enzymatic hydrolysis the next step is fermentation. It is the process of converting glucose (C_6) and other reducing sugars (obtained from the cellulosic/lignocellulosic biomass) using microorganisms such as bacteria or yeast in order to produce bioethanol. Some yeasts and bacteria can metabolize monosaccharides, such as glucose and fructose as well as disaccharides, such as sucrose and maltose (in the absence of oxygen) resulting in the production of bioethanol (C_2H_5OH) and carbon dioxide (CO_2). Efficient utilization of glucose and xylose using fermentation is one of the bottlenecks in the efficient production of bioethanol from LB [17]. The process of Direct Microbial Conversion (DMC) or Consolidated Bioprocessing (CBP) encompasses all the reactions that are needed for the transformation of LCB into bioethanol. This approach varies from other bioconversion studies as it makes use of only a single microbial community that can effectively carry out both production of enzymes and fermentation process (*i.e.* enzyme production; enzymatic hydrolysis as well as fermentation) in a single step. This process holds many advantages since no capital is required for enzyme production and complete substrate utilization takes place for the optimal release of reducing sugars.

2 Bioconversion of lignocellulosic biomass waste into biofuels and other valuable by-products

A wide range of established pretreatment technologies along with abundant availability of LCB waste is a boon that can be used to suffice the energy demands of mankind. LCB can be converted *via* physical, chemical, or biological processes for generating energy and a wide range of products (lactic acid, xylitol, succinic acid, itaconic acid, furfural, *etc.*) at a competitive price. Around 50

Table 1 Comparative assessment of key parameters of various pretreatment methods

Pretreatment Type	Operational conditions	Advantages	Disadvantages	Developmental stage
Milling/Mechanical Comminution	1. Particle size: ~0.003-30 mm	1. Decreases cellulose crystallinity 2. Does not produce inhibitory compounds	1. Requires high energy and power [9,10]	Commercial [11]
Freezing	1. Freezing temperature = ~ -20 °C	1. Utilizes negligible quantity of chemicals 2. Decreased negative effects on the environment	1. Expensive [12]	Laboratory scale
Alkaline	1. Temperature range= 15-210 °C 2. Alkali concentration= 0.5–15% 3. Solid to liquid ratio-1:5–1: 20	1. Low inhibitor formation 2. High xylose and glucose yield 3. Improves accessible surface area 4. Removes hemicelluloses and lignin content	1. The residence time required is high 2. Salts formed are irrevocable and incorporated into the biomass 3. High percentage of water is required for salt removal	Commercial [13]
Acidic	1. Temperature= 130–210 °C 2. Acid=0.2–2.5% (w/w)	1. Removes hemicellulose 2. Yields high glucose content 3. Production of xylose without the use of enzymes	1. Solvents need to be recycled 2. Corrosion issues 3. Inhibitor formation 4. Formation of degradation products	Commercial [13]
Steam explosion	1. Temperature=160–260 °C 2. Pressure=0.7–5 MPa	1. Hemicellulose solubilization and transformation of lignin 2. Higher glucose yield 3. Cost-effective	1. Toxic compound generation 2. Partial degradation of hemicellulose	Commercial [13]
Ammonia Fiber Explosion (AFEX)	For AFEX 1. Weight of ammonia=1-2 kg 2. Temperature=60–100 °C 3. Time=5–30 min For ARP (Ammonia Recycled Percolation) 1. 10–15 wt% aqueous ammonia 2. Temperature=150–170 °C 3. Time=10–12 min	1. Increased accessible surface area 2. Inhibitor concentration formed is low	1. Not efficient for biomass containing high lignin content 2. The cost of ammonia is high	Commercial: Method operationalized to DuPont's cellulosic ethanol biorefinery in the US [13,14]
Biological	1. Depends on the type of fungus, bacteria or enzyme used or their combination (crude or concentrated)	1. Degrades hemicellulose and lignin 2. Consumption of energy is low	1. Rate of hydrolysis/biodegradation is low 2. Costly due to extended pretreatment time [15]	Laboratory scale
Liquid hot water	1. Temperature= 160-220 °C 2. Time= Minutes to hours	1. Chemicals are not required therefore no need for neutralization making this process cost-efficient	1. Water consumption and energy required are high 2. The rise in temperature leads to the generation of inhibitory compounds	Commercial scale [13]

Deep Eutectic Solvents	1. Temperature=60–200 °C 2. Time=0.5–24 h 3. Biomass loading =1:5–1:3	1. Biodegradable 2. Cheap 3. Synthesized easily 4. Biocompatible	1. Economic viability is to be maintained by recycling/reusing the chemicals	Laboratory scale [16]
Organosolv treatment	1. Temperature= 180–200 °C 2. Biomass loading=4:1–10:1 3. pH =2–4 4. Time=30–90 min	1. Disruption of hemicellulose and lignin bonds	1. Expensive process 2. Removal of solvents is necessary	Laboratory scale
Combined pre-treatment	The operational conditions affect the pretreatment process	Usually more effective in comparison to other pretreatment methods	The difficulties vary according to the combination type	-

billion kilograms of total renewable biochemical (apart from biofuels) are produced every year on a global scale [14]. The main aim is to focus on progress in pretreatment technologies paving the way toward improved biofuel and biogas production (Figure 1).

2.1 Production of Biogas

A considerable increase in biogas (methane) production has been reported by anaerobic digestion of LCB after the pretreatment process. 6% w/w NaOH pretreatment of *Asparagus* stem yielded 242.3 (mL/g VS) biogas after 18 days of digestion in comparison to that produced by untreated biomass substrate (175.1 mL/g VS). In another study, the digestion time of corn straw was reduced by 34.6% using yeast and cellulolytic bacteria [18]. The induction of an upgraded biogas system into LCB biorefinery will be able to promote a green economy and induce environmental susceptibility.

2.2 Liquid biofuel production

Liquid biofuels like bioethanol, biobutanol and biodiesel can be generated from LCB. A wide range of research and studies have been carried out to enhance the production of biofuels primarily targeting pretreatment techniques. An ionic liquid [Emim][OAc] pretreatment of spruce and oak sawdust (forest residues) at a low temperature of 45 °C resulted in approximately 2.5-4 times increase in the ethanol yield as compared to the native samples [19]. A few notable LCB biorefineries include the Raizen plant in Brazil and the POET-DSM Advanced Biofuels plant in South Dakota. They generate 8 ml/yr and 80 ml/year of bioethanol from bagasse and corn stover, respectively, by fermentation [20].

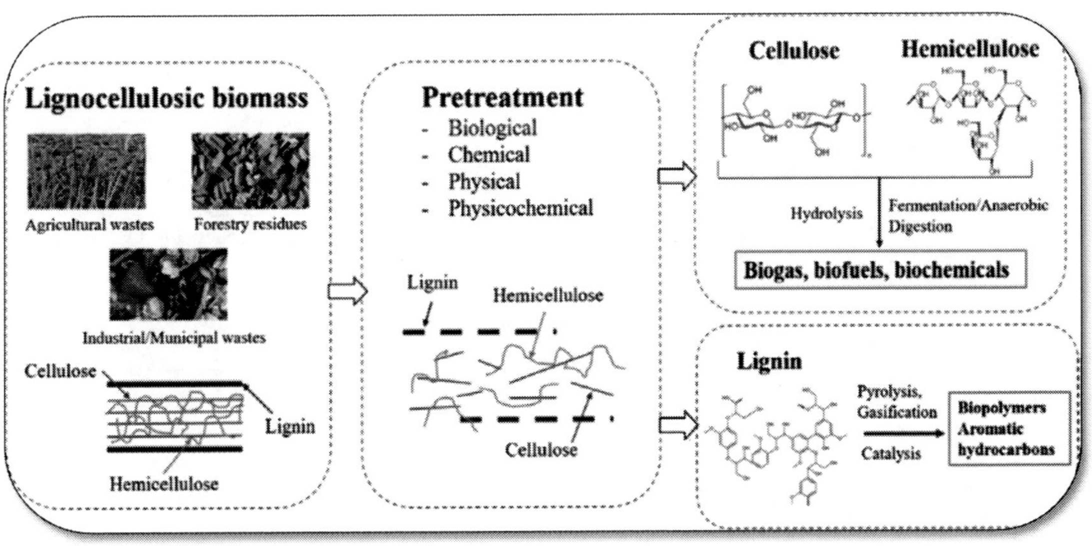

Figure 1. Process of bioconversion of lignocellulosic biomass residues into bio-chemicals and bio-products *via* pretreatment [18]

2.3 Biochemicals and Bioproducts

LCB can be used to generate organic acids (lactic acid, acetic acid), essential chemicals (phenolic compounds, phenyl propanoids), enzymes (pectinase, amylase, and cellulase), and chemicals for biopolymer production. USA and Asia Pacific currently dominate the lactic acid market. The current valuation of lactic acid market (a biochemical) was placed at 0.69 billion USD in 2017 and by the year 2023, it is estimated to reach 2.09 billion USD. Similarly, the production of xylitol, which finds application in chewing gum industry is estimated to reach 130 million USD by the year 2025. In the case of vanillin, the global market is expected to cross 700 million USD approximately [1]. These biobased compounds can be produced using the process of Separate Hydrolysis and Fermentation (SHF) and Simultaneous Saccharification and Fermentation (SSF). The recalcitrance of LCB does play a major role in affecting the production of these bioprocesses as they do in the case of biofuels. This again calls for the development of a suitable pretreatment process keeping in view that the bio-based market is anticipated to remain around 50 billion EURO by the year 2030 [21].

3 Research gaps and future research directions

Due to wide variability in the composition of LCB in our surroundings, it becomes really difficult to optimize a specific type of pretreatment process for optimum recovery of biofuels and other by-products. It varies with the biomass type and the desired end-product. However; these factors are intertwined with the need for environmental sustainability and cost-effectiveness making the process more challenging. Efforts directed at developing advancements in recycling/recovering of enzymes, chemicals and by-products from pretreated hydrolysate could greatly improve cost benefits. In addition, the effects of removal of inhibitors such as acetic acid, furfurals, hydroxymethyl furfurals (HMFs) and phenolics could further improve process economics. Advancement in the detoxification process can also facilitate the commercialization of LCB biorefinery. Combining different pretreatment techniques has shown promising results by synergizing their effects in terms of sugar yields, biofuel production and many various by-products. However, the cost of equipments and chemicals needs to be regulated.

4 Conclusions and future recommendations

The depletion of petroleum reserves and rising energy demands have shifted our research focus towards the potential of LCB waste for the generation of biofuels, biochemicals, biopolymers, food/feed supplements and bioactive compounds *etc*. It is clear that the use of LCB waste can be an optimal solution to our energy demands. Pretreatment being a vital step still requires advancement in terms of cost-cutting, sugar yields and therefore overall process feasibility. Many pretreatment methods consume a large amount of water, energy, chemicals and equipments in addition to release of inhibitory and harmful waste into the environment. This needs to be thoroughly addressed. LCB-based biorefineries could be the 'Zero-waste' based biomass biorefineries but they face a series of challenges, which hinder their advancement towards the production of a sustainable and eco-friendly economy. Full-scale conversion of LCB into fuel, chemicals and by-products is challenged by factors such as sustainability, policy initiatives, social, socio-economic issues and market challenges. On the part of industries and research sectors, a collaborative approach is the need of the hour that can address all these challenges in a holistic way. Suitable integration of pretreatment methods with minimum input of resources and energy will contribute to the complete commercialization of 'zero-waste' biorefinery processes using LCB.

References

[1] N. Singh, R.R. Singhania, P.S. Nigam, C.D. Dong, A.K. Patel, M. Puri, Global status of lignocellulosic biorefinery: Challenges and perspectives, Bioresource Technology, 344 (2022):126415.

[2] N. Raina, P.S. Slathia, P. Sharma, Experimental optimization of thermochemical pretreatment of sal (Shorea robusta) sawdust by Central Composite Design study for bioethanol production by co-fermentation using *Saccharomyces cerevisiae* (MTCC-36) and Pichia stipitis (NCIM-3498), Biomass and Bioenergy, 143(2020): 105819.

[3] N. Raina, P.S. Slathia, P. Sharma, Response surface methodology (RSM) for optimization of thermochemical pretreatment method and enzymatic hydrolysis of deodar sawdust (DS) for bioethanol production using separate hydrolysis and co-fermentation (SHCF), Biomass Conversion and Biorefinery (2020): 1-21.

[4] S. Sachdeva, V.K. Garg, N.K. Labhsetwar, Anita Singh, K.N. Yogalakshmi, Zero Waste Biorefinery: A Comprehensive Outlook, In Zero Waste Biorefinery, Springer, Singapore (2022):3-22.

[5] P. Jusakulvijit, A. Bezama D. Thrän, The Availability and Assessment of Potential Agricultural Residues for the Regional Development of Second-Generation Bioethanol in Thailand, Waste and Biomass Valorization 12(11) (2021):6091-6118.

[6] M. Hidayat, N.A. Aqilah, Adi Winata, Pretreatment of Oil Palm Empty Fruit Bunch Using Caustic Soda Solution for Lignin Isolation,

Journal of Applied Science and Engineering, 25(6) (2022):1025-1030.
[7] W. Rodiahwati, M. Sriariyanun, Lignocellulosic Biomass to Biofuel Production: Integration of Chemical and Extrusion (Screw Press) Pretreatment, King Mongkut's University of Technology North Bangkok International Journal of Applied Science and Technology, 9(4) (2016):289-298.
[8] M.P. Gundupalli, M. Sriariyanun, Recent Trends and Updates for Chemical Pretreatment of Lignocellulosic Biomass, Applied Science and Engineering Progress 16(1) (2023):5842.
[9] M. Raud, K. Orupõld, L. Rocha-Meneses, V. Rooni, O. Träss and T. Kikas, Biomass Pretreatment with the Szego Mill™ for Bioethanol and Biogas Production. Processes, 8(10) (2020):1327.
[10] H. Chandel, P. Kumar, A.K. Chandel, M.L. Verma, Biotechnological advances in biomass pretreatment for bio - renewable production through nanotechnological intervention, Biomass Conversion and Biorefinery (2022):1-23.
[11] J. Park, B. Jones, B. Koo, X. Chen, M. Tucker, J.H. Yu, T. Pschorn, R. Venditti, S. Park, Use of mechanical refining to improve the production of low-cost sugars from lignocellulosic biomass. Bioresource technology, 199 (2016):59-67.
[12] A. Devi, A. Singh, S. Bajar, D. Pant, Z.U. Din, Ethanol from lignocellulosic biomass: An in-depth analysis of pre-treatment methods, fermentation approaches and detoxification processes, Journal of Environmental Chemical Engineering, 9(5) (2021):105798.
[13] B. Kumar, N. Bhardwaj, K. Agrawal, V. Chaturvedi, P. Verma, Current perspective on pretreatment technologies using lignocellulosic biomass: An emerging biorefinery concept. Fuel Processing Technology, 199 (2020):106244.
[14] Amisha Patel, Amita R. Shah. Integrated lignocellulosic biorefinery: Gateway for production of second generation ethanol and value added products, Journal of Bioresources and Bioproducts, 6(2) (2021): 108-128.
[15] S. Rezania, B. Oryani, J. Cho, A. Talaiekhozani, F. Sabbagh, B. Hashemi, P.F. Rupani, A.A. Mohammadi, Different pretreatment technologies of lignocellulosic biomass for bioethanol production: An overview, Energy, 199 (2020):117457.
[16] A. Satlewal, R. Agrawal, S. Bhagia, J. Sangoro, A.J. Ragauskas, Natutal deep eutectic solvents for lignocellulosic biomass pretreatment: Recent developments, challenges and novel opportunities, Biotechnology advances, 36(8) (2018):2032-2050.
[17] F. Wirawan, C.L. Cheng, Y.C. Lo, C.Y. Chen, J.S. Chang, S.Y. Leu, D.J. Lee, Continuous cellulosic bioethanol co-fermentation by immobilized Zymomonas mobilis and suspended Pichia stipitis in a two-stage process, Applied Energy, 266 (2020):114871.
[18] H.P. Vu, L.N. Nguyen, M.T. Vu, M.A.H. Johir, R. McLaughlan, L.D. Nghiem, A comprehensive review on the framework to valorize lignocellulosic biomass as biorefinery feedstocks, Science of the Total Environment, 743 (2020):140630.
[19] R. Alayoubi, N. Mehmood, E. Husson, A. Kouzayha, M. Tabcheh, L. Chaveriat, C.Sarazin, I. Gosselin, Low temperature ionic liquid pretreatment, of lignocellulosic biomass to enhance bioethanol yield, Renewable Energy, 145 (2020):1808-1816.
[20] GHD. Biofuels and transport: an Australian opportunity: Australian Renewable Energy Agency and Clean Energy Finance Corporation 2019.
[21] S.S. Hassan, G.A. Williams, A.K. Jaiswal, Moving towards the second generation of lignocellulosic biorefineries in the EU: Drivers, challenges, and opportunities, Renewable and Sustainable Energy Reviews, 101 (2019):590.

The combined effect of inorganic salt and ionic liquid in pretreatment on enzymatic saccharification of rice straw

Diana Jose[1], *Neelu* Raina[2], *Rajagopal* Deepakkumar[3], *Elizabeth Jayex* Panakkal[1], *Malinee* Sriariyanun[1], and *Tawiwan* Kangsadan[1,*]

[1]Biorefinery and Process Automation Engineering Center, Department of Chemical and Process Engineering, TGGS, King Mongkut's University of Technology North Bangkok, 10800 Bangkok, Thailand
[2]School of Biotechnology, Shri Mata Vaishno Devi University, Katra, J&K-182320, India
[3]Mechanical Engineering, Vellore Institute of Technology, Vellore, Tamil Nadu, India

Abstract. The pretreatment method is one of the challenging steps in the production of biofuel through the biorefinery process that unlocks the recalcitrant nature of lignocellulosic biomass. Ionic liquid pretreatment gained attention for being highly effective to improve the enzymatic saccharification of the biomass, however its high cost hinders its industrial application. In this study, the combined effect of ionic liquid 1-ethyl-3-methylimidazolium acetate (EMIM-Ac) with inorganic salts (NaCl and KCl) was used for the pretreatment of rice straw. Optimization of pretreatment was conducted based on Response Surface Methodology and sugar yields obtained by EMIM-Ac+NaCl (160 °C, 88.7 min, 7.6%wt) and EMIM-Ac+KCl (160 °C, 68.2 min, 12.5%wt) were 670.7 and 392.9 mg/g-biomass, respectively. The effect of combined pretreatment on ethanol production was analyzed after 48h fermentation. The results showed that the ethanol yield from pretreated samples with EMIM-Ac+NaCl (0.72%) and EMIM+KCl (0.76%) was increased by 2.18 and 2.25 fold times, respectively, compared to untreated sample (0.33%). This combined effect of inorganic salts and ionic liquid significantly removed the lignin during pretreatment, while maintaining efficient enzymatic saccharification of rice straw. Thus, this cost-effective combined chemical method may be an alternative strategy for increasing cellulosic ethanol production.

Keyword. Biorefinery, Enzymatic saccharification, Inorganic salt, Ionic liquid, Lignocellulose, Pretreatment

1 Introduction

Due to intensive awareness of global society on the global warming effects, the concept of Bioeconomy, Circular economy and Green economy (BCG economy) has been implemented for manufacturing and industrialization. The use of fossil fuels as raw materials for the production of fuels and chemicals can be substituted by using various types of biomass. Lignocellulose biomass is acknowledged as the most abundant biomass on earth. After the harvesting season, lignocellulose biomass in agricultural waste is burned on fields to clear the land for the next season and air pollution was uncontrolled released into the environment, especially PM2.5 and PM10 [1]. Therefore, it is important to utilize lignocellulose biomass in the controllable system to minimize the negative impact to the environment and it has the potential to produce value-added products (e.g. biofuels, biochemicals, platform chemicals) to gain more benefit to the process [2-5].

To efficiently convert the lignocellulose biomass to biofuels, a multi-step process, including pretreatment, hydrolysis, fermentation and product recovery/separation, has to be conducted [6]. Pretreatment has been proved in many studies to be critical to determine the success of the process, in terms of efficiency and economy by improvement of enzymatic hydrolysis. The functions of pretreatment include i) modifying cellulose fibril to be more vulnerable to enzymatic saccharification, ii) increasing surface area to allow enzyme accessibility, iii) removing hemicellulose and lignin content [7]. Pretreatment methods are mainly classified to be chemical, physical and biological methods. The selection of the appropriate pretreatment method for each type of biomass requires research and preliminary investigation to maximize the process efficiency with less complication and investment.

Among various types of chemical pretreatments, ionic liquids (ILs) have been demonstrated to be green solvents with high efficiency for the improvement of enzymatic saccharifications of lignocellulose biomass [8]. IL is a salt composed of cation and anion, and it mostly has a liquid state at room temperature or below 100 °C. ILs have various properties that are suitable for the biorefining process, such as high polarity, low volatility, low vapor pressure, heat stability, low melting point, and recyclability. The most common ILs derivatives used in lignocellulose pretreatment are imidazolium, choline, and ammonium, pyridinium, etc. [9]. ILs function in fractionation and solubilization of cellulose and lignin, as well as modification of cellulose arrangement to promote the enzyme accessibility to

* Corresponding author: tawiwan.k@tggs.kmutnb.ac.th

biomass during hydrolysis reaction [10-12]. Although IL is an efficient chemical for pretreatment, it has a high cost, which makes it infeasible for industrial applications.

This study, the inorganic salts were formulated with an IL, 1-ethyl-3-methylimidazolium acetate (EMIM-Ac), to reduce the fraction of IL for pretreatment and the cost. Previously, inorganic salts, NaCl, KCl, $MgCl_2$, $NaNO_3$, KNO_3, $Mg(NO_3)_2$, were used in the pretreatment of rice straw biomass, and the yields of reducing sugars were improved for more than 4.7 times when using NaCl and KCl [13]. The optimization of pretreatment with mixed EMIM-Ac and NaCl and KCl was conducted based on Response Surface Methodology (RSM) to maximize the sugar yields obtained from cellulase saccharification. The hydrolysates were then converted to ethanol to assess the potential of biomass for biofuel production.

2 Materials and methods

2.1. Material preparation and chemicals

Rice straw samples were harvested after the harvesting season from a local rice field located in the central part of Thailand. The size of the sample was reduced by hand cutting to about 1-cm long pieces and dried in a hot air oven at 60 °C until the dried weight was obtained. To obtain a consistent sample size, dried rice straw was ground by blender and sieved through a 20-mesh aluminium sieve. The prepared biomass was kept in a sealed plastic bag until used. EMIM-Ac and CelluClast 1.5L were purchased from Sigma-Aldrich. Other chemicals, including inorganic salts, DNS, and glucose were purchased Ajax Finechem.

2.2. Optimization of pretreatment

The EMIM-Ac and inorganic salts (NaCl and KCl) were pre-mixed with the ratio of 1:10 by weight. The pretreatment experiment was designed based on Box-Behnken by using Design Expert Software (Version 7.0.0, Stat-Ease) [14-15] to target to vary the three pretreatment factors, including pretreatment temperature (X_1: 100-160 °C), pretreatment time (X_2: 60-180 min) and biomass loading ratio (X_3: 5-15%wt). A total of 17 experimental runs (Table 1) were designed and each pretreatment run was conducted. Then, each pretreated sample was enzymatic hydrolyzed and the reducing sugar yield was analyzed. The influences of pretreatment factors and interacting factors were evaluated based on ANOVA. The mathematical model to represent the correlation of pretreatment factors and sugar yields was constructed and the reliability of the model was assessed based on the R^2 value and p-value.

To pretreat the rice straw sample, the pre-mixed solution (EMIM-Ac and inorganic salts) was added to rice straw in different ratios of biomass loading (5-15%wt). Then, the mixture was placed in a hot air oven at the targeted pretreatment temperature and time. After pretreatment, the sample was cooled down to room temperature. Then, the solid fraction of the sample was collected by centrifugation at 8,000 xg for 10 min. The pretreated sample was washed with distilled water three times to remove IL and salt residues. The washed solid sample was subjected to enzymatic saccharification.

2.3. Enzymatic saccharification

The pretreated solid sample was loaded at the loading ratio of 2.5%wt in 4 ml of hydrolysis buffer (containing 20 FPU/g-biomass, 40 ul of 2% sodium azide in 50 mM sodium citrate buffer, pH 4.8). The mixture was placed in a shaking incubator at 45 °C, 200 rpm, for 72 h. After 72 h, the progress of the reaction was stopped by heating in the waterbath at 100 °C for 5 min. The supernatant fraction was collected from rice straw hydrolysate by centrifugation at 8,000 xg for 10 min [16-17]. The reducing sugar concentration in the hydrolysate was measured by using a DNS assay. All measurements were repeated three times.

2.4. Ethanol fermentation

The liquid fraction of hydrolysate (without the addition of sodium azide) was subjected to fermentation for ethanol production using yeast, *Saccharomyces cerevisiae* TISTR 5606 [18-19]. For each experiment, 5% of yeast inoculum (at the concentration of about 10^8 cells/ml) was added to 19 ml of rice straw hydrolysate supplemented with 1%w/v glucose and 1% w/v yeast extract [18, 19]. The sample was incubated in a shaking incubator at 30 °C, 150 rpm for 72 h. Then, the supernatant fraction of yeast culture was separated by centrifugation at 8,000 xg for 10 min. The ethanol concentration in the supernatant fraction was measured by using our published protocol based on spectrophotometry [20]. All measurements were conducted in triplicates to confirm the reliability of the analysis.

2.5. FT-IR analysis

The chemical modification of rice straw by IL-inorganic salt pretreatment was assessed by using Fourier transform infrared (FT-IR) spectrometry. The pretreated and untreated samples were finely ground before analysis. The absorption spectra were scanned with a resolution of 4 cm^{-1} in the range of 400–1800 cm^{-1}. The spectral data were developed by Spectrum 2.00 software (Perkin Elmer, USA).

3 Results and discussion

3.1. Optimization of pretreatment by EMIM-Ac and inorganic salts

The pretreatment experiments were designed based on RSM by varying 3 pretreatment factors, including pretreatment temperature, pretreatment time and biomass loading ratio. After pretreatment, biomass was enzymatically hydrolyzed and reducing sugars were quantitated. In this study, the ranges of reducing sugar yields were between 86.7 (run 10) and 526.2 (run 7) mg/g-biomass, which was about 6 fold time difference (Table 1). This range suggested the importance of optimization to improve the pretreatment efficiency.

The effects of each pretreatment factor and interacting factor on reducing sugar yields were analyzed

https://doi.org/10.1051/e3sconf/202235501002

Table 1. Experimental design of pretreatment by EMIM-Ac and inorganic salts based on RSM.

Run	Pretreatment condition			Reducing sugar yield (mg/g-biomass)	
	Temperature (°C)	Time (min)	Biomass loading (%wt)	EMIM-Ac+NaCl	EMIM-Ac+KCl
1	130	120	10	407.1	283.3
2	160	60	10	758.2	451.0
3	160	120	15	761.9	421.8
4	130	180	15	420.8	214.3
5	130	120	10	378.2	203.0
6	160	120	5	618.3	237.4
7	160	180	10	663.7	526.2
8	130	60	15	282.3	142.4
9	130	120	10	490.7	212.4
10	100	120	15	277.1	84.8
11	130	120	10	387.7	151.4
12	130	180	5	461.0	265.8
13	100	60	10	171.4	86.7
14	100	180	10	191.6	117.4
15	130	60	5	300.8	222.3
16	100	120	5	183.1	128.8
17	130	120	10	448.6	227.0

based on ANOVA analysis and mathematical models were generated. The model significance levels of both EMIM-Ac+NaCl and EMIM-Ac+KCl pretreatments were calculated to be P-value <0.0001. The correlation efficiency (R^2) values are 0.8864 and 0.8811, respectively, advocating the reliability of the generated models [21]. The mathematic models obtained from both EMIM-Ac+NaCl and EMIM-Ac+KCl were presented as shown in eq 1 and eq 2, respectively.

Reducing sugar yield (mg/0.1g-biomass) = -8.10267 + 0.10307 x temperature (eq 1)

Reducing sugar yield (mg/0.1g-biomass) = -5.32690 + 0.063467 x temperature (eq 2)

The impacts of pretreatment factors on reducing sugar yields were visualized in 3D models to observe the impacts of two pretreatment factors at a time (Figure 1). It could be seen that the pretreatment factor was the significant parameter for reducing sugars. The increasing of pretreatment temperature improves the reducing sugar yields. However, in this study, the pretreatment time and biomass loading ratio were not significant factors influencing sugar yields, which may be the result of an inappropriate range of pretreatment factors.

Based on the RSM model, the optimal pretreatment conditions to maximize the sugar yields were predicted for both EMIM-Ac+NaCl (160 °C, 88.7 min, 7.6%wt) and EMIM-Ac+KCl (160 °C, 68.2 min, 12.5%wt) and the optimal sugar yields were predicted to be 670.7, and 392.9 mg/g-biomass, respectively (Table 2). To verify the reliability of the predicted mathematic models obtained from RSM, the pretreatments at optimal conditions were conducted and the reducing sugar yields were measured and compared with the yields obtained from the predicted model. The sugar yields obtained from the pretreatment by EMIM-Ac+NaCl and EMIM-Ac+KCl at the optimal condition were 670.7 and 392.9 mg/g-biomass, respectively (Table 2), which were only 0.04% and 1.73% difference compared to the predicted

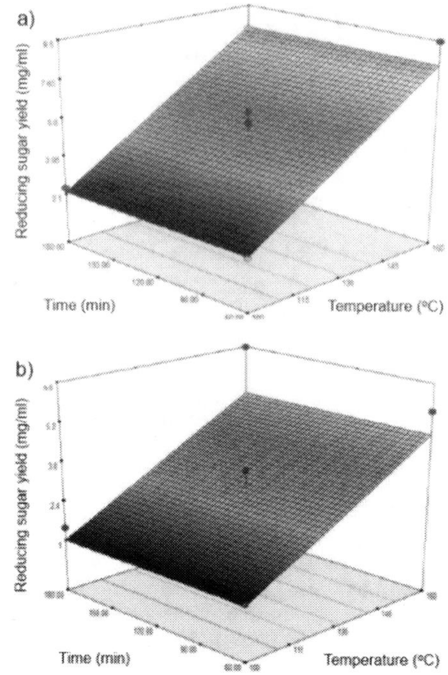

Figure 1. 3D models representing the correlation of pretreatment factors on reducing sugar obtained from a) EMIM-Ac+NaCl pretreated sample and b) EMIM-Ac+KCl pretreated sample.

yields. This result proves the validity of the RSM model of pretreatment. It should be noted that both pretreatments improved the reducing sugar yields by 2.24 and 1.32 fold times, compared to the untreated sample (298.1 mg/g-biomass).

Table 2. Predicted pretreatment conditions and reducing sugar yields obtained from the RSM model.

Chemical	Pretreatment condition			Reducing sugar yield (mg/g-biomass)	
	Temperature (°C)	Time (min)	Biomass loading (%wt)	Predicted value	Experimental value
EMIM-Ac-NaCl	160	88.7	7.6	671.0	670.7
EMIM-Ac-KCl	160	68.2	12.5	386.2	392.9

3.2. Microscopic chemical analysis of pretreated rice straw

To evaluate the effect of pretreatment on lignocellulose biomass, the arrangements of chemical bonds in biomass were analyzed by using FT-IR spectrometry. The spectra peaks that are corresponding to the chemical bonds in cellulose, hemicellulose and lignin and the linkages between these compounds were selectively analyzed (Figure 2). The spectral peaks at 900 cm^{-1} and 1060 cm^{-1} representing β-glucosidic bond and C=O stretching bond in cellulose fibrils were increased in pretreated samples, suggesting enriched cellulose contents [22]. The height of the peak at 1315 cm^{-1}, assigned intermolecular hydrogen bond in amorphous cellulose, was also increased in pretreated samples, suggesting the enhancement of enzymatic saccharification. On the other hand, the peak height at 1627 cm^{-1}, representing C=C bond in an aromatic skeleton of lignin [23-24], was reduced in pretreated samples, suggesting the removal of lignin from biomass during pretreatment.

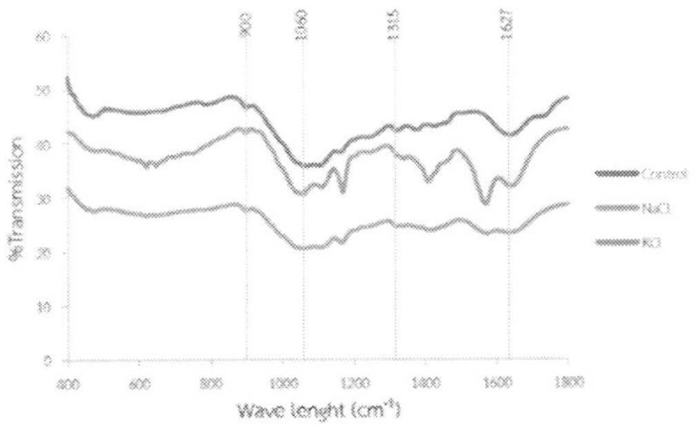

Figure 2. FT-IR spectra observed from untreated and pretreated rice straw

3.3. Effect of pretreatment by combined EMIM-Ac and inorganic salts on ethanol production

To evaluate the efficiency of pretreatment of lignocellulose for application in the biorefining process, the hydrolysates obtained from untreated and pretreated biomass were converted to ethanol by fermentation with S. cerevisiae TISTR 5606. Within 48 h fermentation time, the ethanol concentrations in fermented hydrolysates were analyzed. The results showed that the ethanol yields at 48 h obtained from pretreated samples by EMIM-Ac+NaCl (0.72%) and EMIM+KCl (0.76%) were increased by 2.18 and 2.25 fold times, respectively, compared to untreated sample (0.33%) (Figure 3).

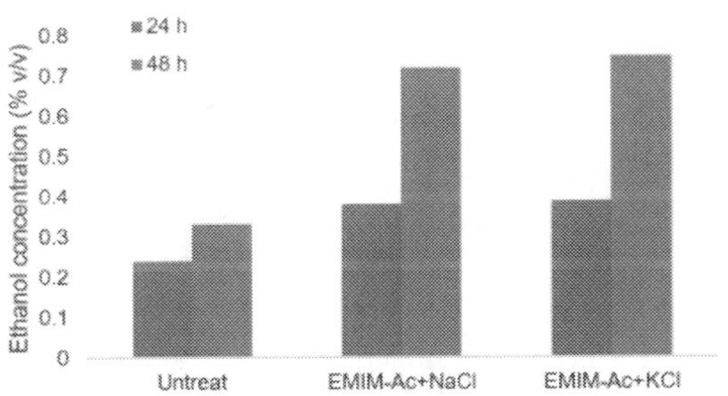

Figure 3. Ethanol production from untreated and pretreated rice straw by *S.cerevisiae* TISTR5606

It should be noted that the increases in ethanol yields by EMIM-Ac+NaCl and EMIM-Ac+KCl were similar, while the increase in sugar yield caused by EMIM-Ac+NaCl pretreatment was higher compared to EMIM-Ac+KCl (Table 2). This trend for the improvement of sugar yield production by NaCl and KCl was also demonstrated in the previous study that NaCl was the better chemical for rice straw pretreatment compared to KCl [13]. Here, when these two inorganic salts were formulated with EMIM-Ac, a similar effect on sugar yield was observed. It could be hypothesized that Na^+ has high ionic strength compared to K^+, which could interfere with the arrangements of intermolecular bonding of cellulose fibrils and improve the disintegration of cellulose during saccharification. However, although the sugar yield obtained from EMIM-Ac+NaCl pretreated sample was higher than EMIM-Ac+KCl pretreated sample, the ethanol yields of both samples were similar. In previous studies, the inhibitory effects of NaCl and KCl on cellulase enzyme were observed during enzymatic saccharification [25]. Testing with the same concentration, the enzyme activities were reduced with a higher degree by KCl compared to NaCl. So, it is a plausible clause for ethanol production trend observed in this study.

4 Conclusion

The synergistic effect of inorganic salt and IL on the enzymatic saccharification of rice straw was studied. Optimization study using RSM generated higher yields of reducing sugars using rice straw biomass. Increased intermolecular H-bonding and reduction in the peak positions (C=C) in regions representing aromatic lignin structures suggest decreased recalcitrance by lignocellulosic biomass structure. Na^+ ions having higher ionic strength than K^+ gave higher reducing sugar yields using EMIM-Ac+NaCl in comparison to that obtained by pretreatment using EMIM-Ac+KCl. Ethanol yield of pretreated samples increased by using EMIM-Ac+NaCl (0.72%) and EMIM+KCl (0.76%) by 2.18 and 2.25 fold, respectively. The present work thus suggests a combined pretreatment study targeted at an enhanced yield of reducing sugars and ethanol using rice straw.

Acknowledgements

The authors would like to thank King Mongkut's University of Technology North Bangkok (Grant Contract No. KMUTNB-FF-65-37, KMUTNB-PHD-66-03) for financial support during this work.

References

1. T. Ruensodsai, M. Sriariyanun, Sustainable Development and Progress of Lignocellulose Conversion to Platform Chemicals, The Journal of King Mongkut's University of Technology North Bangkok, 32(4) (2022):2465-4698
2. Y.S. Cheng, P. Mutrakulcharoen, S. Chuetor, K. Cheenkachorn, P. Tantayotai, E.J. Panakkal, M. Sriariyanun, Recent situation and progress in biorefining process of lignocellulosic biomass: Toward green economy, Applied Science and Engineering Progress, 13(4) (2020):299-311.
3. P. Rachamontree, T. Douzou, K. Cheenkachorn, M. Sriariyanun, K. Rattanaporn, Furfural: A Sustainable Platform Chemical and Fuel, Applied Science and Engineering Progress, 13(1) (2020):3-10.
4. E.J. Panakkal, N. Kitiborwornkul, M. Sriariyanun, J. Ratanapoompinyo, P. Yasurin, S. Asavasanti, W.

Rodiahwati, P. Tantayotai, Production of Food Flavouring Agents by Enzymatic Reaction and Microbial Fermentation, Applied Science and Engineering Progress, 14(3) (2021):297-312.

5. M. Sriariyanun, J.H. Heitz, P. Yasurin, S. Asavasanti, P. Tantayotai, Itaconic Acid: A Promising and Sustainable Platform Chemical, Applied Science and Engineering Progress, 12(2) (2019):75-82.
6. M. Sriariyanun, P. Tantayotai, Y.S. Cheng, P. Pornwongthong, S. Chuetor, K. Cheenkachorn, Progress in Development of Biorefining Process: Toward Platform Chemical-Derived Polymeric Materials, Value-Added Biocomposites. CRC Press, (2021):189-212.
7. M.N.F. Norrahim, R.A. Ilyas, N.M. Nurazzi, M.S.A. Rani, M.S.N. Atikah, S.S. Shazleen, Chemical Pretreatment of Lignocellulosic Biomass for the Production of Bioproducts: An Overview, Applied Science and Engineering Progress, 14(4) (2021):588-605.
8. M.P. Gundupalli, K. Rattanaporn, S. Chuetor, W. Rodiahwati, M. Sriariyanun, Biocomposite Production from Ionic Liquids (IL)-Assisted Processes Using Biodegradable Biomass, Value-Added Biocomposites. CRC Press, (2021):213-249.
9. M.P. Gundupalli, K. Rattanaporn, S. Chuetor, W. Rodiahwati, M. Sriariyanun, Ionic liquid assisted pretreatment to improve cellulose fractionation of lignocellulosic biomass, Ionic Liquid-Based Technologies for Environmental Sustainability. Elsevier, (2022):75-99.
10. M. Sriariyanun, N. Kitiborwornkul, P. Tantayotai, K. Rattanaporn, P.L. Show, One-pot ionic liquid-mediated bioprocess for pretreatment and enzymatic hydrolysis of rice straw with ionic liquid-tolerance bacterial cellulose, Bioengineering, 9(1) (2022):17.
11. S. Chuetor, E.J. Panakkal, T. Ruensodsai, K. Cheenkachorn, S. Kirdponpattara, Y. S. Cheng, M. Sriariyanun, Improvement of Enzymatic Saccharification and Ethanol Production from Rice Straw Using Recycled Ionic Liquid: The Effect of Anti-Solvent Mixture, Bioengineering, 9(3) (2022): 115.
12. M.P. Gundupalli, K. Bano, T.K. Panda, M. Sriariyanun, D. Bhattacharyya, Understanding the effect of low-concentrated protic ionic liquids (PILs) on coconut (*Cocos nucifera*) residues, Biomass Conversion and Biorefinery, (2022):1-17.
13. P. Suwannabun, K. Cheenkachorn, M. Prongjit, A. Tawai, M. Sriariyanun, Pretreatment of rice straw by inorganic salts and 1-ethyl-3-methylimdazolium acetate for biofuel production, 2019 2nd Asia conference on energy and environment engineering (ACEEE) IEEE, (2019):12-15.
14. E.J. Panakkal, M. Sriariyanun, J. Ratanapoompinyo, P. Yasurin, K. Cheenkachorn, W. Rodiahwati, P. Tantayotai, Influence of sulfuric acid pretreatment and inhibitor of sugarcane bagasse on the production of fermentable sugar and ethanol, Applied Science and Engineering Progress, 15(1) (2022):5238.
15. E.J. Panakkal, . Cheenkachorn, M. Paulraj Gundupalli, N. Kitiborwornkul, M. Sriariyanun, Impact of sulfuric acid pretreatment of durian peel on the production of fermentable sugar and ethanol, Journal of the Indian Chemical Society, 98(12) (2021):100264.
16. K. Cheenkachorn, T. Douzou, S. Roddecha, P. Tantayotai, M. Sriariyanun, Enzymatic saccharification of rice straw under influence of recycled ionic liquid pretreatments, Energy Procedia, 100(2016):160-165.
17. P. Tantayotai, P. Rachmontree, W. Rodiahwati, K. Rattanaporn, M. Sriariyanun, Production of ionic liquid-tolerant cellulase produced by microbial consortium and its application in biofuel production, Energy Procedia, 100(2016):155-159.
18. M.P. Gundupalli, Y.S. Cheng, S. Chuetor, D. Bhattacharyya, M. Sriariyanun, Effect of dewaxing on saccharification and ethanol production from different lignocellulosic biomass, Bioresource Technology, 339(2021):125596.
19. M.P. Gundupalli, S. Chuetor, K. Cheenkachorn, K. Rattanaporn, P.L. Show, Y.S. Cheng, M. Sriariyanun, Interferences of Waxes on Enzymatic Saccharification and Ethanol Production from Lignocellulose Biomass, Bioengineering 8(11) (2021):171.
20. M. Sriariyanun, P. Mutrakulcharoen, S. Tepaamorndech, K. Rattanaporn, A rapid spectrophotometric method for quantitative determination of ethanol in fermentation products, Oriental Journal of Chemistry 35(2) (2019):744.
21. L.K. Akula, R.K. Oruganti, D. Bhattacharyya, K.K. Kurilla, Treatment of marigold flower processing wastewater using a sequential biological-electrochemical process, Applied Science and Engineering Progress, 14(3) (2021):525-542.
22. M.P. Gundupalli, S.T.A. Sahithi, E.P. Jayex, S. Asavasanti, P. Yasurin, Y.S. Cheng, M. Sriariyanun, Combined effect of hot water and deep eutectic solvent (DES) pretreatment on a lignocellulosic biomass mixture for improved saccharification efficiency, Bioresource Technology Reports, 17(2022):100986.
23. M.P. Gundupalli, P. Tantayotai, E. J. Panakkal, S. Chuetor, S. Kirdponpattara, A. S. S. Thomas, B. K. Sharma, M. Sriariyanun, Hydrothermal pretreatment optimization and deep eutectic solvent pretreatment of lignocellulosic biomass: An integrated approach, Bioresource Technology Reports, 17(2022):100957.
24. M.P. Gundupalli, ST A. Sahithi, Y. S. Cheng, P. Tantayotai, M. Sriariyanun, Differential effects of inorganic salts on cellulase kinetics in enzymatic saccharification of cellulose and lignocellulosic biomass, Bioprocess and Biosystems Engineering, 44(11) (2021):2331-2344.

Effects of oxygen and water vapour on adhesion of scale grown on ferritic stainless steels in solid oxide fuel cells

Thanasak Nilsonthi*

High Temperature Corrosion Research Centre and Department of Materials and Production Technology Engineering, Faculty of Engineering, King Mongkut's University of Technology North Bangkok, 1518, Pracharat 1 Road, Wongsawang, Bangsue, Bangkok, 10800, Thailand

Abstract. The adhesion of thermal oxide scale grown at 800°C on ferritic stainless steel AISI 441 proposed as SOFCs interconnectors was investigated. The influence of oxidising atmosphere was considered by 20% O_2-N_2 or 20% H_2O-N_2 as the representative cathode and anode atmospheres, respectively. A tensile test in an SEM chamber was used to assess the adhesion behaviour. The spallation of oxide scale on AISI 441 has to important as the candidate of interconnectors in SOFCs. The adhesion energy was considered as strain energy stored in the oxide, which was calculated at the first spallation of scale. The adhesion energy of scale on steel substrate was shown in the range of 173-214 J m^{-2}. The adhesion energy of scale on steel oxidised in water vapour was higher than that in oxygen. This might be due to the presence of a silica layer at the scale-steel interface of steel oxidised in water vapour. The qualitative assessment of scale adhesion in terms of strain initiating the first spallation suggested the good adhesion behaviour of scale grown in the anode atmosphere.

Keyword. SOFC interconnectors, Stainless steel, Oxide scale, Adhesion, Water vapour, Oxygen

1 Introduction

Among the types of fuel cells that are currently under development: proton-exchange membrane fuel cells (PEMFCs), phosphoric acid fuel cells (PAFCs), molten carbonate fuel cells (MCFCs), alkaline fuel cells (AFCs), and solid oxide fuel cells (SOFCs). The SOFCs are gaining increased attention. SOFCs have several advantages over other types of fuel cells due to their high operating temperature (600-1000°C) and solid materials. Since they are operated at high temperatures, the SOFCs have a better ability to tolerate the presence of impurities. Also the component is solid, there is no need for electrolyte loss maintenance and electrode corrosion is eliminated [1]. Solid oxide fuel cells are electrochemical conversion device that generates energy by oxidising a fuel. Solid oxide fuel cells use a solid oxide material as the electrolyte.

A solid oxide electrolyte is used in SOFCs to transfer negative oxygen ions from the cathode to the anode. When multiple cells are stacked in a series, an interconnector is used to connect the fuel-side anode and the air-side cathode. It also enables electron transmission from one cell to the next and the external circuit. Ceramic materials, such as doped lanthanum chromite have been proposed as interconnectors for standard SOFCs operating at approximately 1000°C [2-4]. The interconnector must be extremely stable because it is exposed to both the oxidising and reducing sides of the cell at high temperature. As interconnect materials, ceramics have shown to be more effective than metals. However, when compared to metals, these ceramic interconnect materials are quite expensive. As lower temperature (600-800°C) SOFCs are developed, steel-based alloys are becoming more promising. Stainless steels are considered promising interconnector materials [5-8]. Mostly because of their gas-tightness and high electrical conductivity.

In comparison to ceramic materials, economic advantages such as low cost, availability, and processability are also important. Ferritic materials are chosen from among the numerous varieties of stainless steel due to their low coefficient of thermal expansion. However, when stainless steel is used at SOFCs temperatures such as 800°C the oxide scale forms as chromia on the steel substrate [9-11]. This scale is semi-conductor and reduces the electrical conductivity [12, 13]. At high temperatures, the increasing oxide scale can cause spallation resulting in an air gap between the substrate and the oxide, which perform as an insulator [14]. This work has been interested in industrial ferritic grade (AISI 441). This steel has 18% Cr content and is suitable for usage in high temperature applications. Good adhesion is critical to avoid oxide-steel decohesion, which could operate as a local infinite electrical resistor inhibiting current flow through the cell. The scale formation and adhesion on metallic substrates are influenced not only by the metal chemical compositions [15, 16], but also by their roughness and polishing direction. The adhesion of oxide scale on metallic alloys is normally determined by observing the oxide-steel interface qualitatively or directly measuring spallation in service. Only a few testing methods, such as the indentation test [17], have been developed to quantify the interfacial fracture of scale, although they are not entirely precise. Alternatively, the inverted blister test [18, 19] can

* Corresponding author: thanasak.n@eng.kmutnb.ac.th

https://doi.org/10.1051/e3sconf/202235501003

be used although thin metal specimens are required. The tensile test is also developed to assess scale adherence on stainless steel [20-22]. For this study, the *in situ* tensile tests are employed to measure scale adhesion on a steel substrate. This test allows for continuous observation of oxide scale spallation, while increasing applied strain. The formation of oxide scale on steel isothermally oxidised in 20% O_2-N or 20% H_2O-N_2 as representative cathode and anode atmospheres respectively is investigated. The adhesion of scale in terms of adhesion energy is also studied.

2 Materials and methods

2.1 Materials and oxidation procedures

The study used an AISI 441 sheets of ferritic stainless steel grade, which is available from ARCELOR with a thickness of 2 mm. Table 1 shows the chemical composition of the steel.

Table 1. Chemical composition of the AISI 441 steel (wt.%).

Element	wt.%
Cr	17.83
Ni	0.10
C	0.01
Mn	0.24
Si	0.60
Ti	0.13
Nb	0.55
Fe	Bal.

A sample is cut from AISI 441 sheets and polished with SiC paper up to 1200 grade. Then, the sample is cleaned by an ultrasonic cleaning method using ethanol cleaner, followed by drying in air. The sample is oxidised in a horizontal furnace with 20% O_2-N_2 or 20% H_2O-N_2 as simulated SOFCs atmosphere at 800°C for 100 hours. Oxidation in water vapour (20% H_2O) is calculated by the Clausius-Clapeyron equation as follows,

$$ln\frac{P}{1\,atm} = \frac{-\Delta H^{l\to v}}{R}\left(\frac{1}{T} - \frac{1}{373}\right) \quad (1)$$

where P is the vapour pressure (0.2 atm), ΔH is the enthalpy change of reaction (40,893 J mol^{-1}), R is the gas constant (8.314 J mol^{-1} K^{-1}), and T is the temperature in Kelvin. Using equation (1), the temperature at 20% of the vapour pressure of water is 332.43 K (59.43°C).

The scale thickness can be calculated by average mass gain as pure chromia assumed. The following equation is commonly used to estimate scale thickness.

$$Scale\ thickness\ (\mu m) = \frac{\Delta m}{A}\cdot\frac{1}{M_O}\cdot M_{FeO}\cdot\frac{1}{\rho_{FeO}}\cdot 10^4 \quad (2)$$

where Δm is the mass change before and after oxidation (g), A is an area (cm^2), M_O is the molecular weight of oxygen (16 g mol^{-1}), M_{FeO} is the molecular weight of iron oxide (72 g mol^{-1}), and ρ_{FeO} is the density of the iron oxide (5.745 g cm^{-3}).

2.2 Determination of oxide scale adhesion energy

The determination of adhesion energy by the tensile test is extensively described in previous papers [20-23]. Electro-erosion is used to prepare specimens with a special shape that fit into the tensile machine sitting in a scanning electron microscope chamber as shown in Figure 1.

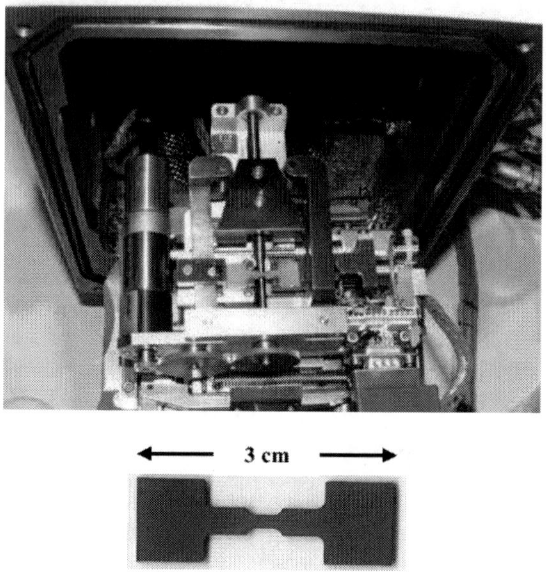

Figure 1. In situ tensile machine and typical shape of steel.

During tensile loading, stress and strain of the sample are continuously recorded. The strain rate is 7×10^{-5} s^{-1} and the test is conducted at room temperature. The scale morphology and scale failure during the tensile test is observed by scanning electron microscope (SEM-EDS, JEOL JSM-6400). The oxide surface is observed by backscatter electrons mode and secondary electrons mode with a magnification of 200×, 500×, 1500×, and 2000×. The adhesion energy of the oxide scale is identified as interfacial fracture energy which is derived from stress and strain at the first spallation. Due to the small size machine and sample, the elongation during the test is far from specimen central. Equation (3) is used to evaluate the actual strain.

$$\varepsilon_{actual} = \varepsilon_{experiment} - \left[\sigma\left(\frac{1}{E_{experiment}} - \frac{1}{E_{metal}}\right)\right] \quad (3)$$

where ε_{actual} is actual metal strain, $\varepsilon_{experiment}$ is an experimental strain of the metal, σ is metal stress (GPa), $E_{experment}$ is experimental Young's modulus of the metal (GPa), and E_{metal} is Young's modulus of the metal (GPa).

https://doi.org/10.1051/e3sconf/202235501003

The adhesion energy of oxide scale grown on metal have been published [20, 21]. As is widely accepted [24], it is assumed that the scale-metal interface has complete adhesion. The scale-metal interfacial fracture is called as scale spallation. The mechanical adhesion energy at the initial spallation can be determined using the following equation.

$$G = W \cdot \vartheta \quad (4)$$

where G is the mechanical adhesion energy (J m^{-2}), W is the strain energy per unit volume accumulated in the oxide scale until the first spallation (J m^{-3}), and ϑ is the scale thickness (m). The strain energy can be calculated using the stress-strain curve in loading and transverse directions until the first spallation. The strain energy is calculated by the following equation as described in the literature [20, 21].

$$W = \int \sigma \cdot d\varepsilon = \sum W_{fx} + \sum W_{fy} \quad (5)$$

3 Results and discussion

The average mass gain was used to calculate the thickness of the oxide scale, assuming pure chromia formation. The thickness of the oxide scale was 1.67 um for steel oxidised in 20% O_2-N_2, which was higher than 1.09 um for steel oxidised in 20% H_2O-N_2. Scanning electron microscopy was used to examine the surface morphology of steel isothermally oxidised in oxygen and water vapour as shown in Figure 2.

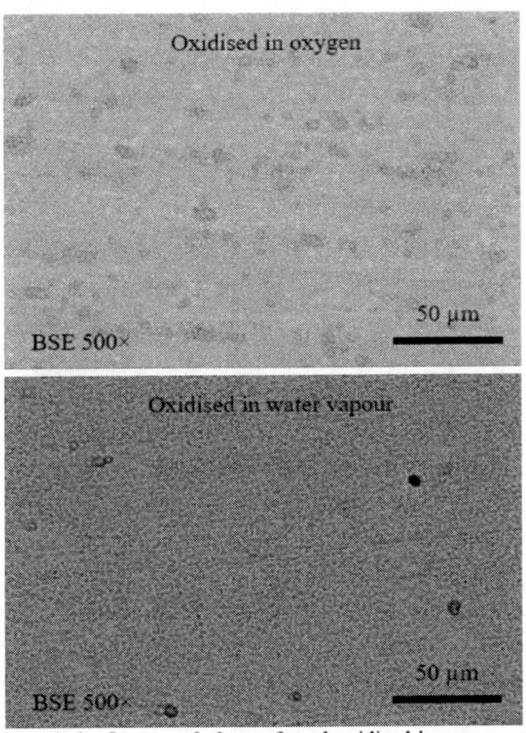

Figure 2. Surface morphology of steel oxidised in oxygen and water vapour.

It was seen that the surface of steel oxidised in water vapour seems to be smoother. Figure 3 illustrates the surface of steel oxidised in oxygen with a magnification of 2000× by secondary electrons mode was shown. An energy-dispersive spectroscopy equipped with a scanning electron microscope was applied to characterise the steel surface. Peaks of Fe, Cr, O, C and Ti were detected. This indicated the existence of oxide containing Cr and Ti at the surface. Meanwhile, Figure 4 illustrates the surface of steel oxidised in water vapour with a magnification of 1500× by secondary electrons mode was shown. Peaks of Fe, Cr, O, C and Ti were observed. The predominance of Cr and Ti was presented on both steels. The scale consisted of major Cr_2O_3-rich oxide and dispersion of TiO_2 as reported [11].

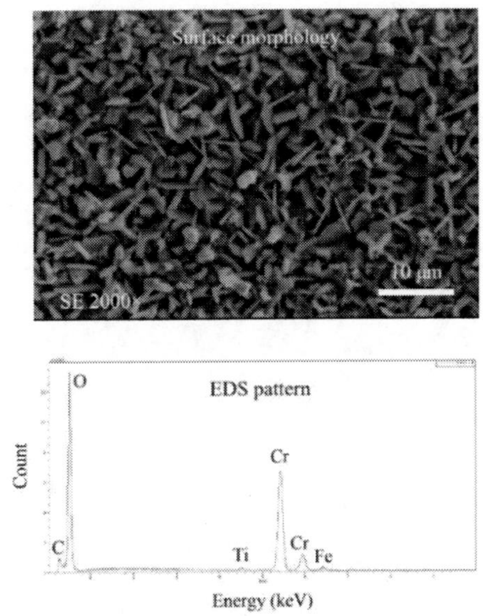

Figure 3. Surface morphology of steel oxidised in oxygen magnified of 2000× and EDS pattern on the surface of steel oxidised in oxygen.

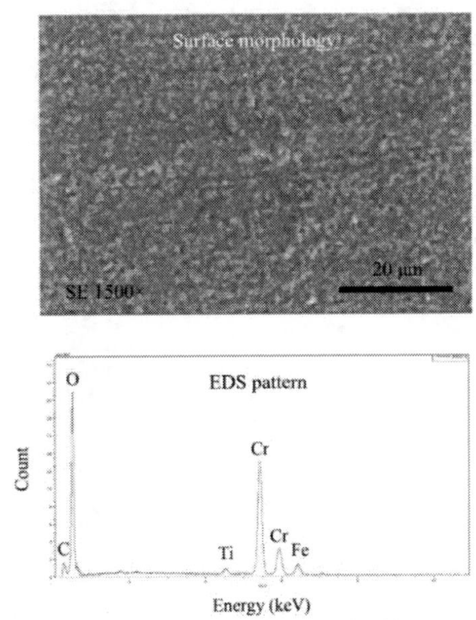

Figure 4. Surface morphology of steel oxidised in water vapour magnified of 1500× and EDS pattern on the surface of steel oxidised in water vapour.

For a tensile test, scale transverse cracks perpendicular to the tensile loading was first observed, followed by local scale spallation. Figure 5 depicts the evolution of scale failure during the tensile test. In these images, steel oxidised in 20% H_2O-N_2 significantly showed less spallation. However, it can be noted that the through-scale transverse cracks perpendicular to the tensile loading was first observed, followed by local scale spallation. This indicated that the steel oxidised in H_2O-N_2 was smoother and less defective. The force of spallation was increased with increasing strain. It should be also noted that the transverse crack on steel oxidised in water vapour was evidently increased by strain. Figure 6 presents the strain initiating the first spallation of the steel oxidised in H_2O-N_2 was 5.50%. This value was higher than that of the steel oxidised in O_2-N_2 which was 2.70%.

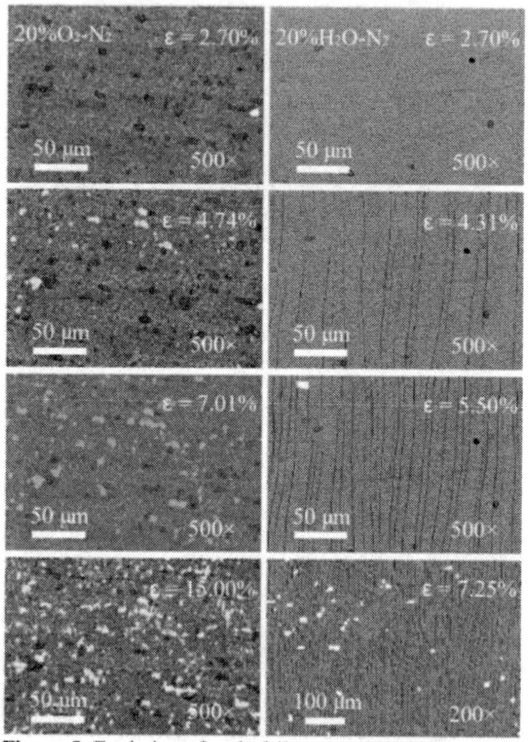

Figure 5. Evolution of scale failure on steel oxidised at 800°C during 100 h in 20%O_2-N_2 (left) and 20%H_2O-N_2 (right).

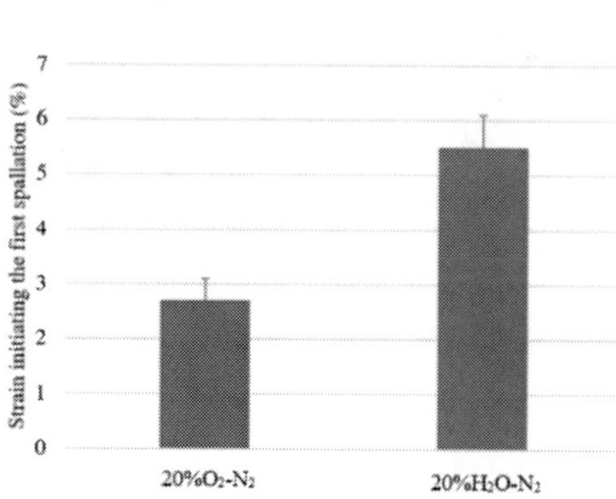

Figure 6. Strain initiating the first spallation of scale on steel oxidised at 800°C during 100 h in 20%O_2-N_2 and in 20%H_2O-N_2.

The fraction of surface area during the tensile test was recorded. The spallation ratio was obtained from these results and plotted in Figure 7. It was observed that the strain at the first strain of steel oxidised in 20% O_2-N_2 was lower than that of the steel oxidised in 20% H_2O-N_2. Moreover, the spallation ratio of steel oxidised in 20% O_2-N_2 was obviously increased with increasing strain. It can be noted that the spallation ratio of steel oxidised in 20% H_2O-N_2 trends to be identical for initiating strain, while rapidly increased with increasing strain for steel oxidised in 20% O_2-N_2.

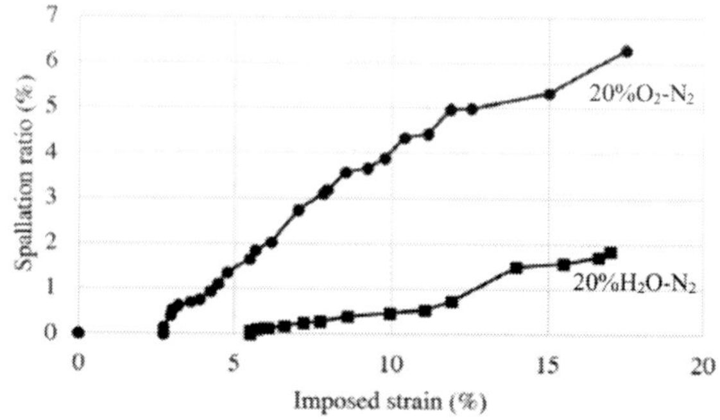

Figure 7. Spallation ratio of oxide scale on steel oxidised at 800°C during 100 h in 20%O_2-N_2 and in 20%H_2O-N_2.

The strain energy was considered as residual stress. Due to the mismatch between the oxide and steel thermal expansion coefficients, it was widely known that oxide growth and cooling on steel accumulates strain in the oxide. The oxide scale was accumulated by residual compressive stress due to oxide growth and cools down with the value of 0.2 GPa. The values of strain energy for chromia scale of 103.47 MJ m^{-3} for steel oxidised in 20% O_2-N_2 and 196.65 MJ m^{-3} for steel oxidised in 20% H_2O-N_2 were shown in Figure 8. It was observed that the strain energy of steel oxidised in 20% H_2O-N_2 was higher than that of the steel oxidised in 20% O_2-N_2.

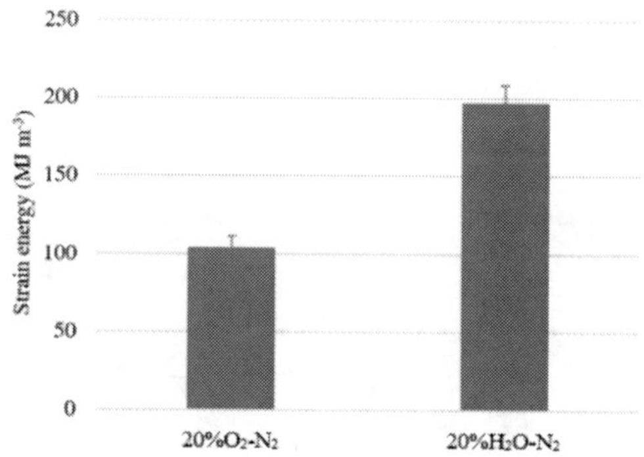

Figure 8. Strain energy of oxide scale formed in 20%O_2-N_2 and 20%H_2O-N_2 at 800°C during 100 h.

The calculated adhesion energy of steel oxidised in 20%O_2-N_2 was 172.79 J m^{-2} which was 2.70% of the strain initiating the first spallation. This value was lower than that of the steel oxidised in 20% H_2O-N_2 which was 214.34 J m^{-2} at 5.50% by strain initiating the first spallation. The mechanical adhesion energy was shown in Figure 9.

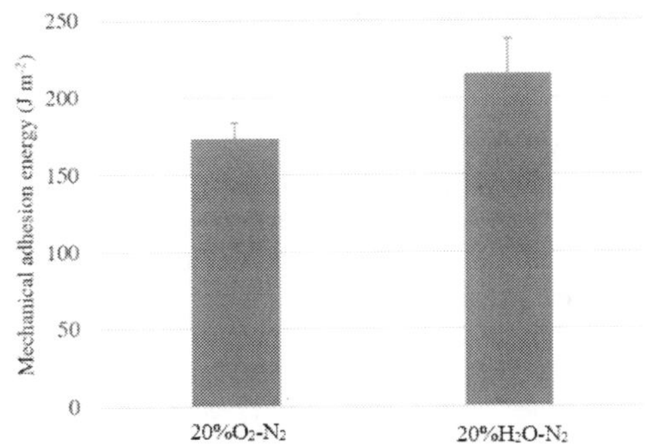

Figure 9. Mechanical adhesion energy of oxide scale formed in 20%O_2-N_2 and 20%H_2O-N_2 at 800°C during 100 h.

Figure 10 and Figure 11 shows tensile specimen in the SEM chamber before and after the tensile loading of steel oxidised at 800°C for 100 hours in 20%O_2-N_2 and 20%H_2O-N_2 respectively. The image was captured using backscatter electrons at a magnification of 30×. It was seen that the spallation of oxide scale revealed global scale failure, particularly on specimen centre. During tensile loading, the maximum stress was subjected in the region of gage length and it was due to the sudden change in the cross-section near the gage length that leads to high stress concentration. The stress greatly affected the global area of the specimen. The scale failure on the specimen was depending on the oxide scale formation. The spallation of scale can be seen more in steel oxidised in oxygen and less in steel oxidised in water vapour. This was due to lower adhesion at the scale-steel interface when the scale oxidised in oxygen.

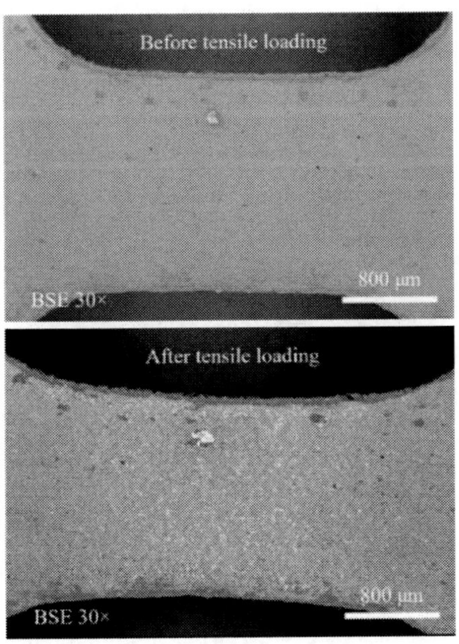

Figure 10. Tensile specimen in SEM chamber before and after tensile loading of steel oxidised in 20%O_2-N_2 at 800°C during 100 h.

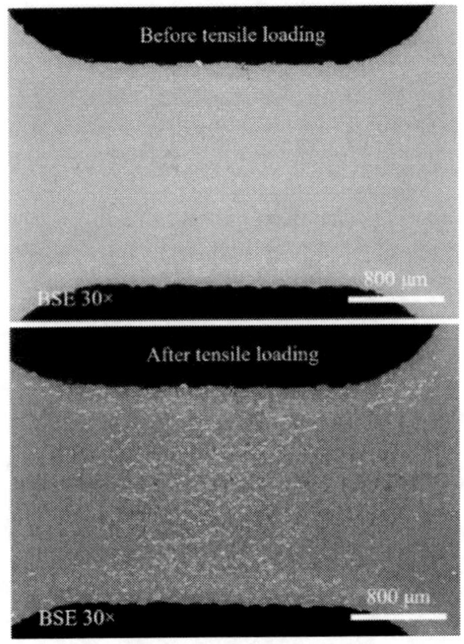

Figure 11. Tensile specimen in SEM chamber before and after tensile loading of steel oxidised in 20%H_2O-N_2 at 800°C during 100 h.

Figure 12 and Figure 13 shows the specimen after the tensile test of steel oxidised at 800°C for 100 hours in 20%O_2-N_2 and 20%H_2O-N_2 respectively. It was shown that the area of scale spalled out after the tensile test. An EDS equipped with SEM was applied to characterise the steel substrate. In the case of steel oxidised in O_2, peaks of Fe, Cr and C were detected. In the case of steel oxidised in H_2O, peaks of Fe, Cr, Si, Ti, C and O were detected.

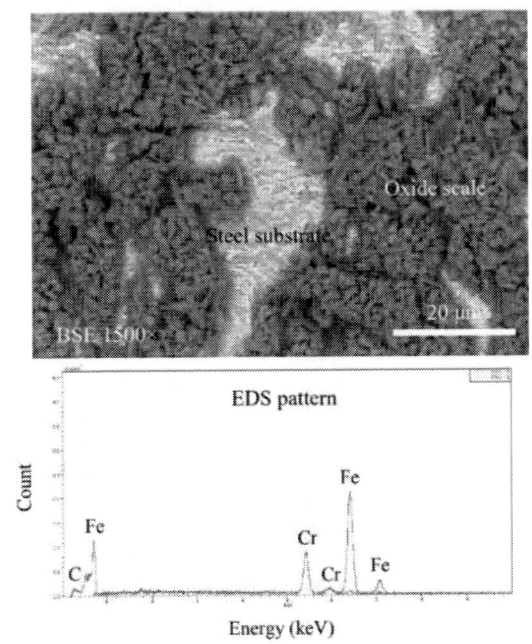

Figure 12. Oxide scale and substrate of steel oxidised in 20%O_2-N_2 at 800°C during 100 h after straining by 27.14% and EDS pattern on metal substrate observed as the white area.

Importantly, this indicated the silica layer formation at the scale-steel interface. This was due to the relatively high Si content (0.60%) in the AISI 441 stainless steel, which corresponded to promoting Si activity in the steel substrate and allowing the formation of an insulating interfacial silica layer. This result was shown to reduce an oxide thickness and exhibited to promote scale adherence. The qualitative assessment of scale adhesion was summarised in Table 2.

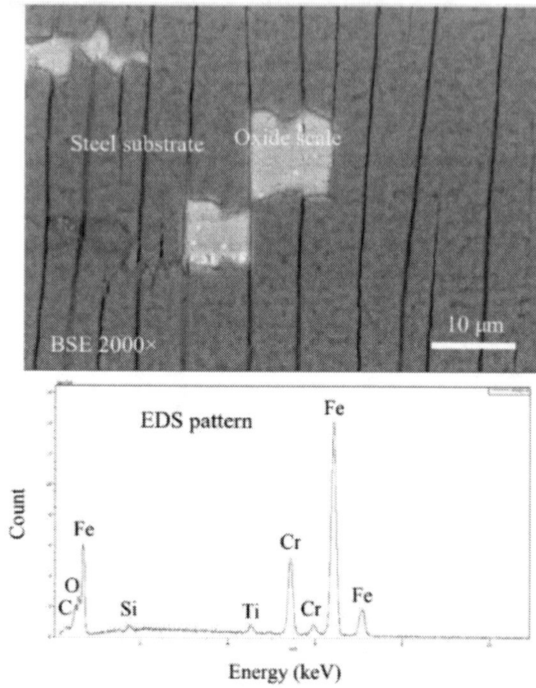

Figure 13. Oxide scale and substrate of steel oxidised in 20%H_2O-N_2 at 800°C during 100 h after straining by 7.65% and EDS pattern on metal substrate observed as the white area.

Table 2. The conclusion of the experiment results.

Scheme	O_2-N_2	H_2O-N_2
Thickness (μm)	1.67	1.09
Strain at first crack (%)	-	3.82
Strain at first spall (%)	2.70	5.50
Strain energy (MJ m^{-3})	103.47	196.65
Adhesion energy (J m^{-2})	172.79	214.34

The qualitative assessment of scale adhesion in terms of strain initiating the first spallation confirmed the good adhesion behaviour of scale grown in water vapour (anode atmosphere). With this value, it was possible to compare results from the literature [25]. It was investigated that the adhesion of thermal oxide scale grown at 800°C in 2% H_2O in H_2 on AISI 441 ferritic stainless steel proposed as interconnectors in solid oxide fuel cells. The adhesion energies were shown to lie in the range 10-100 J m^{-2}. Adhesion values exhibited decreasing with increasing oxide thickness and good adhesion of oxide grown in water vapour compared to oxygen.

For this study, the adhesion energy at the strain initiating the first spallation was found in the range 173-214 J m^{-2}. As a matter of discussion, the good adhesion of scale grown in water vapour compared to oxygen. This could be the result of hydrogen-containing species diffusing inwards in the oxide during scale growth, leading to a scale-steel interface of better growth direction. Another idea was that the oxide containing Si observed at the scale-steel interface during steel oxidised in a water vapour atmosphere. This oxide could be promoted scale adhesion on a steel substrate. Another point of interest was the higher scale thickness of steel oxidised in oxygen compared to water vapour. The adhesion of steel oxidised in oxygen should be less. This might be due to forming voids at the scale-steel interface during metals diffuse to the external scale. The thicker scale had higher voids. Then the adhesion should be worse.

The significance of this study was to focus on the adhesion of oxide scale on the ferritic stainless steel AISI 441. Comparison between oxidation in oxygen and water vapour. Because of high-temperature operation, the oxide scale that grows on SOFCs can cause spallation resulting an insulating air gap between the oxide and the substrate. It could inhibit current flow through the cell. Adhesion results should confirm whether the atmosphere could grow oxide scale adhesive proper to be used as SOFCs interconnectors. The materials used as components have thermally challenged that need to be solved. For improving materials as shown in the literature [26], a concentrated solar power plant (CSP) was another future sustainable energy source like fuel cells which produce electrical energy by chemical reactions. The CSP was the most efficient, reliable, and cost-effective. However, the corrosion was still violent for most materials at high temperature, thus the selection of the material was important. The ferritic stainless steel AISI 409 can be improved by the Al slurry coating technique.

4 Conclusion

The ferritic stainless steel AISI 441 on a candidate to be used as interconnectors in SOFCs was isothermally oxidised in oxygen (cathode atmosphere) and water vapour (anode atmosphere). The tensile test was used to determine adhesion energy. From the results, the following conclusions could be drawn:

4.1 The thickness of steel oxidised in 20%H_2O-N_2 was shown more lowly than steel oxidised in 20%O_2-N_2. This might be due to the present silica layer as a barrier at scale-steel interface.

4.2 Tensile loading could be used to assess the adhesion behaviour of oxide formed in the H_2O-N_2 atmosphere, giving through-scale transverse cracking followed by scale spallation, which was unseen when oxidised in O_2-N_2.

4.3 The adhesion energy at the strain initiating the first spallation was found in the range 173-214 J m^{-2}. Adhesion energy was higher for scale grown in an H_2O-N_2 atmosphere due to the possible formation of an oxide containing Si at the scale-steel interface during scale growth, which promotes scale adhesion.

4.4 Regarding oxidation and oxide scale adhesion, ferritic stainless steel AISI 441 oxidised in water vapour seems to be better than oxidised in an oxygen atmosphere. This was due to lower scale thickness and formation of silica layer on steel oxidised in water vapour.

4.5 The present work studied the oxidation and adhesion of ferritic stainless steel AISI 441. The additional study of ferritic stainless steel AISI 444 should be carried out for comparison.

Acknowledgements

This research was funded by National Science, Research and Innovation Fund (NSRF), and King Mongkut's University of Technology North Bangkok with Contract no. KMUTNB-FF-65-09. Ugine & ALZ (Arcelor-Mittal Group) for the provision of the AISI 441 (F18TNb) stainless steel for the study.

References

1. Fiseha M. Guangul, Girma T. Chala, A comparative study between the seven types of fuel cells, Applied Science and Engineering Progress, 13, 3, (2020): 185-194.
2. W.Z. Zhu, S.C. Deevi, Development of interconnect materials for solid oxide fuel cells, Materials Science and Engineering A, 384, (2003): 227-243
3. C. Déportes, M. Duclot, P. Fabry, J. Fouletier, A. Hammou, M. Kleitz, E. Siebert, J.L. Souquet, *Electrochimie des solides* (Presses Universitaires de Grenoble, France, 1994)
4. L. Antoni, Materials for solid oxide fuel cells: the challenge of their stability, Materials Science Forum, 461-464, (2004): 1073-1090
5. W.J. Quadakkers, J. Piron-Abellan, V. Shemet, L. Singheiser, Metallic interconnectors for solid oxide fuel cells-a review, Materials at High Temperatures, 20, 2 (2003): 115-127
6. Z. Yang, G. Xia, K.D. Meinhardt, K.S. Weil, J.W. Stevenson, Chemical stability of glass seal interfaces in intermediate temperature solid oxide fuel cells, Journal of Materials Engineering and Performance, 13, (2004): 327-334
7. W.Z. Zhu, S.C. Deevi, Development of interconnect materials for solid oxide fuel cells, Materials Science and Engineering A, 384, (2003): 227-243
8. J.W. Fergus, Metallic interconnects for solid oxide fuel cells, Materials Science and Engineering A, 397, (2005): 271-283
9. P. Promdirek, G. Lothongkum, S. Chandra-ambhorn, Y. Wouters, A. Galerie, Materials and Corrosion, 62 (2011): 616-622
10. W. Wongpromrat, H. Thaikan, W. Chandra-ambhorn, S. Chandra-ambhorn, Chromium vaporisation from AISI 441 stainless steel oxidised in humidified oxygen, Oxidation of Metals, 79, 5-6, (2013): 529-540
11. P. Promdirek, G. Lothongkum, S. Chandra-Ambhorn, Y. Wouters, A. Galerie, Oxidation kinetics of AISI 441 ferritic stainless steel at high temperatures in CO_2 atmosphere, Oxidation of Metals, 81, 3-4, (2014): 315-329
12. T. Brylewski, M. Nanko, T. Maruyama, K. Przybylski, Application of Fe–16Cr ferritic alloy to interconnector for a solid oxide fuel cell, Solid State Ionics, 142, (2001): 131-150
13. N. Sakai, T. Horita, Y. P. Xiong, K. Yamaji, H. Kishimoto, M.l E. Brito, H. Yokokawa, T. Maruyama, Structure and transport property of manganese-chromium-iron oxide as a main compound in oxide scales of alloy interconnects for SOFCs, Solid State Ionics, 176 (2005): 681-686
14. S Chandra-Ambhorn, F. Roussel-Dherbey, F. Toscan, Y. Wouters, A. Galerie, Determination of mechanical adhesion energy of thermal oxide scales on AISI 430Ti alloy using tensile test, Materials Science and Technology, 23, 4, (2007): 497-501
15. T. Nilsonthi, S. Chandra-ambhorn, Y. Wouters, A. Galerie, Adhesion of thermal oxide scales on hot-rolled conventional and recycled steels Oxidation of Metals, 79, (2013): 325-335
16. S. Chandra-ambhorn, T. Nilsonthi, Y. Wouters, A. Galerie, Oxidation of simulated recycled steels with 0.23 and 1.03 wt.% Si in Ar–20%H_2O at 900°C, Corrosion Science, 87, (2014): 101-110
17. Q. Ma, J.L. Beuth, F. S. Pettit, G. H. Meier, M.J. Stiger, *Use of Indentation Fracture Tests to Investigate Toughness Loss Mechanisms in Thermal Barrier Coating Systems* (ASM International/TMS, Materials Solutions, Pittsburgh, PA, September, 2005)
18. J. Mougin, M. Dupeux, A. Galerie, L. Antoni, Inverted blister test to measure adhesion energy of thermal oxide scales on metals or alloys, Materials Science and Technology, 18, 10, (2002): 1217-1220
19. J. Mougin, M. Dupeux, L. Antoni, A. Galerie, Adhesion of thermal oxide scales grown on ferritic stainless steels measured using the inverted blister test, Materials Science and Engineering A, 359, (2003): 44-51
20. A. Galerie, F. Toscan, E. N'Dah, K. Przybylski, Y. Wouters, M. Dupeux, Measuring adhesion of Cr_2O_3 and Al_2O_3 scales on Fe-based alloys, Materials Science Forum, 461-464, (2004): 631-638
21. F. Toscan, L. Antoni, Y. Wouters, M. Dupeux, A. Galerie, Oxidation kinetics and scale spallation of iron-chromium alloys with different titanium contents, Materials Science Forum, 461-464, (2004): 705-712
22. S. Chandra-ambhorm, F. Roussel-Dherbey, F. Toscan, Y. Wouters, A. Galerie, M. Dupeux, Determination of mechanical adhesion energy of

thermal oxide scales on AISI 430Ti alloy using tensile test, Materials Science and Technology, 23, 4, (2007): 497-501

23. S. Chandra-ambhorn, T. Nilsonthi, Y. Madi, A. Galerie, Application of the micro-tensile testing to investigate the adhesion of thermal oxide scales grown on AISI 441 stainless steel sheet oxidised in air and water vapour, Key Engineering Materials, 410-411, (2009): 187-193
24. H.E. Evans, Stress effects in high temperature oxidation of metals, International Materials Reviews, 40, 1, (1995): 1-40
25. S. Chandra-Ambhorn, Y. Wouters, L. Antoni, F. Toscan, A. Galerie, Adhesion of oxide scales grown on ferritic stainless steels in solid oxide fuel cells temperature and atmosphere conditions, Journal of Power Sources, 171, (2007): 688-695
26. P. Kongkaoroptham, M. Boonpensin, T. Siripongsakul, P. Promdirek, Corrosion behavior of AISI 409 stainless steel with Al slurry coating in molten salt, Applied Science and Engineering Progress, 15, 1, (2022): 1-7.

* Corresponding author: thanasak.n@eng.kmutnb.ac.th

Dilute inorganic acid pretreatment of mixed residues of Cocos nucifera (coconut) for recovery of reducing sugar: optimization studies

Marttin Gundupalli Paulraj[1,*], *Malinee* Sriariyanun[2], and *Debraj* Bhattacharyya[1]

[1] Department of Chemical and Process Engineering, TGGS, King Mongkut's University of Technology North Bangkok, Thailand
[2] Department of Civil Engineering, Indian Institute of Technology Hyderabad, Sangareddy, Telangana

Abstract. Inorganic acids, such as sulphuric acid, hydrochloric acid, and nitric acid are widely used for the pretreatment of lignocellulosic biomass for bioenergy production. In this study, the effect of different acids on the recovery of reducing sugar from coconut residues (coir and pith) mixed in different ratios was studied. The pretreatment conditions for different acids were optimized using response surface methodology (RSM). The independent variables, such as biomass ratio, time and acid concentration were considered for the optimization studies with reducing sugar as the dependent variable. The maximum recovery of reducing sugar (45%) from mixed biomass was observed during nitric acid (NA) pretreatment. The recovery of reducing sugar was lower for hydrochloric acid (HA) and sulphuric acid (SA). The lower yield was attributed to the possible formation of sugar degradation compounds during acid pretreatment. Therefore, NA pretreatment was found suitable for mixed biomass compared to other acids. Further studies are required to understand the effect of NA pretreatment through a detailed study of liquid hydrolysate and the introduction of the saccharification process. Mixed biomass benefits the biorefinery industries for sustainable bioenergy production.

Keyword. Inorganic acid, pretreatment, reducing sugar, optimization, coconut

1 Introduction

Lignocellulosic biomass is the most abundant and promising alternative renewable resource for the production of bioenergy and valuable chemical by-products [1]. Lignocellulose is a complex structure with three polymer layers: cellulose, hemicellulose, and lignin. Lignocellulosic biomass includes agricultural residues, energy crops, wood residues and other waste. Coconut coir and pith residues of *Cocos nucifera* (Coconut) is an agricultural waste generated in higher amounts in India [2]. The annual coconut coir and pith production in India is 280,000 and 50,000 metric tonnes, respectively. Most of the generated coir and pith is considered waste and is incinerated as part of waste management. In this study, the residues of coconut were utilized for the recovery of sugar during pretreatment. Coconut residues are rich substrates for the cellulose and lignin content [3]. Lignin and hemicellulose hinder the enzymes during the saccharification process and microbial attack. Similarly, bioenergy production from lignocellulosic biomass requires three essential steps: pretreatment, saccharification, and fermentation. Pretreatment is a process of breaking the inter-and intra-hydrogen linkages present in lignin and hemicellulose structure [4]. The pretreatment process usually is performed by physical, chemical, or biological processes. Pretreatment methods under the chemical process include acid, alkaline, ionic liquid, Organosolv, Ammonia fibre expansion (AFEX) and deep eutectic solvent (DES) [5].

The current study has focused on acid pre-treatment in a dilute state to break bonds between hemicellulose and lignin to recover reducing sugar. Inorganic acids such as sulfuric acid, hydrochloric acid, and nitric acid were used to extract reducing sugar from coconut residue. Inorganic acids readily dissociate, and the hydrogen bonds are cleaved in the presence of free H^+ ions. Different inorganic acids have different effects during the pretreatment process. Therefore, the effect of different acids on the recovery of reducing sugar from coconut residues was studied. Several studies have reported the effect of acid during pretreatment using single biomass. Zhang *et al.* studied the effect of different acids under mild conditions for selective saccharification by a fast pyrolysis process. Phosphoric acid showed improved results compared to other inorganic acids [6]. Different acids have different effects on the recovery of reducing sugars. Pretreatment with acids leads to the production of inhibitory compounds such as furfural and 5-hydroxymethylfurfural (drawback). However, the percentage of such compounds varies depending on the acidity of the various acids and pretreatment conditions (time, temperature, type of acid and its concentration). The pretreatment conditions for various acids were optimized

* Corresponding author: mpgundupalli@gmail.com

using the response surface methodology (RSM). The processing of one biomass does not give a stable production of bioenergy. Single biomass is not always available, so mixed biomass offers an alternative solution for the sustainable production of bioenergy. In our previous published study, the effect of different acids on coconut coir for recovery of maximum reducing sugar was investigated [7].

In this study, attempts have been made to investigate the effects of different acids on different proportions of coconut coir and pith in different ratios. In addition, the pretreatment conditions such as biomass ratio, time and acid concentration were optimized using RSM. The present study investigated the potential of different acids to recover maximum reducing sugar from mixed biomass. Reducing sugar was considered the response/dependent variable for the generation of a second-order model to determine the optimum condition.

2 Materials and Methods

2.1 Collection and Characterization of Coconut residues

Coconut residues were collected separately from a coconut processing facility in the central part of Hyderabad, Telangana, India. The size of the pith sample was reduced to 1 mm using a food processor after drying in a hot air oven at 65°C for 48 h. The coir sample was soaked in liquid N2 and grounded to powder (size 1 mm) using a mortar and pestle. The samples were characterized for cellulose, hemicellulose and lignin by following the NREL standard protocol [8].

2.2 Central Composite Design – Design of Experiment

The pretreatment of coconut residues was carried out using inorganic acids such as sulphuric acid (SA), hydrochloric acid (HA), and nitric acid (NA). The pretreatment conditions such as biomass ratio (X_1), time (X_2, min) and acid concentration (X_3, w/w) were optimized using response surface methodology. The optimum conditions were determined and considered maximum reducing sugar yield after pretreatment from mixed biomass. The central compound design (CCD) was considered for three independent variables, resulting in 20 experimental runs. The biomass ratio was calculated from the expression shown in Eq (1).

$$\text{Biomass ratio} = \frac{\text{Coconut coir, g}}{\text{Coconut pith, g}} \quad (1)$$

The CCD matrix with 20 runs includes 8 full factorial runs, 6 axial runs, and 6 central runs. The acid pretreatment was carried out for individual experimental runs to generate a second-order model for different acids. The general second-order model expression for three factors is shown in Eq (2).

$$Y = \beta_o + \sum_{i=1}^{3} \beta_i x_i + \sum_{i=1}^{3} \beta_{ii} x_i^2 + \sum \sum_{i<j=3}^{3} \beta_{ij} x_i x_j \quad (2)$$

Where Y is the response, x_i and x_j are the independent variables, β_o, β_i, β_{ii}, and β_{ij} is the constant coefficient, i[th] linear coefficient, quadratic coefficient, and ij[th] interaction coefficient, respectively. The experimental data were analysed statistically and processed using Statistica (7.0).

2.3 Inorganic acid pretreatment

Inorganic acid pretreatment of mixture biomass (coir/pith) using sulphuric acid (SA), hydrochloric acid (HA), and nitric acid (NA). The acid pretreatment was performed in a laboratory autoclave following the conditions for each experimental run summarized in Table 1. The pretreatment temperature was fixed at 121°C to avoid the formation of sugar degradation compounds. The solids and liquid hydrolysate were separated by filtration after pretreatment. The pH of the liquid hydrolysate was neutralized to 7 and analysed for reducing sugar following the DNS protocol [9]. The yield of reducing sugar was calculated based on the expression shown in Eq. (3)

$$Y = \frac{RS, mg/mL}{Total\ sugar\ in\ biomass, mg/mL} \times 100 \quad (3)$$

$$RS, mg/mL = \frac{A}{2.615} \times D \quad (3.1)$$

where RS - reducing sugar mg/mL, A – absorbance (575 nm), D - dilution factor, 2.615 - slope value (glucose standard curve)

Table 1. Central composite design (CCD) matrix for inorganic acid pretreatment of coconut residues

Run no	Biomass ratio (X_1)	Time, min (X_2)	Acid concentration, % w/w (X_3)
1	0.47	20.1	0.70
2	0.47	20.1	1.30
3	0.47	49.9	0.70
4	0.47	49.9	1.30
5	2.11	20.1	0.70
6	2.11	20.1	1.30
7	2.11	49.9	0.70
8	2.11	49.9	1.30
9	0.25	35.0	1.00
10	4.00	35.0	1.00
11	1.00	10.0	1.00
12	1.00	60.0	1.00
13	1.00	35.0	0.50
14	1.00	35.0	1.50
15	1.00	35.0	1.00
16	1.00	35.0	1.00
17	1.00	35.0	1.00
18	1.00	35.0	1.00

Run no	Biomass ratio (X_1)	Time, min (X_2)	Acid concentration, % w/w (X_3)
19	1.00	35.0	1.00
20	1.00	35.0	1.00

3 Results and Discussion

3.1 Acid pretreatment: Optimization studies

The characterization of coconut coir and pith was performed, and the chemical composition was shown in Figure 1. The glucose content was higher in coir compared to the pith. Acid pretreatment using SA, HA, and NA was carried out for the coconut residues following the experimental condition displayed in Table 1. The reducing sugar for individual experimental runs were determined. The reducing sugar for SA, HA, and NA varied from 13-43%, 28-47%, and 18-56%, respectively. It was observed that the maximum reducing sugar yield was observed for nitric acid pretreatment.

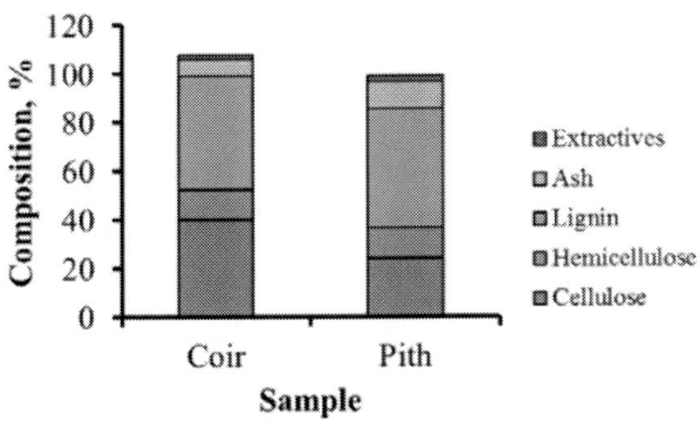

Figure 1. Chemical composition of coconut coir and pith

The second order model for SA, HA, and NA was generated by multiple regression analysis of the experimental data obtained for different experimental runs. The second order model for SA, HA, and NA was shown in Eq (4), Eq (5), and Eq (6), respectively

$$Y_{SA} = -38.28 - 4.88X_1 + 1.762X_2 + 57.22X_3 - 1.03X_{11} - 0.021X_{22} - 33.58X_{33} - 0.0789X_{12} + 13.09X_{13} + 0.1928X_{23} \quad (4)$$

$$Y_{SA} = -17.83 + 9.93X_1 + 1.17X_2 + 37.96X_3 - 0.567X_{11} - 0.011X_{22} - 13.08X_{33} - 0.095X_{12} - 1.21X_{13} - 0.0013X_{23} \quad (5)$$

$$Y_{NA} = -57.3 + 21.92X_1 + 0.75X_2 + 92.1X_3 - 5.23X_{11} + 0.0026X_{22} - 33.46X_{33} + 0.0037X_{12} + 2.07X_{13} - 0.445X_{23} \quad (6)$$

The generated models were tested for their adequacy and significance by (a) coefficient of determination (R^2), (b) analysis of variance (ANOVA), (c) lack of fit, and (d) residual plots. The value of R^2 is dependent upon the variance between the observed data versus predicted data. A low variance between these data shows that the model is statistically significant. A second-order model with R^2 greater than 0.8 is significant with observed data closer to the predicted data [10]. The R^2 value for SA, HA, and NA was greater than 0.95 showing the model was statistically significant. The predicted and observed data plot for SA, HA, and NA was shown in Figure 2.

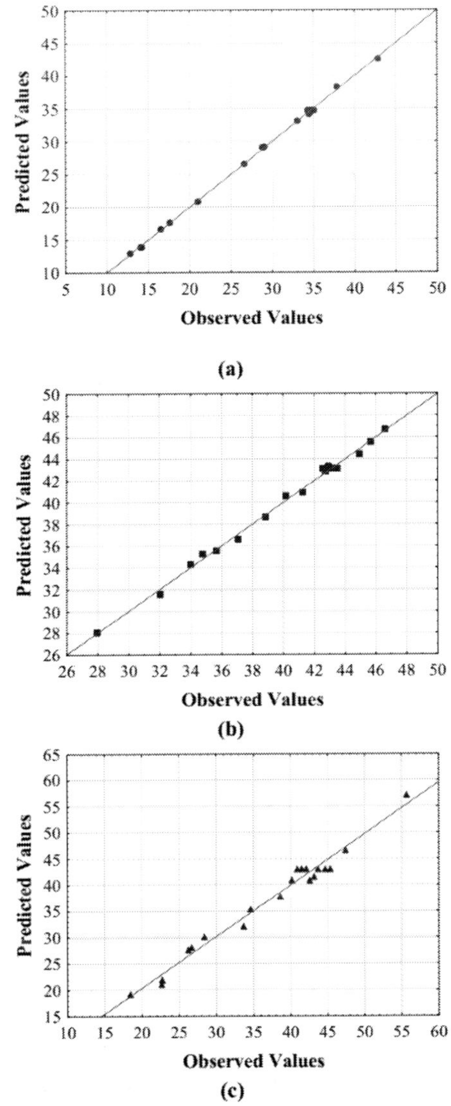

Fig 2. Predicted and observed data plots for the different second-order models. (a) Sulphuric acid (SA); (b) Hydrochloric acid (HA); (c) Nitric acid (NA)

The ANOVA performed for SA, HA, and NA second-order model showed that the model was significant (Table 2). The significance of the model was depicted by a p-value < 0.05. Furthermore, the model's significance and adequacy were examined through the insignificance of lack-of-fit (p-value > 0.05). Lack of fit usually describes the relationship between the dependent and independent variables. It was observed that the lack of fit for all the models was insignificant. Therefore, it is clear that the second-order models were significant. Further significance and adequacy of the second-order model were investigated through residual plots. The residual values are the difference between the predicted and observed values.

Table 2. Analysis of variance (ANOVA) for different second order model

Parameter	SS	df	MS	F	P
Sulphuric acid (SA)					
X_1	19.07	1	19.07	264.63	0.00
X_{11}	23.03	1	23.03	319.67	0.00
X_2	275.37	1	275.37	3822.06	0.00
X_{22}	314.20	1	314.20	4361.02	0.00
X_3	719.87	1	719.87	9991.70	0.00
X_{33}	132.96	1	132.96	1845.47	0.00
X_{12}	13.69	1	13.69	190.06	0.00
X_{13}	150.61	1	150.61	2090.52	0.00
X_{23}	5.81	1	5.81	80.65	0.00
Lack of Fit	0.78	5	0.16	2.17	0.21
Pure Error	0.36	5	0.07		
Total SS	1600.9	19			
Hydrochloric acid (HA)					
X_1	150.67	1	150.67	1354.87	0.00
X_{11}	7.17	1	7.17	64.47	0.00
X_2	98.49	1	98.49	885.66	0.00
X_{22}	89.07	1	89.07	800.91	0.00
X_3	101.89	1	101.89	916.24	0.00
X_{33}	19.26	1	19.26	173.18	0.00
X_{12}	19.90	1	19.90	178.96	0.00
X_{13}	1.29	1	1.29	11.57	0.02
X_{23}	0.00	1	0.00	0.00	0.96
Lack of Fit	1.64	5	0.33	2.95	0.13
Pure Error	0.56	5	0.11		
Total SS	477.62	19			
Nitric acid (NA)					
X_1	59.65	1	59.65	18.33	0.01
X_{11}	610.55	1	610.55	187.60	0.00
X_2	757.57	1	757.57	232.78	0.00
X_{22}	4.67	1	4.67	1.44	0.28
X_3	235.30	1	235.30	72.30	0.00
X_{33}	126.70	1	126.70	38.93	0.00
X_{12}	0.08	1	0.08	0.02	0.89
X_{13}	3.40	1	3.40	1.05	0.35
X_{23}	29.82	1	29.82	9.16	0.03
Lack of Fit	23.06	5	4.61	1.42	0.36
Pure Error	16.27	5	3.25		
Total SS	1843.7	19			

X_1, X_2, and X_3 – Main terms; X_{11}, X_{22}, and X_{33} – Quadratic terms; X_{12}, X_{13}, and X_{23} – Interaction terms; SS – sum of squares; df – degree of freedom; MS – mean sum of square

The residual plots for SA, HA, and NA second-order models were plotted and shown in **Figure 3**. The residual graph is plotted with residual values and predicted values on a vertical and horizontal axis. The model is considered significant when the residual values are scattered randomly without forming any patterns. The residual plots showed that the model was significant since the residual values were scattered randomly. The overall observation of the models showed that the SA, HA, and NA models were significant for the selected levels.

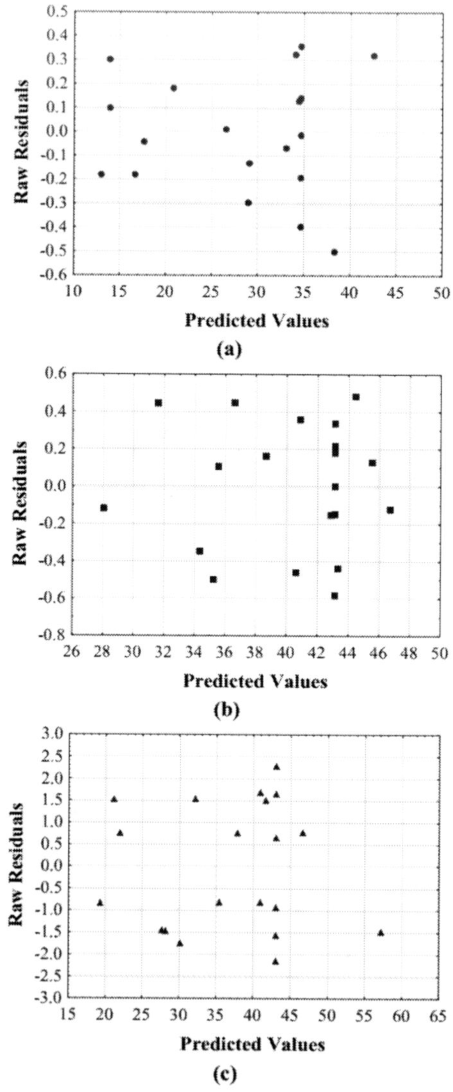

Fig 3. Residual plots for a different second-order model. (a) Sulphuric acid (SA); (b) Hydrochloric acid (HA); (c) Nitric acid (NA)

3.1.1 Interaction effect of biomass ratio and time

The interaction effect of biomass ratio and time for all the models were shown as a three-dimensional surface plot with the contour in Figure 4. Two variables were varied, with the third variable condition fixed at a center

https://doi.org/10.1051/e3sconf/202235501004

run. It was observed that the model predicted reducing sugar yield in the range of 6-30%, 8-40%, and 2-60% for the SA, HA, and NA models, respectively. The reducing sugar recovery from mixed biomass of coconut residues increased as the time increased for SA and HA models. On the other hand, the reducing sugar yield increased for the NA model when the biomass ratio levels were in the mid-range and higher reaction time. Therefore, it is understood that the recovery of reducing sugar from mixed biomass is more dependent on reaction time followed by biomass ratio.

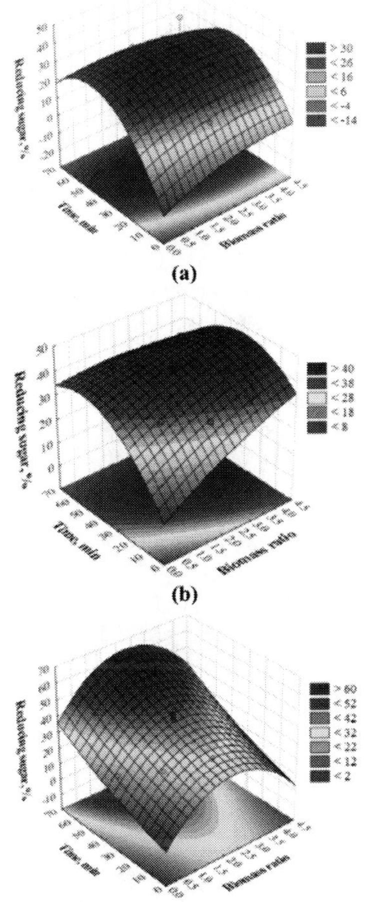

Figure 4. Three-dimensional response surface plot for time versus biomass ratio. (a) Sulphuric acid (SA); (b) Hydrochloric acid (HA); (c) Nitric acid (NA)

3.1.2 Interaction effect of Biomass ratio and acid concentration

The interaction effect of biomass ratio and acid concentration for all the models was shown as a three-dimensional surface plot with the contour in Figure 5. The model predicted reducing sugar yield in the range of 6-50%, 23-45%, and 6-40% for the SA, HA, and NA models, respectively. The reducing sugar recovery from mixed biomass of coconut residues after SA and HA pretreatment increased for the conditions of acid concentration and biomass ratio at the far-end range. On the other hand, the reducing sugar yield increased for the NA model when the biomass ratio and acid concentration levels were in the mid-range. Therefore, it was understood that the recovery of reducing sugar from mixed biomass increases when the biomass ratio and acid concentration interact.

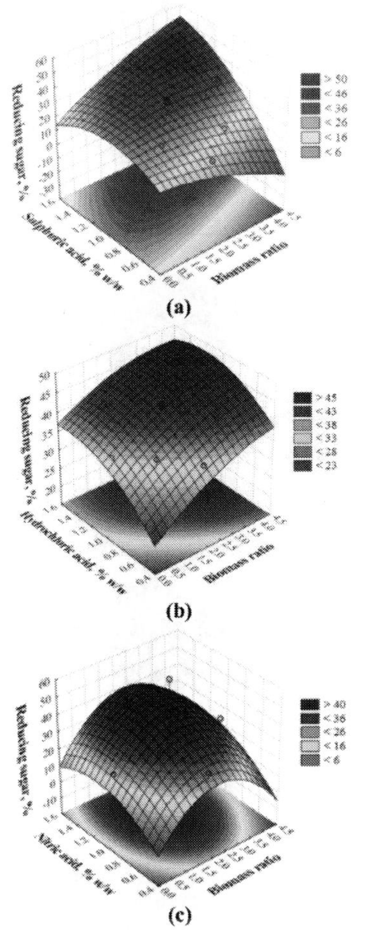

Figure 5. Three-dimensional response surface plot for acid concentration versus biomass ratio. (a) Sulphuric acid (SA); (b) Hydrochloric acid (HA); (c) Nitric acid (NA)

3.1.3 Interaction effect of acid concentration and time

The interaction effect of acid concentration and reaction time for all the models were shown as a three-dimensional surface plot with the contour in Figure 6. The model predicted reducing sugar yield in the range of 6-40%, 13-45%, and 6-60% for the SA, HA, and NA models, respectively. The reducing sugar recovery from mixed biomass of coconut residues after SA and HA pretreatment increased at mid-range pretreatment conditions. On the other hand, the reducing sugar yield increased for the NA model when the acid concentration and time parameters were maintained at mid and far-range, respectively. Therefore, the recovery of reducing sugar from mixed biomass increased when the reaction time and acid concentration interacted for SA and HA models. However, the interaction between acid concentration and reaction time was insignificant under initial conditions. The far-end range of parameters resulted in an increase in reducing sugar yield due to interaction.

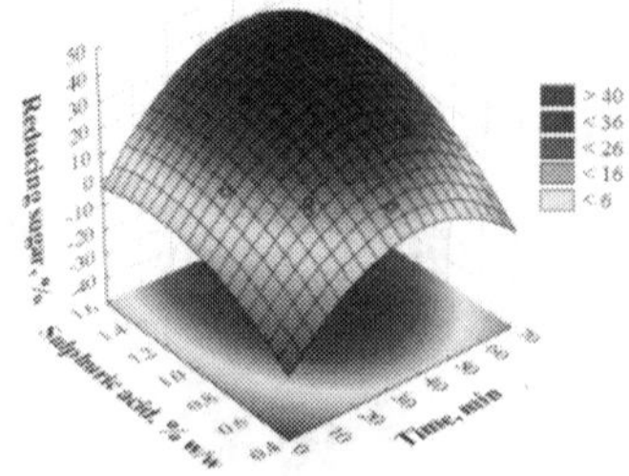

(a)

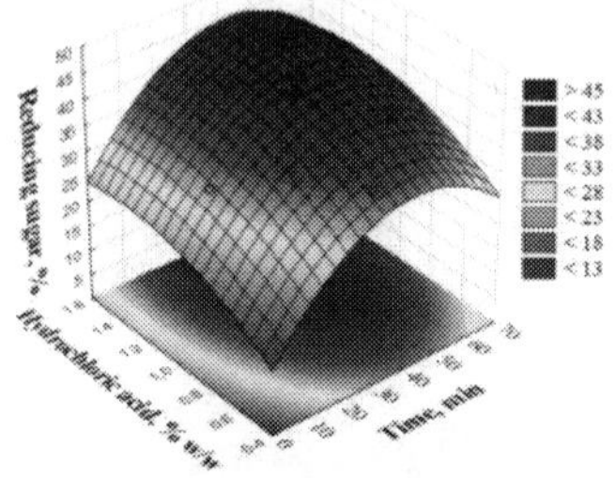

(b)

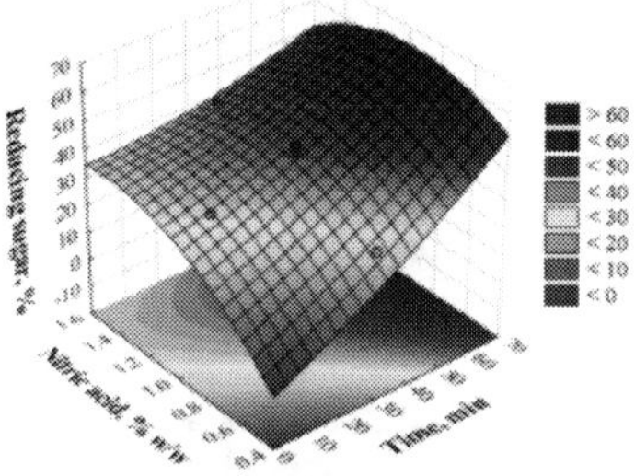

(c)

Figure 6. Three-dimensional response surface plot for time versus acid concentration. (a) Sulphuric acid (SA); (b) Hydrochloric acid (HA); (c) Nitric acid (NA)

3.2 Validation of the optimum conditions

The optimum conditions for the acid pretreatment were determined by the steepest ascent method to achieve maximum reducing sugar. The optimum conditions for SA, HA, and NA pretreatment are summarized in Table 3. The optimum conditions were validated by performing the studies using the optimum conditions thrice. Repetition of the studies using optimum conditions lowers the variance between the experimental data. It was observed that the optimum conditions for three acid pretreatments were almost similar. However, the maximum reducing sugar was obtained for nitric acid pretreatment. The maximum reducing sugar yield of 45% was observed for NA pretreatment, followed by HA (41.44%) and SA (36.55%) pretreatment. Different acids had a different effect on mixed biomass on recovery of reducing sugar. The variation in reducing sugar yield could be attributed to the strength of the acids. The pKa of sulphuric acid, hydrochloric acid, and nitric acid is -10, -5.9, and -1.5, respectively. Among the selected acid, nitric acid is the weakest due to its lower acidity strength.

Table 3. Optimum conditions determined for acid pretreatment of mixed biomass.

Parameter	SA			HA			NA		
	O_P	Y_P	Y_O	O_P	Y_P	Y_O	O_P	Y_P	Y_O
Biomass ratio	2.125	34.7%	36.55±1.92%	2.05	43.02%	41.44±0.73	2	43.13%	45.11±2.2%
Time, min	35			40			30		
Acid concentration w/w	1			1.05			1.1		

The lower yield of reducing sugar for sulphuric acid and hydrochloric acid was ascribed to the formation of sugar degradation compounds at 120°C. Furfural and 5-hydroxymethylfurfural are formed due to pretreatment at harsh conditions, such as longer reaction time, higher temperature, acid concentration, and so on [11]. During pretreatment, sulphuric acid and hydrochloric acid could degrade the sugar to degradation compounds, such as furfural and 5-hydroxymethylfurfural [12]. Since nitric acid has lower acid strength, the loss of sugar during pretreatment was lower. A study reported a maximum reducing sugar from sugarcane bagasse after sulphuric acid pretreatment compared to hydrochloric acid and nitric acid [13]. The results in this study were contradictory to the previously reported studies. However, the present study was performed for mixed biomass, and the selected biomass has higher lignin content than sugarcane bagasse. Another study reported the influence of sulphuric acid pretreatment on sugarcane bagasse for the recovery of reducing sugar and ethanol production [14]. Another study reported that acids with free H^+ ions break the inter-and intra-hydrogen bonds in lignin and hemicellulose [15]. Therefore, it was understood that the strength of acids affects the recovery of reducing sugar from lignocellulosic biomass. Furthermore, a detailed study is required to recover complete sugar from mixed biomass through the saccharification process.

4 Conclusion

The effect of different inorganic acids on the recovery of reducing sugar from mixed biomass (coir/pith) was studied. A reducing sugar recovery of 45%, 41.44%, and 36.55% was obtained from nitric acid, hydrochloric acid, and sulphuric acid pretreatment, respectively. The recovery of reducing sugar increased in NA>HA>SA. The results showed that different acids with similar optimum conditions showed different recovery of reducing sugar. Furthermore, a detailed study is necessary to analyse the liquid hydrolysate for furan compounds and introduce the saccharification process to completely recover reducing sugar.

Acknowledgements

This work was supported by the Department of Science and Technology under the FIST program and the Ministry of Education, Government of India and King Mongkut's University of Technology North Bangkok (KMUTNB-FF-65-37).

References

[1] L. Zhao, Z.F. Sun, C.C. Zhang, et al., Advances in pretreatment of lignocellulosic biomass for bioenergy production: Challenges and perspectives, Bioresource Technology. 343, (2022): 126123

[2] M.P. Gundupalli, H. Kajiura, T. Ishimizu, and D. Bhattacharyya, Alkaline hydrolysis of coconut pith: process optimization, enzymatic saccharification, and nitrobenzene oxidation of Kraft lignin, Biomass Conversion and Biorefinery. 12, 7, (2022): 2349–2367

[3] M.P. Gundupalli and D. Bhattacharyya, Sequential acid hydrolysis and enzymatic saccharification of coconut coir for recovering reducing sugar: Process evaluation and optimization, Bioresource Technology Reports. 6, (2019): 70–80

[4] N. El-Ahmady, S. Deraz, and A. Khalil, Bioethanol Production from Lignocellulosic Feedstocks Based on Enzymatic Hydrolysis: Current Status and Recent Developments, Biotechnology (Faisalabad). 13, 1, (2013): 1–21

[5] W.Y. Cheah, R. Sankaran, P.L. Show, et al., Pretreatment methods for lignocellulosic biofuels production: current advances, challenges and future prospects, Biofuel Research Journal. 7, 1, (2020): 1115–1127

[6] F. Zhang, L. Xu, F. Xu, and L. Jiang, Different acid pretreatments at room temperature boost selective saccharification of lignocellulose via fast pyrolysis, Cellulose. 28, 1, (2021): 81–90

[7] M. Paulraj Gundupalli and D. Bhattacharyya, Effect of different mineral acids on coconut coir for recovery of reducing Sugar: Process optimization using response surface Methodology (RSM), Materials Today: Proceedings. (2021):

[8] A. Sluiter, B. Hames, R. Ruiz, et al., NREL/TP-510-42618 analytical procedure - Determination of structural carbohydrates and lignin in Biomass, Laboratory Analytical Procedure (LAP). April 2008, (2012): 17

[9] G.L. Miller, Use of Dinitrosalicylic Acid Reagent for Determination of Reducing Sugar, Analytical Chemistry. 31, 3, (1959): 426–428

[10] M.P. Gundupalli, N. Senthilkumar, and D. Bhattacharyya, Process Optimization for Recovery of Reducing Sugar from Coconut Pith Using Sequential Hydrothermal Pretreatment and Enzymatic Saccharification, International Journal of Chemical Engineering and Applications. 9, 6, (2018): 94–199

[11] P. Rachamontree, T. Douzou, K. Cheenkachorn, M. Sriariyanun, and K. Rattanaporn, Furfural: A Sustainable Platform Chemical and Fuel, Applied Science and Engineering Progress. 13, 1, (2020):3-10

[12] P.T. Pienkos and M. Zhang, Role of pretreatment and conditioning processes on toxicity of lignocellulosic biomass hydrolysates, (2009)

[13] R. Sindhu, M. Kuttiraja, P. Binod, K.U. Janu, R.K. Sukumaran, and A. Pandey, Dilute acid pretreatment and enzymatic saccharification of sugarcane tops for bioethanol production, Bioresource Technology. 102, 23, (2011): 10915–10921

[14] E.J. Panakkal, M. Sriariyanun, J. Ratanapoompinyo, et al., Influence of Sulfuric Acid Pretreatment and Inhibitor of Sugarcane Bagasse on the Production of Fermentable Sugar and Ethanol, Applied Science and Engineering Progress. 15, 1, (2021): 5238

[15] J. Satarn, W. Lamamorphanth, and K. Kamwilaisak, Acid Hydrolysis from Corn Stover for Reducing Sugar, Advanced Materials Research. 931–932, (2014): 1608–1613

* Corresponding author: mpgundupalli@gmail.com

A comparative study of the anticorrosive response of *Tinospora cordifolia* stem extract for Al and Cu in biodiesel-based fuels

Kumar Amgain[1], *Bhesh Nath* Subedi[1], *Susan* Joshi[1], and *Jagadeesh* Bhattarai[1,*]

[1]Central Department of Chemistry, Tribhuvan University, Kirtipur 44619, Nepal

Abstract. The anticorrosive effect of methanol extract of *Tinospora cordifolia* stem for Al and Cu metals in pure biodiesel and its 10% blend with petrodiesel was investigated at 25±2 °C in a closed system by corrosion, inhibition efficiency, adsorption, and electrochemical tests. The corrosion inhibiting action of the plant extract for Cu in the biodiesel was more efficient than in the blend, while the extract showed more anticorrosive behavior of Al metal in the blend than in the biodiesel. Adsorption of the plant extract on the metal surface conformed to the Langmuir adsorption model. The plant extract functioned as a mixed-type corrosion inhibitor for both the metals in both biodiesel and its blend based on the experimental results. Outcomes of the study confirm the suggestive evidence to formulate the green extract-based biodiesel additives to enhance the anti-corrosive response for the Al and Cu metal parts of the vehicle engine. Nepal-origin *Tinospora cordifolia* stem extract could be used as an anticorrosive agent to control the corrosion of Al and Cu metals in biodiesel-based fuels.

Keyword. Biodiesel, Corrosion inhibitor, Immersion test, Plant extract, Polarization

1 Introduction

The European countries put forwarded rules of the renewable energy directive II for targeting to share at least 32% biofuels in the transport system by 2030 [1]. Biodiesel and its blends are one sort of biofuel, which are proven alternative renewable fuel that substitutes petroleum diesel. Because biodiesel-based fuels (BDBFs) have close properties with petroleum diesel [2]. They are eco-friendly [3, 4], and perform almost the same behavior as that of non-renewable petroleum fuels [5]. The utilization of such greener biodiesel-based fuels in vehicle engines becomes advantageous because they significantly diminish the toxic gas emissions, and are biodegradable, economical sources of renewable energy [6]. The BDBFs produce about 4.5 times of energy units than the counter fossil fuels [7] and have shown superlubricity behavior [8].

Biodiesel has been mandated for its usage in a blending range of up to 20% with petroleum diesel all over the world [9], although there are some challenges to making such greener BDBFs into reality. In comparison to petroleum diesel, the BDBFs are more corrosive to the engine parts of transportation means and the fuel storage systems [10, 11], which are mostly fabricated by steel, cast iron, aluminum, copper so on [12]. There is a growing trend to replace steel and cast iron with Al and Cu metals, and their alloys in vehicle engines and fuel storage systems in recent decades [13].

Excellent mechanical properties, good corrosion resistance, and adequate ductility of high-strength Al alloys led to significant attention in various industrial sectors [14], including aviation and automotive industries to replace steel components with much lighter parts of aluminum-based materials [15]. The Al and its alloys are usually utilized in different parts of the vehicle engine like pistons, cylinder heads, and engine blocks [16], whereas pumps, injectors, starters, ignition coils, and heat exchangers are made of Cu and its alloys [17].

The anti-corrosive nature of petroleum diesel decreases by increasing the biodiesel concentration in the BDBFs, mostly due to fostering microorganism activities in the biodiesel fuels [18]. Besides, researchers corroborated that the BDBFs reported more corrosive than the petroleum diesel on uncoated gray cast iron [19], and AA 3003 aluminum meal [20], mostly due to a marked increase in water, total acidity, electrical conductivity, density, and viscosity so on of the BDBF as compared to the petroleum diesel after immersion tests. From the electrochemical tests, the corrosion-resistance of carbon steel was reported the highest (at least one order of magnitude) in comparison to Al as well as the Cu in biodiesel, and the Cu metal showed the lowest corrosion-resistant properties than carbon steel and Al metals in biodiesel [21]. In most cases, the blending of biodiesel with 80% petroleum diesel or more has been practiced to improve its physical, chemical, and anticorrosive properties [22].

For the last decades, it becomes one of the effective techniques for minimizing the corrosion rates of different structural materials by mixing sufficient amounts of plant-based green inhibitors in aggressive electrolytes [23-25], except biofuels. Few studies demonstrated that biodiesel and its blends with a small addition of the plant-based green extracts have sufficient capability and possibilities to inhibit the corrosion

* Corresponding author: bhattarai_05@yahoo.com

problems of metals in the biofuels [26, 27]. One more research step is projected in this study to perform the anticorrosive response of methanol extract of *Tinospora cordifolia* stem (METcS) as a green inhibitor in pure biodiesel (BD-100) and its 10% blend (BDB-10) to control the corrosive nature of these biofuels towards Al and Cu metals using the immersion and electrochemical tests.

Tinospora cordifolia (known as GURJO in Nepali and heart-leaved moonseed in English), belonging to the Menispermaceae family, is one of the largest deciduous climbing shrubs and is distributed throughout Nepal, including other South Asian countries. It is used in folk and the Ayurvedic system of medicine by the local people of Nepal [28]. The stem extract of the Nepal-origin *T. cordifolia* plant reported various secondary plant metabolites like alkaloids, glycosides, coumarins, triterpenes, phenols, saponins, and tannins based on the phytochemical screening tests only [29]. The presence of the phytomolecules having O, N, S heteroatoms, multiple bonds, and aromatic ring compounds of the METcS will act as efficient green-based additives in different biofuels, such as biodiesel-based fuels (BDBF), to enhance the anticorrosive response to the aluminum, copper and their alloys.

However, it has not been reported so far the important aspects of this plant species as a corrosion inhibitor source to extend the lifetime and corrosion-resistant behavior of different automotive engine metals in the BDBF, although very few studies have been reported the corrosion inhibiting efficacy of *T. cordifolia* [30, 31], and *T. crispa* [32] extracts for steel in acidic electrolytes. In such conditions, the main goal of this study is to explore the METcS as green inhibiting additives to the BD-100 and BDB-10 fuels on a commercial scale to control their corrosive nature towards Al and Cu metals. The output of the experimental results would be helpful to manufacturers and policymakers for the mass production of automotive engine benign biodiesel and its blend fuels in Nepal.

2 Materials and methods

A rectangular-shaped (i.e., 4 cm × 3 cm × 0.4 cm) Al and Cu plats were mechanically polished to make a mirror shine surface, using 200-1500 grit numbers SiC paper in ethanol, rinsed by acetone, and air-dried to obtain consistent results from immersion, inhibition and electrochemical tests. The test electrolytes to accomplish this research were pure biodiesel (BD-100), which was produced at a laboratory scale from a cooked-oil residue (COR) as per the trans-esterification method, as described elsewhere [33]. Besides, a 10% blend (BDB-10) of the biodiesel with commercially available petroleum diesel (PetD) was also prepared. Then, some physical properties of the BD-100 synthesized from the COR were assessed using ASTM standards and compared with the properties of the PetD. Besides, gas chromatographic (GC), and FTIR spectroscopic techniques were carried out using instruments of a GC-2010 model (Shimadzu, Japan) and IRTacer-100 model (Shimadzu, Japan), respectively, for the quality analysis of both the BD-100 and PetD.

The METcS extract from stem powder of the *T. cordifolia* plant was acquired by soxhlet extraction using the rotatory evaporation technique, as explained elsewhere [34]. Then, the methanol extract was dried and stored at 4 °C before its use as biofuel additives. The dried solid plant extract obtained was named the METcS. The required amount of the stored METcS was weighted and added in both BD-100 and BDB-10 fuels separately to prepare their 500, 1000, 1500, and 2000 ppm electrolytic solutions.

The weight lost (Δw) of each Al and Cu specimen after immersion time (t) in the BD-100 and BDB-10 electrolytes without and with METcS concentrations was used for estimating their corrosion rate (CR) using a formula, as depicted in equation (1), where Fc (= 87600) is a conversion factor [35]. Also, the CR of both metals in a commercially available PetD was estimated for comparative study.

After the estimation of the CR of each Al or Cu specimen in different test biofuels without and with the METcS, the degree of surface coverage (θ) and % corrosion inhibition efficiency ($CIE_{\%}$) were calculated using equations (2) and (3), respectively [36]. The authors assumed herein the calculated θ is based on the METcS entirely controls the dissolution of Al or Cu from the covered surfaces [37].

$$CR\,(mm/yr) = \frac{Fc \times \Delta w\,(g)}{area\ (cm^2) \times density\ (g/cm^3) \times t\,(hr)} \quad (1)$$

$$\theta = \frac{CR_{without} - CR_{with\ METcS}}{CR_{without}} \quad (2)$$

$$CIE_{\%} = \frac{CR_{without} - CR_{with\ METcS}}{CR_{without}} \times 100 \quad (3)$$

Besides, the protective layer formation mechanism on the corroded Al and Cu surfaces after exposing these metals in BDBF electrolytes with the METcS was investigated using Langmuir and Temkin adsorption isotherm models. The inhibiting action of the METcS to the corroded Al and Cu metals was explained from the calculated adsorption equilibrium constant (K_{ads}) using the Langmuir adsorption model [38], as given in equation (4), where the C_{METcS} is the used concentration of the METcS.

In addition, the Temkin adsorption model [39] could be used to explain the adsorption mechanism of inhibition action of phytomolecules of the METcS onto the corroded metal surfaces, as shown in equation (5). In equation (5), the Temkin equilibrium constant (K_T) and constant B (= $RT/\Delta H_{ads}$) provide the information about binding energy, and change of heat (ΔH_{ads}) for an adsorption experiment, respectively [40]. The graphical plot of θ aginst $\log C_{METcS}$ shows a line that obeys the Temkin adsorption model [39].

$$\frac{C_{METcS}}{\theta} = \frac{1}{K_{ads}} + C_{METcS} \quad (4)$$

$$\theta = B \log K_T + B \log C_{METcS} \quad (5)$$

The anodic polarization curve was recorded from the rest of the potential value observed after a half-hour immersion to understand the passivity and electrochemical behavior of Al and Cu metals in both the BD-100 and BDB-10 biofuels containing different concentrations of the METcS at 25±2 °C, as elucidated in the previous work [41]. For this purpose, a potentiostat/galvanostat (Hokuto Denki HA-151, Japan) was used by adjusting the 15 mV/min scan rate. Before the electrochemical measurements, about 0.1 g/L $LiClO_4$ supporting electrolyte was added to each BD-100 and BDB-10 test electrolyte to increase their conductivity, as described elsewhere [42]. A saturated $Hg/Hg_2Cl_2/Pt$, Al or Cu coupon, and graphite rod were used as a reference, working, and auxiliary electrodes, respectively.

3 Results and discussion

The experimental data on the physical properties of the cooked-oil residue (COR) biodiesel are found within the limits of the ASTM D6751-14 standard of commercial biodiesel [43], as summarized in Table 1. Based on the observed physical properties, virtually pure and contamination-free biodiesel (i.e., BD-100) was successfully synthesized from the COR using transesterification. Hence, the BD-100 and its bend could be used in a diesel engine as a renewable biofuel by improving its corrosive response.

Table 1. Physical properties of the COR biodiesel.

Properties	COR biodiesel	Commercial biodiesel [43]
Appearance	Yellow	Yellow
Odor	Mild	–
Water and sediment (%)	Trace	0.05
Physical state	Liquid	Liquid
Boiling point (°C)	234	182-338
Kinematic viscosity (mm^2/s)	4.94	1.9 to 6.00
Specific gravity at 25 °C	0.91	0.86 to 0.90

Furthermore, the quality analysis of the BD-100 and its blend becomes important for their successful use as a surrogate of petroleum diesel on a commercial scale. From the qualitative GC analysis, six peaks were recorded in the chromatogram of the COR biodiesel (not shown), and their percentage area, retention time, and name of the FAMEs are listed (Table 2), based on the library search report. No other peaks except the six peaks of the FAME were observed in the Gas Chromatography (GC), indicating the absence of the byproducts like glycerol and other impurities in the synthesized COR biodiesel.

Table 2. Lists of the constituents of the COR biodiesel identified from the gas chromatograph.

Rt* (min.)	Area (%)	Name of FAMEs
52.86	8.92	Hexadecanoic acid methyl ester
58.19	50.93	9,12-Octadecadienoic acid methyl ester
58.38	33.84	11-Octadecenoic acid methyl ester
58.51	1.39	16-Octadecenoic acid methyl ester
58.58	1.19	9-Octadecenoic acid methyl ester
59.11	3.73	Cyclopentanetridecanoic acid methyl ester

**Retention time*

However, only the GC is not an adequate method for the characterization of the biodiesel blend with petroleum diesel, mostly due to the jeopardy of the overlapping signals of the complex spectra of long chains of petroleum carbon [44]. Thus, the FTIR spectroscopic method becomes more relevant to evaluate the FAME contents and quality of the biodiesel and its blends. The results of the non-existence of broadband around the 3100-3500 cm^{-1} region confirmed the absence of moisture, free fatty acid, and even glycerol in the synthesized COR biodiesel (Table 3). Three FTIR peaks in the 3008-2853 cm^{-1} range can be credited to the stretching vibration of the C–H bonds of the $-CH_2$ group, which are in agreement with the FTIR of edible oil [45].

Table 3. Comparative discussion of the FTIR spectra between the synthesized biodiesel and its raw cooked-oil residue.

Cooked-oil Residue (COR)		*COR Biodiesel (BD-100)*	
Vibration (cm^{-1})	Spectral assignment	Vibration (cm^{-1})	Spectral assignment
3008	C–H stretching; $C{=}CH_2$	3008	C–H stretching; $C{=}CH_2$
2923	C–H asymmetrical stretching; $-CH_2$	2923	C–H asymmetrical stretching; $-CH_2$
2853	C–H symmetrical stretching; $-CH_2$	2853	C–H symmetrical stretching; $-CH_2$
1744	C=O stretching; ester	1741	C=O stretching; $-OCH_3$
1460	C–H bending; $-CH_2$	1458	C–H bending; $-CH_2$
–		1435	Deformation vibration, $-OCH_3$
1377	C–H bending; $-CH_2$	1366	C–H bending; $-OCH_3$ (biodiesel)
1236	C–H bending; $-CH_3$	1229	C–O stretching of a typical ester
–		1215	C–O stretching; $-OCH_3$
–		1205	C–O stretching; FAMEs
–		1169	C–O stretching; FAMEs
1159	C–O stretching; C–O–C (Average)	–	
1097		–	
1032	C–O stretching; ester	1020	C=O bending; $-OCH_3$
968	C–H deformation; unsaturated fatty acid	–	
–		880	
721	C–H rock bending; $(-CH_2)_n$, n≥4	721	C–H rock bending; $(-CH_2)_n$, n≥4

It demonstrated the strongest peak at 1744 cm^{-1} among all the others absorption bands of the specimen, which is the characteristic of the ester C=O stretching vibration. However, a shifting of the peak to 1741 cm^{-1} with strong absorption intensity was ascribed to the C=O group of the methyl ester (Table 3). The C=O stretch of the ester absorbs light at 1740±5 cm^{-1} in the FTIR spectra, which is employed in the analysis of biodiesel content as the peak maximum [46].

Biodiesel has several characteristic FTIR absorption bands at 1435 cm^{-1}, 1365 cm^{-1}, and multiple peaks in the 1229-1169 cm^{-1} range [47]. A narrow absorption band of 1435 cm^{-1} moves along from the 1460 cm^{-1} of the raw cooking oil. The band at 1435 cm^{-1} was contemplated to the deformational vibration of the $-OCH_3$ group of the biodiesel. The band at 1377 cm^{-1} was attributed to the O-CH_2 of the raw COR, and this band was diminished with appearing as a strong band at 1366 cm^{-1} in the spectra of the COR biodiesel (Table 3), mostly due to the C−H bending vibration of O−CH_3 group.

The average energy of the triglycerides is no longer present at 1159 cm^{-1} after the synthesis of biodiesel from the raw COR, and this band was separated into multiple peaks in the region between 1229-1169 cm^{-1}. They assign to the C−O stretching and bending of the $-OCH_3$ group of the COR biodiesel. Very weak bands at 1032 cm^{-1} and 968 cm^{-1} were observed in the raw COR, which may be due to the C-O stretching vibrations, and the C-H out-of-plane deformation of unsaturated fatty acids, respectively [48], which disappeared in the biodiesel sample. The 721 cm^{-1} band in both the raw COR and synthesized biodiesel samples remarked to rock-bending vibration of the $(-CH_2)_n$ groups for $n \geq 4$, as ascribed elsewhere [47]. The summaries of these FTIR peak assignments are condensed in Table 3.

The corrosive response of two biodiesel-based biofuels (i.e., BD-100 and BDB-10), including PetD fossil fuel to Al and Cu metals is described from the estimated CR of the metals after immersion for about 16-189 days in 3 types of vehicle fuels at 25±2 °C, and the results are summed up in Table 4. It is easy to perceive from the tabulated data that both metals showed a lower CR in PetD than BD-100. Such corrosion behavior was reported more or less in the literature for the Al, Cu, cast iron, and carbon steel in palm biodiesel [49], partially due to increasing the total acid number (TAN) values of the biodiesel at the end of immersion for four months [50].

A unique elemental dissimilarity between the BDBFs and PetD is due to the presence of oxygen atoms in the biofuels, which make them more polar and able to create hydrogen bonds with water. This property of the BDBFs enhanced to increase the corrosion rate of metals than in the PetD. Besides, the blending of the biodiesel with petroleum diesel decreases the anticorrosive response to Cu metal, although an anomalous corrosion trend is observed for Al metal in BDB-10.

Table 4. The corrosion rate of Al and Cu metals in the BDBFs, including the PetD at 25±2 °C in a closed vessel at different periods.

Immersion time	CR of Al metal (mm/yr) in			CR of Cu metal (mm/yr) in		
(hr)	*BD-100*	*BDB-10*	*PetD*	*BD-100*	*BDB-10*	*PetD*
384	0.00697	0.00216	0.00291	0.00787	0.00330	0.00068
768	0.00531	0.00208	0.00301	0.00793	0.00334	0.00074
1536	0.00287	0.00155	0.00159	0.00799	0.00316	0.00065
2184	0.00228	0.00156	0.00138	0.00907	0.00331	0.00054
2976	0.00191	0.00117	0.00113	0.01121	0.00352	0.00055
4520	0.00104	0.00101	0.00089	0.01021	0.00334	0.00042

A high corrosive level of biodiesel (BD-100) to Cu should be due to low oxidative stability in comparison to Al, which is also in agreement with the earlier outcomes [51]. The oxidative stability determines the fuel quality and essentially influences the biodiesel stability [22]. It is reported that the addition of plant-based natural anti-oxidative phytomolecules in biodiesel and its blends enhanced their oxidative stability for long period uses or storage in vehicle engines [52]. The CR of the Al metal in the BDB-10 is slightly lower than in PetD up to the initial 24 days of exposure, and it becomes almost the same until about 189 days of exposure. In this circumstance, it becomes a fascinating topic to perform experimental works for knowing the effectiveness of different concentrations of METcS to increase the anticorrosive response of Al and Cu metals in BD-100 and BDB-10 fuels.

For this purpose, the corrosion-resistant behavior of Al and Cu metals was studied in BD-100 and BDB-10 in the absence and presence of 500, 1000, 1500, or 2000 ppm METcS concentrations. The anticorrosive behavior of Al metal increased significantly (i.e., nearly one order of magnitude) by increasing the concentrations of the METcS in BD-100 and BDB-10 biofuels, as shown in Figure 1(a), and 1(b), respectively. However, the corrosive effect of the different concentrations of the plant extract in both the BD-100 and BDB-10 fuels to Cu metal seems to be different and is less than that of Al metal.

The corrosion rate of Cu in each BD-100 or BDB-10 fuel with 500 ppm METcS is lower than in other concentrations, which is slightly increased with increasing the concentrations of the METcS from 1000 ppm, although their corrosive nature to the Cu is lower than in BD-100 and BDB-10 without the extract, as shown in Figure 1(c) and 1(d), respectively. These results revealed two important findings; firstly the METcS could be used as a good green-based corrosion inhibitor to increase the corrosion resistance properties of different vehicular engine metals like Al and Cu in pure biodiesel (BD-100) obtained from the COR, and its blend (BDB-10) at 25 °C in a closed system. The addition of 500 ppm METcS in both the biofuels seems to be a sufficient concentration of the METcS to improve

* Corresponding author: bhattarai_05@yahoo.com

https://doi.org/10.1051/e3sconf/202235501005

the corrosion-resistant properties of Cu in both the DBBFs at room temperature.

Besides, Table 5 shows the result of the calculated % corrosion inhibition efficiency ($CIE_{\%}$) of the METcS for Al and Cu metals in both the BDBFs (i.e., BD-100 and BDB-10) fuels. The $CIE_{\%}$ increases by increasing the METcS amounts (i.e., 1000-2000 ppm) in BDB-10 in the case of Al metal to maintain the optimal $CIE_{\%}$ (i.e., 86.1%), despite the additions of 1000-2000 ppm METcS in the BD-100 enhanced to increase the $CIE_{\%}$ slightly to attain the optimal value of about 78.8%.

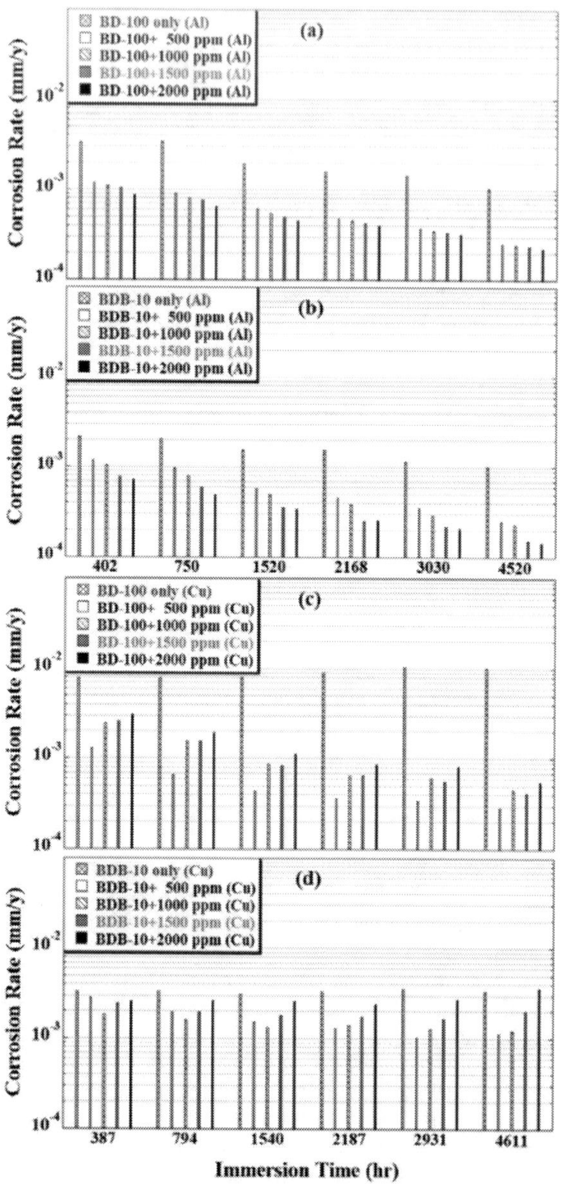

Figure 1. Effect of different concentrations of METcS on the corrosion rates of Al in (a) BD-100, (b) BDB-10, and Cu in (c) BD-100, and (d) BDB-10 at 25±2 °C and closed condition, against the exposure time.

In contrast, the $CIE_{\%}$ of 1000-2000 ppm extract in BD-100 fuel to Cu metal is decreased, and especially the decreasing trend of the $CIE_{\%}$ by increasing of 1000-2000 ppm METcS in the BDB-10 is significantly higher than in BD-100 for the Cu metal (Table 5). Consequently, the addition of 500 ppm METcS in BD-100 and BDB-10 shows the optimal $CIE_{\%}$ of about 97.3% and 66.2%, respectively, compared to Cu.

Results disclosed that the corrosion inhibition action on Cu metal significantly decreases with the addition of more than 500 ppm METcS in the B10. Such behavior of the significant acceleration of the corrosion rate of Cu metal in BDB-10 with the additions of 1000-2000 ppm METcS is more, probably due to the formation of an unstable surface film on Cu at a high amount of the METcS in BDB-10. At large, the corrosion inhibition study, based on the adsorption isotherm model, reported that the fractional surface coverage (θ) was increased by increasing the inhibitor amounts to a fixed optimal concentration, mostly due to the saturation of the corroded metal surface sites according to the adsorption model [53].

Overall, the order of the inhibiting efficiencies of METcS in both the BDBFs for Al and Cu metals could be arranged as; $CIE_{\%\text{ for Cu in BD-100+METcS}} > CIE_{\%\text{ for Al in BDB-10+METcS}} > CIE_{\%\text{ for Al in BD-100+METcS}} > CIE_{\%\text{ for Cu in BDB-10+METcS}}$. Therefore, the methanol fraction of the stem extract of *T. cordifolia* (i.e., METcS) would be efficient green-based materials for the formulation of the BDBF additives for controlling their corrosive nature to Al and Cu.

The Langmuir and Temkin adsorption models have been used to approximate the thermodynamic parameters to understand the adsorption mechanism of the corrosion inhibitors [54]. In the Langmuir model, the linear relationship between C_{METcS}/θ and C_{METcS} for Al and Cu metals in both the BDBFs with METcS concentration is displayed in Figure 2(a). Also, a linear form of the Temkin adsorption model was tested for comparative elucidation, as illustrated in Figure 2(b). The Langmuir adsorption model should be the best-fitted, based on the experimentally calculated linear correlation coefficient (R^2 ~1) and slope values (Figure 2a). However, the R^2 ranges from 0.8414 to 0.9748 for the Temkin equation.

These adsorption properties suggest the homogeneous distribution of adsorbed sites onto the corroded Al and Cu surfaces to form a corrosion-preventing monolayer film, although the Al metal in BD-100 shows a disparity in the Langmuir slope value. It deviates to some degree from the unity, which indicates that there is not a uniform explanation of the assumption of monolayer adsorption of the phytomolecules of the METcS onto the Al surface in BD-100. The Langmuir adsorption model suggested the interaction between the adsorbate and adsorbent, not between the adsorbate-adsorbate interactions [55]. A similar consequence was reported in the literature on the surface of mild steel specimens to control their corrosion in aggressive acids [56, 57].

Moreover, an endothermic type of the adsorption reaction enhances the formation of a protective passive film on the corroded Al surface in both BD-100 and BDB-10 biofuels, which is supported by the slope value (i.e., B>0) of the Temkin adsorption equation [58]. Such reaction mechanism obeyed to inhibit the corrosion of Al in both the BDBFs with 500-2000 ppm METcS. However, the corrosion inhibiting the efficiency of Cu metal is decreased in BD-100 with more than 500 ppm METcS, mostly due to the existence of an exothermic type of adsorption reaction, which is not beneficial for further protection of the passive film of Cu metal in the presence of 1000-2000 ppm METcS in both the BDBFs. Such results from the Temkin model indicated that the intermediate concentration range of the METcS should be 500 ppm or less, not more than it for the corrosion control of Cu in the BDBFs.

Table 5. The corrosion rate (CR) and % corrosion inhibition efficiency ($CIE_{\%}$) of Al and Cu metals in the BD-100 and BDB-10 with different concentrations of the METcS at 25±2 °C in a closed system.

Concentration of METcS (ppm)	*Al metal in DBFs*				*Cu metal in BDBFs*			
	CR_{BD-100} (mm/y)	CIE_{BD-100} (%)	CR_{BDB-10} (mm/y)	CIE_{BDB-10} (%)	CR_{BD-100} (mm/y)	CIE_{BD-100} (%)	CR_{BDB-10} (mm/y)	CIE_{BDB-10} (%)
0	0.00104	–	0.00101	–	0.01021	–	0.00334	–
500	0.00026	75.0	0.00032	68.3	0.00028	**97.3**	0.00113	**66.2**
1000	0.00024	76.9	0.00023	77.2	0.00045	95.6	0.00121	63.8
1500	0.00023	77.9	0.00015	85.1	0.00042	95.9	0.00199	40.4
2000	0.00022	**78.8**	0.00014	**86.1**	0.00055	94.6	0.00359	-00.1

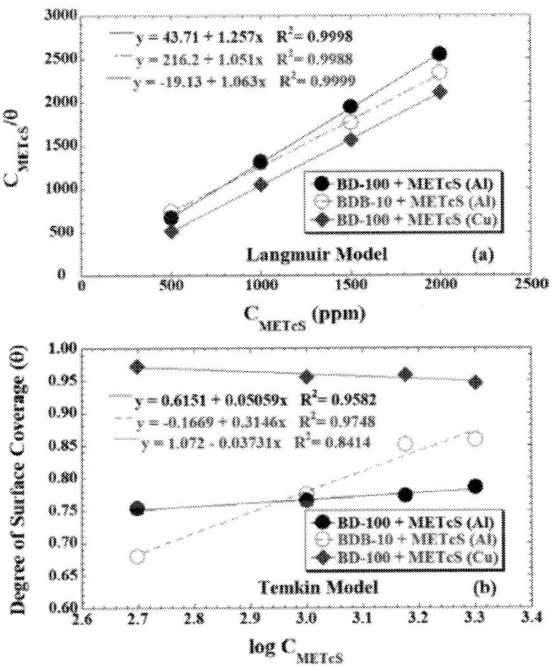

Figure 2. The Langmuir (a), and Temkin (b) adsorption isotherm plots for Al and Cu metals after exposure for 4520 h in both the BDBFs with different concentrations of METcS at 25±2 °C in a closed system.

Consequently, the corrosion-resistant of Cu decreased by increasing the METcS concentration, more than 500 ppm in both the BDBFs, as shown in Figure 1(c), and 1(d) also, although a high surface coverage (θ) is obtained for Cu metal in BD-100 with METcS than for Al in both the BD-100 and BDB-10 electrolytes (Table 5). In general, the Temkin adsorption isotherm model is assumed to be valid only at the intermediate concentration of adsorbate [59]. It reported a heterogeneous surface by which the adsorption heat changes (ΔH_{ads}) differ linearly from the θ by the adsorbate [60]. Consequently, the estimated ΔH_{ads} is decreased linearly with increasing the θ (Figure 2b).

The main purpose of the METcS addition in BDBFs is to increase the anticorrosive response and thus the corrosion-resistant properties of the vehicle engine Al and Cu metals. It presumed that the METcS inhibitors with electronegative heteroatoms including oxygen-, nitrogen-, sulfur, and the multiple bonded aromatic ring compounds like tinocordiside, isocolumbin, berberine, magnoflorine, *N*-formylannonain, yangambin, 11-hydroxymustakone so on are reported in the literature [61, 62]. Thus, adsorption of these phytoconstituents on the surfaces of the corroded Al and Cu enhances to fabrication of highly stable and less reactive layers, which blocks anodic/cathodic or both sides, and hence prevents significantly the corrosion rates of the metals in the presence of the inhibitors [63]. The degree of such corrosion-preventing layer is mostly in proportionate to the inhibitor concentrations, and it affects the physicochemical characteristics as described above in the adsorption models [64].

Laboratory-based electrochemical methods have been used to look into the corrosion inhibiting behavior of the METcS. The open-circuit potential (OCP) and potentiodynamic polarization measurements were also carried out in this study to know a better understanding of the passivation and stability of the corrosion preventive layers, which formed on the surfaces of Al and Cu metals in BD-100 and BDB-10 fuels without and with METcS concentrations. Figure 3(a) and 3(b) show the changes in the OCP and anodic current density (i_a) of the Al and Cu metals in BD-100 and BDB-10 without and with 500, 1000, 1500, and 2000 ppm METcS at 25±2 °C.

The shifting of the OCP in a noble (more positive) direction and the depletion of the i_a of Al metal is seen with the addition of 500 ppm METcS in the BD-100, while the i_a of Al is increased and shifted the corrosion potential in the less noble direction with the addition of 1000 ppm METcS or more in the BD-100, as noticed on the left-hand side of Figure 3(a). The electrochemical results revealed the fact that the addition of 500 ppm METcS in the BD-100 is enough to inhibit the anodic sides by utilizing an indiscernible adsorption film, which lessens the dissolution reaction of Al metal significantly [65]. Furthermore, additions of 1000-2000 ppm METcS control the corrosion rate of the Al metal in BD-100 by

 https://doi.org/10.1051/e3sconf/202235501005

forming the adsorbed layers on both anodic and cathodic sites, which safeguards further corrosion reactions.

On the other hand, as shown on the right-hand side of Figure 3(a), the i_a of the Cu is increased with METcS additions in BD-100. Besides, the Cu metal shows passive-transpassive dissolution. The dissolution of the passive films formed on the Cu metal could be explained based on the transpassive potential shifting near to the OCP of Cu in pure BD-100 without the METcS inhibitor. The transpassive potential was shifted to the cathodic direction with the addition of 1000 ppm METcS or more in BD-100, as revealed from the anodic polarization curves of the Cu metal. A similar passivation behavior is observed in the anodic polarization curves of the Cu metal in BDB-10 with 500-1500 ppm METcS. Hence, it can be assumed that the METcS functions as a cathodic-anodic inhibitor to enhance the corrosion inhibition efficiency of Al and Cu metals in BD-100, and also the Cu metal in BDB-10.

The cathodic-anodic (i.e., mixed-type) inhibitors are layer-forming compounds that are required to construct an exterior passive layer by hindering both the anodic and cathodic sites of the corroded metals [66]. However, the i_a of Al metal in BDS-10 is usually decreased with increasing the concentrations of the METcS without a remarkable OCP change, as depicted on the left-hand side of Figure 3(b), although the corrosion speed of the Al decreases with the addition of 500-2000 ppm METcS in BDB-10 (Figure 1b). Thereby, the METcS functions as an anodic-type corrosion inhibitor for Al metal even in BDB-10 biofuel.

A comparable consequence of the METcS on the strength and solidity of the anodic passive films of Al and Cu metals is studied in both the BD-100 and BDB-10 biofuels, as shown in Figure 4. A stronger anodic film is formed on both the metals in the 10% biodiesel blend (i.e., BDB-10) than in pure biodiesel (BD-100) in the presence of different concentrations of the plant extract, except 500 ppm METcS for the Al metal, as clearly noticed in Figure 4(a).

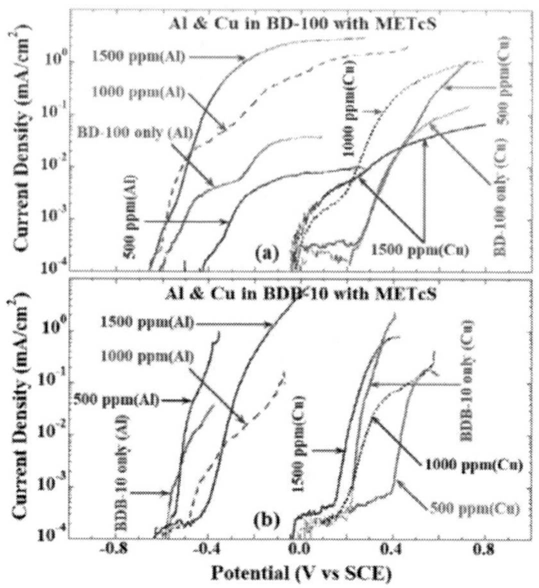

Figure 3. The effects of different concentrations of the METcS inhibitor on the i_a and OCP characteristics of both the Al and Cu metals in (a) BD-100 and (b) BDB-10 at 25±2 °C

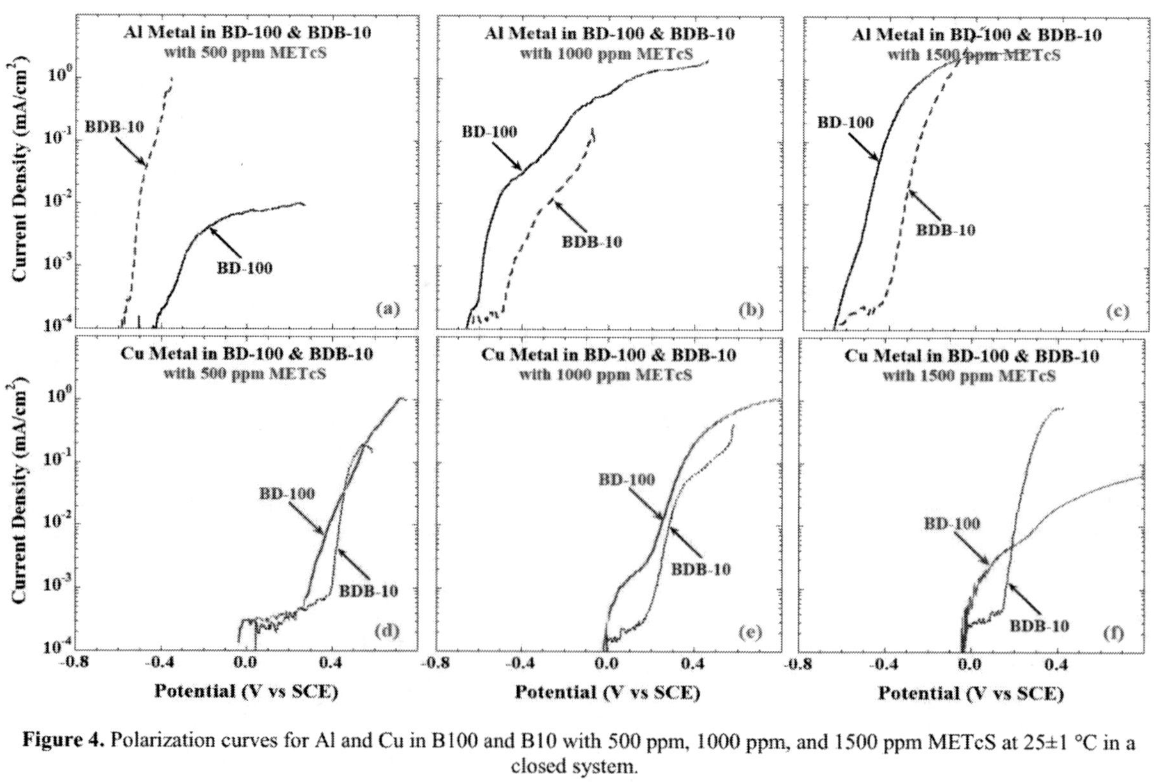

Figure 4. Polarization curves for Al and Cu in B100 and B10 with 500 ppm, 1000 ppm, and 1500 ppm METcS at 25±1 °C in a closed system.

Both the Al and Cu metals show better stability of the anodic passive films formed in the BDB-10 than in BD-100 with different concentrations of the METcS. These results are in close agreement with the experimentally calculated corrosion rate from the weight loss measurement, as discussed above in Figure 1.

4 Conclusions

The anticorrosive effects of the plant-based METcS concentrations for the Al and Cu were reported first time after exposure for more than six months in two varieties of the BDBFs (i.e., BD-100, BDB-10) at 25±2 °C and corked condition, using corrosion, inhibition efficiency, adsorption isotherm, and anodic polarization tests. The novel outcomes of the present work are enumerated as follows;

- The corrosion inhibitory action of the METcS extract on the Al metal increased by increasing its 500-2000 ppm concentrations in BDBFs, although the METcS shows the anomaly behavior to the Cu metal by increasing the concentrations more than 500 ppm BDBFs.
- The outcomes of the electrochemical and corrosion tests suggest the METcS performed as a mixed-type inhibitor for Al and Cu metals in the BDBFs.
- Thermodynamic estimations based on the best fitted Langmuir model alongside the Temkin adsorption models clearly show the adsorption of the majority of METcS phytochemicals on the Al and Cu surfaces without interactions between the METcS phytochemical molecules themselves, which also support the experimental outcomes of the corrosion-resistant and electrochemical polarization tests.
- The results of the adsorption studies also argue for the formation of a corrosion protective layer on the metal surface.
- The green-based METcS could be utilized as a source of biofuel additives on a commercial scale to improve the long-term anticorrosive response of the biodiesel fuels to Al- and Cu-based vehicle engine parts or storage systems.
- More extracts of other plant species can offer as an anticorrosive biofuel additive.

Acknowledgments

This research work was partially supported under the NAST Research Grant No.: 2-2075/76 of Nepal Academy of Science and Technology, Khumaltar, Lalitpur.

References

[1] EU, *Renewable Energy Directive*, (European Commission: Directive 2009/28/EC, revised in 2018) https://energy.ec.europa.eu/topics/renewable-energy/renewable-energy-directive-targets-and-rules/renewable-energy-directive_en (Accessed on 20 June 2022)

[2] D. Singh, D. Sharma, S.L. Soni, S. Sharma, P.K. Sharma, A. Jhalani, A review on feedstocks, production processes, and yield for different generations of biodiesel, *Fuel*, 262 (2020): 116553

[3] B.K. Selvan, S. Das, M. Chandrasekar, R. Girija, S.J. Vennison, N. Jaya, N. Rajamohan, Utilization of biodiesel blended fuel in a diesel engine–combustion engine performance and emission characteristics study, *Fuel*, 311 (2022): 122621

[4] R.R. Kumal, J. Liu, A. Gharpure, R.L. Vander Wal, J.K. Kinsey, B. Giannelli, J.J. Swanson, Impact of biofuel blends on black carbon emissions from a gas turbine engine, *Energy & Fuels*, 34(4) (2020): 4958-4966

[5] S. Dharma, H.C. Ong, H.H. Masjuki, A.H. Sebayang, A.S. Silitonga, An overview of engine durability and compatibility using biodiesel-bioethanol-diesel blends in compression-ignition engines, *Energy Conversion and Management*, 128 (2016): 66-81

[6] H.M. Mahmudul, F.Y. Hagos, R. Mamat, A.A. Adam, W.F.W. Ishak, R. Alenezi, Production, characterization and performance of biodiesel as an alternative fuel in diesel engines – a review, *Renewable and Sustainable Energy Review*, 72 (2017): 497-509

[7] *Biodiesel Benefits and Considerations*, https://afdc.energy.gov/fuels/biodiesel_benefits.html (Accessed on April 3, 2022)

[8] M. Mofijur, F. Kusumo, I.M.R. Fattah, H.M. Mahmudul, M.G. Rasul, A.H. Shamsuddin, T.M.I. Mahlia, Resource recovery from waste coffee grounds using ultrasonic-assisted technology for bioenergy production, *Energies*, 13(7) (2020): 1770

[9] I.S.A. Manaf, N.H. Embong, S.N.M. Khazaai, M.H.A. Rahim, M.M. Yusoff, K.T. Lee, G.P. Maniam, A review for key challenges of the development of biodiesel industry, *Energy Conversion and Management*, 185 (2019): 508-517

[10] P. Katuwal, R. Regmi, S. Joshi, J. Bhattarai, Assessment on the effective green-based Nepal origin plants extract as a corrosion inhibitor for mild steel in bioethanol and its blend, *European Journal of Advanced Chemistry Research*, 15 (2020): 1-13

[11] A. Shehzad, A. Ahmed, M.M. Quazi, M. Jamshaid, S.M. Ashrafur Rahman, M.H. Hassan, H.M.A. Javed, Current research and development status of corrosion behavior of automotive materials in biofuels, *Energies*, 14 (2021): 1440

[12] B.N. Subedi, K. Amgain, S. Joshi, J. Bhattarai, Green approach to corrosion inhibition effect of *Vitex negundo* leaf extract on aluminium and copper metals in biodiesel and its blend, *International Journal of Corrosion and Scale Inhibitor*, 8(3) (2019):744-759

[13] M. Tisza, I. Czinege, Comparative study of the application of steels and aluminium in lightweight production of automotive parts, *International Journal of Lightweight Materials and Manufacture*, 1(4) (2018): 229-238

* Corresponding author: bhattarai_05@yahoo.com

[14] C. Berlanga-Labari, M.V. Biezma-Moraleda, P.J. Rivero, Corrosion of cast aluminum alloys: a review, *Metals,* 10(10) (2020): 1384

[15] S.V. Sajadifar, E. Scharifi, T. Wegener, M. Krochmal, S. Lotz, K. Steinhoff, T. Niendorf, On the low-cycle fatigue behavior of thermo-mechanically processed, high-strength aluminum alloys, *International Journal of Fatigue*, 156 (2022): 106676

[16] M.F.X. Wagner, Light-weight aluminum-based alloys- from fundamental science to engineering applications, *Metals*, 8(4) (2018): 260-264

[17] B. Jones, G. Mead, P. Steevens, M. Timanus, The effect of E20 on metals used in automotive fuel system components, *Report No.: 2-22-2008* (St. Paul, USA: Minnesota Department of Agriculture; 2008).

[18] Y.M. Pusparizkita, A. Harimawan, H. Devianto, T. Setiadi, Effect of bacillus megaterium biofilm and its metabolites at various concentration biodiesel on the corrosion of carbon steel storage tank, *Biointerface Research in Applied Chemistry*, 12(4) (2022): 5698-5708

[19] A. Amaya, O. Piamba, J. Olaya, Corrosiveness of palm biodiesel in gray cast iron coated by thermo-reactive diffusion vanadium carbide coating, *Coatings*, 9(2) (2019): 135

[20] M. Soares, L.O. Berbel, C. Vieira, D.C.S. Oliszeski, C.B. Furstenberger, E.P. Banczek, Study of corrosion of AA 3003 aluminum in biodiesel, diesel, ethanol and gasoline media, *Material Science Forum*, 1012 (2020): 407-411

[21] C.I. Rocabruno-Valdes, J.A. Hernandez, A.U. Juantorena, E.G. Arenas, R. Lopez-Sesenes, V.M. Salinas-Bravo, J.G. Gonzalez-Rodriguez, An electrochemical study of the corrosion behavior of metals in canola biodiesel, *Corrosion Engineering, Science and Technology,* 53(2) (2018): 152-162

[22] L. Longanesi, A.P. Pereira, N. Johnston, C.J. Chuck, Oxidative stability of biodiesel: recent insights, *Biofuels, Bioproducts and Biorefining,* 16(1) (2022): 265-289

[23] M. Somai, A. Giri, A. Roka, J. Bhattarai, Comparative studies on the anti-corrosive action of waterproofing agent and plant extract to steel rebar, *Macromolecular Symposia*, in press (2022).

[24] S. Bouazama, J. Costat, J.M. Desjobertb, A. BenAli, A. Guenbou, M. Tabyaoui, Influence of *Lavandula dentata* essential oil on the corrosion inhibition of carbon steel in 1 M HCl solution, *International Journal of Corrosion and Scale Inhibitor*, 8(1) (2019): 25-41

[25] J. Bhattarai, M. Rana, M.R. Bhattarai, S. Joshi, Effect of green corrosion inhibitor of *Callistemon* plant extract on the corrosion behavior of mild steel in NaCl and HCl solutions, In *Proceedings of CORCON 2016* (Paper No. MI-17, 2016, New Delhi, India: NIGIS/NACE, p. 11)

[26] F.D. Fernandes, L.M. Ferreira, M.L.C.P. da Silva, Application of *Psidium guajava* L leaf extract as a green corrosion inhibitor in biodiesel: biofilm formation and encrustation, *Applied Surface Science Advances*, 6 (2021): 100185

[27] P. Katuwal, K.R. Gaire, J. Bhattarai, Study on the effects of ethylenediamine and plant extract as a corrosion inhibitor for mild steel passivation in bioethanol, In *Proceedings of CORCON-2018* (Paper No. MCI-35, Jaipur, India: NIGIS/ NACE, 2018, p. 9)

[28] R.M. Kunwar, K.P. Shrestha, R.W. Bussmann, Traditional herbal medicine in far West Nepal: a pharmacological appraisal, *Journal of Ethnobiology and Ethnomedicine*, 6 (2010): 35

[29] T. Shrestha, J. Lamichhane, Assessment of phytochemicals, antimicrobial, antioxidant and cytotoxicity activity of methanolic extract of *Tinospora cordifolia* (Gurjo), *Nepal Journal of Biotechnology,* 9(1) (2021): 18-23

[30] M. Rana, S. Joshi, J. Bhattarai, Extract of different plants of Nepalese origin as green corrosion inhibitor for mild steel in 0.5 M NaCl solution, *Asian Journal of Chemistry*, 29(5) (2017): 1130-1134

[31] A. Saxena, D. Prasad, K.K. Thakur, J. Kaur, PDP, EIS, and surface studies of the low-carbon steel by the extract of *Tinospora cordifolia*: a green approach to the corrosion inhibition, *Arabian Journal of Science and Engineering*, 46 (2021): 425-436

[32] M.H. Hussin, M.J. Kassim, N.N. Razali, N.H. Dahon, D. Nasshorudin, The effect of *Tinospora crispa* extracts as a natural mild steel corrosion inhibitor in 1 M HCl solution, *Arabian Journal of Chemistry*, 9 (2016): S616-S624

[33] R. Guzatto, Defferrari D, Reiznautt QB, Cadore IR, Samios D. Trans-esterification double step process modification for ethyl ester biodiesel production from vegetable and waste oil. *Fuel*, 92(1) (2021): 197-203

[34] J. Bhattarai, M. Somai, N. Acharya, A. Giri, A. Roka, N.R. Phulara, Study on the effects of green-based plant extracts and water-proofers as anti-corrosion agents for steel-reinforced concrete slabs, *E3S Web of Conferences*, 302 (2021): 02018

[35] J. Bhattarai, Study on the synergism of corrosion-resistant W-xNb alloys by angle-resolved X-ray photoelectron spectroscopy, *Ceylon Journal of Science*, 50(4) (2021): 513-520

[36] D.B. Pokharel, D.B. Subedi, D. VK, J. Bhattarai, Effects of tungstate and nitrite ions as corrosion inhibitor for Cr-10Zr-10W alloy in 0.5 M NaCl solution, *International Journal of Metal and Alloys*, 5(1) (2019): 11-19

[37] Y.I. Kuznetsov, N.N. Andreev, S.S. Vesely, Why we reject papers with calculations of inhibitor adsorption based on data on protective effects? *International Journal of Corrosion and Scale Inhibitor*, 4(2) (2015): 108-109

[38] I. Langmuir, The constitution and fundamental properties of solids and liquids: part I-Solid, *Journal of the American Chemical Society*, 38(11) (1916): 2221-2295

[39] M.I. Temkin, Adsorption equilibrium and the kinetics of processes on non-homogeneous surfaces and in the interaction between adsorbed molecules, *Zhurnal Fiziche-Skoi Khimii*, 15 (1941): 296-332
[40] F.T. Kamga, Modeling adsorption mechanism of paraquat onto Ayous (*Triplochiton scleroxylon*) wood sawdust, *Applied Water Science*, 9 (2019): 1
[41] D.B. Subedi, D.B. Pokharel, J. Bhattarai, Assessment on the effects of sodium salts of tungstate and nitrite as green inhibitor for the corrosion of Cr–5Ni–53W alloy in 0.5 M NaCl solution, *International Journal of Metallurgy and Alloys*, 6(1) (2020): 25-36
[42] M.A. Deyab, Corrosion inhibition of aluminum in biodiesel by ethanol extracts of Rosemary leaves, *Journal of the Taiwan Institute of Chemical Engineers*, 58 (2016): 536-541
[43] ASTM D6751-14, *Standard specification for biodiesel fuel blend stock (B100) for middle distillate fuels* (West Conshohocken, USA: ASTM International, 2014)
[44] C.R. Whetstine, GC-MS analysis of synthesized biodiesel, In *Forensic Science Master's Projects-1*, 2020
[45] A.S. Luna, A.P. Da Silva, J. Ferre, R. Boque, Classification of edible oils and modeling of their physico-chemical properties by chemometric methods using mid-IR spectroscopy, *Spectrochimi Acta A: Molecular and Biomolecular Spectroscopy*, 100 (2013): 109-114
[46] A.S. Silitonga, A.H. Shamsuddin, T.M.I. Mahlia, J. Milano, F. Kusumo, J. Siswantoro, ... H.C. Ong, Biodiesel synthesis from *Ceiba pentandra* oil by microwave irradiation-assisted transesterification: ELM modeling and optimization, *Renewable Energy*, 146 (2020): 1278-1291
[47] A.S. Luna, J.S. de Gois, Application of chemometric methods coupled with vibrational spectroscopy for the discrimination of plant cultivars and to predict physicochemical properties using R, *Comprehensive Analytical Chemistry*, 80 (2018): 165-194
[48] J.Y. Chen, H. Zhang, J. Ma, T. Tuchiya, Y. Miao, Determination of the degree of degradation of frying rapeseed oil using Fourier-transform infrared spectroscopy combined with partial least-squares regression, *International Journal of Analytical Chemistry*, 2015 (2015): 185367
[49] A.T. Hoang, M. Tabatabaei, M. Aghbashlo, A review of the effect of biodiesel on the corrosion behavior of metals/alloys in diesel engines, *Energy Sources Part A: Recovery, Utilization, and Environmental Effects*, 42(23) (2020): 2923-2943
[50] M. Fazal, A. Haseeb, H. Masjuki, Degradation of automotive materials in palm biodiesel, *Energy*, 40(1) (2012): 76-83
[51] C.L. Kugelmeier, M.R. Monteiro, R. da Silva, S.E. Kuri, V.L. Sordi, C.A. Della Rovere, Corrosion behavior of carbon steel, stainless steel, aluminum and copper upon exposure to biodiesel blended with petrodiesel, *Energy*, 226 (2021): 120344
[52] A. Devi, V.K. Das, D. Deka, Ginger extract as a nature-based robust additive and its influence on the oxidation stability of biodiesel synthesized from non-edible oil, *Fuel*, 187 (2017): 306-314
[53] M. Mobin, I. Ahmad, M. Basik, M. Murmu, P. Banerjee, Experimental and theoretical assessment of almond gum as an economically and environmentally viable corrosion inhibitor for mild steel in 1 M HCl, *Sustainable Chemistry and Pharmacy*, 18 (2020): 100337
[54] M.N. Al-Ghouti, D.A. Da'ana, Guidelines for the use and interpretation of adsorption isotherm models: a review, *Journal of Hazardous Materials*, 393 (2020): 122383
[55] N. Ayawei, A.N. Ebelegi, D. Wankasi, Modelling and interpretation of adsorption isotherms, *Journal of Chemistry*, 2017 (2017):3039817
[56] P. Magrati, D.B. Subedi, D.B. Pokharel, J. Bhattarai, Appraisal of different inorganic inhibitors action on the corrosion control mechanism of mild steel in HNO_3 solution, *Journal of Nepal Chemical Society*, 41(1) (2020): 64-73
[57] D. Vk, J. Bhattarai, Effect of sodium tungstate as a green corrosion inhibitor on the passivation behavior of mild steel in aggressive media, *International Journal of Applied Science and Biotechnology*, 4(2) (2016): 183-190
[58] F. Batool, J. Akbar, S. Iqbal, S. Noreen, S.N.A. Bukhari, Study of isothermal, kinetic, and thermodynamic parameters for adsorption of cadmium: An overview of linear and nonlinear approach and error analysis, *Bioinorganic Chemistry and Applications*, 2018 (2018): 3463724
[59] S.L. Lam, R. Ballinger, C. Forsberg, Modeling and predicting total hydrogen adsorption in nanoporous carbon materials for advanced nuclear systems, *Journal of Nuclear Materials*, 511 (2018): 328-340
[60] C. Aharoni, M. Ungarish, Kinetics of activated chemisorption Part 2: theoretical models, *Journal of the Chemical Society, Faraday Transactions 1: Physical Chemistry in Condensed Phases*, 73 (1977): 456-464
[61] V. Sagar, A.H.S. Kumar, Efficacy of natural compounds from *Tinospora cordifolia* against SARS-CoV-2 protease, surface glycoprotein and RNA polymerase, *Biology, Engineering, Medicine and Science Reports*, 6(1) (2020): 6-8
[62] E. Okon, W. Kukula-Koch, A. Jarzab, M. Halasa, A. Stepulak, A. Wawruszak, Advances in chemistry and bioactivity of magnoflorine and magnoflorine-containing extracts, *International Journal of Molecular Sciences*, 21(4) (2020): 1330
[63] T.J. Harvey, F.C. Walsh, A.H. Nahle, A review of inhibitors for the corrosion of transition metals in aqueous acids, *Journal of Molecular Liquids*, 266 (2018): 160-175
[64] D.B. Pokharel, D.B. Subedi, J. Bhattarai, Study the effect of sodium nitrite as a green inhibitor for the sputter-deposited tungsten-based ternary alloys in 0.5 M NaCl solution, *Bibechana* 12 (2015): 1-12
[65] I.A.W. Ma, S. Ammar, S.S.A. Kumar, K. Ramesh, S. Ramesh, A concise review on corrosion inhibitors: types, mechanisms and electrochemical

evaluation studies, *Journal of Coating Technology and Research*, 19 (2022): 241-268

[66] D.B. Subedi, D.B. Pokharel, J. Bhattarai, Study the corrosion inhibition mechanism of sputter-deposited W-42Cr-5Ni and Cr-10Zr-10W alloys by sodium nitrite as a green inhibitor in 0.5 M NaCl and 1 M NaOH solutions, *International Journal of Applied Science and Biotechnology*, 2(4) (2014): 537-543

* Corresponding author: bhattarai_05@yahoo.com

N and S Dual-Doped Mesoporous Carbon Nanostructure as a High Performance and Durable Metal Free Oxygen Reduction Reaction Electrocatalyst

Apichat Saejio[1], *Nattawan* Pitipuech[1], *Kultida* Kongpunyo[2], *Nutsuda* Buntao[2], *Kittimaporn* Nernprom[2], *Khemika* Boonkor[2], *Kitisak* Wichienwat[2], *Noppavan* Chanunpanich[3,4], *Narong* Chanlek[5], *Sangaraju* Shanmugam[6] and *Kriangsak* Ketpang[2,3*]

[1]Faculty of Engineering and Technology, King Mongkut's University of Technology North Bangkok Rayong Campus, Thailand
[2]Faculty of Science, Energy and Environment, King Mongkut's University of Technology North Bangkok Rayong Campus, Thailand
[3]Integrated Chemistry Research Center for Sustainable Technology (ICRT), Science and Technology Research Institute, King Mongkut's University of Technology North Bangkok, Thailand
[4]Department of Industrial Chemistry, Faculty of Applied Science, King Mongkut's University of Technology North Bangkok, Thailand
[5]Synchrotron Light Research Institute (Public Organization), Nakhon Ratchasima, Thailand
[6]Department of Energy Science & Engineering, Daegu Gyeongbuk Institute of Science & Technology (DGIST), Republic of Korea

Abstract. Discovering a high performance, durable, and cost-effective oxygen reduction reaction (ORR) electrocatalyst is a key strategy for widespread use of the high efficiency and environmentally friendly fuel cell and metal-air battery technologies. Herein, we fabricate a high performance and durable metal free N and S dual-doped mesoporous carbon nanostructure (NS-VXC) ORR catalyst using solid state thermolysis at 700 °C for 1 h. The fabricated catalyst exhibits nanocarbon aggregated chain-like morphology with a high surface area and mesoporous structure. The amount of N and S dopants embedded in mesoporous carbon nanostructure is found to be 3.2 and 1.1%, respectively which significantly attribute to the synergistic effect of spin and charge density leading to not only superior ORR performance but excellent durability in the alkaline environment as well. Rotating ring disk electrode analysis reveals the codoped NS-VXC catalyst possesses a direct 4-electron transfer number pathway with extremely low peroxide intermediate content. Compared to the benchmark Pt/C catalyst, the fabricated NS-VXC catalyst generated 10 mV ORR performance outperform and negligible performance degradation after the 10,000 ORR cycling test. These results suggest that an innovative solid state thermolysis methodology can be a powerful nanomaterial fabrication technique to generate high performance and excellent durability electrocatalyst for green energy applications.

Keyword. N and S codoped carbon, Metal free ORR catalyst, Oxygen reduction reaction, Fuel cell

1 Introduction

The electrochemical oxygen reduction reaction (ORR) is the key reaction directly determining the efficiency of the environmental compatibility energy conversion and storage technologies such as fuel cells and metal-air batteries [1]. Because of its sluggish kinetic, a substantial amount of catalyst is essentially required to kinetically dissociate the bonding of an oxygen molecule in the ORR process [1]. Currently, the precious platinum and/or platinum alloy nanoparticles supported on carbon (Pt/C) have been the most efficient and widely used as the major catalyst for reducing the overpotential of the ORR in both acids and alkaline media [2-4]. However, the nanoparticle Pt/C catalysts have several technical drawbacks such as high cost, CO poisoning, methanol crossover, and poor durability which significantly impede the current commercialization of fuel cell and metal-air technologies [5-6]. Thus, numerous innovative approaches have been intensively focused on the discovery of the high performance, excellent durability, and low-cost non-precious ORR electrocatalyst to replace the precious Pt/C catalyst [7-8].

Recently, carbon composite nanomaterials especially heteroatom doped carbon nanostructures have potentially shown as promising non-precious metal free ORRcatalyst due to their low cost, tunable surface chemistry, and facile electron transfer mobility [8]. Among heteroatoms dopants, nitrogen (N), similar in size to carbon, is the most widely investigated as the efficient dopant to promote the ORR [8]. It has been proposed that incorporation of a nitrogen atom into a carbon framework (NC) induces charge redistribution resulting in a slightly positive charge on carbon atoms adjacent to the doped nitrogen atom owing to the intrinsically higher electron negativity of nitrogen relative to carbon atoms [8]. The positively charged carbon atoms are capable of changing the chemisorption mode of an oxygen molecule from the

* Corresponding author: kriangsak.k@sciee.kmutnb.ac.th

https://doi.org/10.1051/e3sconf/202235501006

usual end-on adsorption to side-on adsorption [8]. The diatomic oxygen adsorption can effectively weakening the O-O bond, facilitating oxygen reduction at the active nitrogen-doped carbon sites with lower ORR overpotential [8]. In this regard, various nanostructures of independent nitrogen-doped carbon (NC) catalysts have been studied as a possible replacement for Pt catalysts including N-graphene, N-CNTs, N-mesoporous carbon and N-nanoparticle carbon [8-9]. An excellent example of a nitrogen-doped carbon catalyst has been described by Gong et al. [9]. The nitrogen-doped vertically aligned carbon nanotubes (VA-NCNTs) electrocatalysts were prepared by pyrolysis of iron phthalocyanine under an NH_3 atmosphere and the obtained metal-free catalysts were subsequently investigated the ORR performance under alkaline media [9]. It was found that the metal-free VA-NCNTs promoted a direct 4-electron transfer ORR process with superior ORR activity relative to the precious Pt/C catalyst. In addition, the VA-NCNTs catalysts were much more stable to the CO poisoning and the methanol crossover in comparison with the precious Pt/C catalyst [9]. However, the number and specification of ORR active sites located in the prepared nitrogen-doped carbon electrocatalysts are preferentially altered when various kinds of carbon and nitrogen precursors are thermally carbonized at high temperatures, leading to unpredictable ORR performance.

Density functional theory (DFT) computation demonstrates that the overpotential of the ORR process dramatically decreases by additional co-incorporating sulfur (S) atom into the structure of NC electrocatalyst [10-11]. S atom possesses a similar electronegativity value as the C atom. Functionalizing a single S atom into a carbon structure contributes to the active positive charge on the doped S element owing to the mismatch of the outermost orbital of the S and C atoms [11]. Interestingly, simultaneous doping of S and N atoms into carbon nanostructure results in not only enhancing the active positive charge density but also substantially improving the spin density of the hexagonal C atom next to a doped N atom [11]. The synergistic effect of both asymmetrical spin and charge density contributes to substantially increasing the number of active C atoms, yielding superior ORR performance [11]. Experimentally, S and N atoms have been dually doped into carbon nanosheet [10, 12-13], carbon nanotube [14-15], three-dimensional carbon foam [16], two-dimensional graphitic sheet [17] by using high temperature pyrolyzing various kinds of carbon, nitrogen and sulfur resources under inert and/or the toxic ammonia gases. However, the ORR performance of the reported S and N codoped nanocarbon catalyst is still less competitive with the benchmark Pt/C catalyst [10-16].

Herein, we have developed a simple, cost-effective, and scalable method for fabricating a high performance and durable N and S codoped mesoporous carbon nanostructure (NS-VXC). The fabrication of the NS-VXC product was accomplished by solid state thermolysis a complex containing commercial carbon nanoparticle (VXC 72R), potassium thiocyanate (KSCN) and 2,2-bipyridine (BPD) at relatively low temperature under ambient atmosphere without applying any gas. The morphology and porous structure of the obtained products were systematically characterized by using a scanning electron microscope (SEM) and Brunauer–Emmett–Teller (BET) method, respectively. The amount of nitrogen and sulfur functional groups decorated in carbon structure was verified by the X-ray photoelectron spectroscope (XPS) technique. Then, the product was employed as a non-precious metal free catalyst to electrochemically promote the ORR in alkaline media using rotating ring disk electrode (RRDE) measurement. A novelty of this report is the synthesis method of the dual-doped N and S into porous carbon nanoparticles under solid state thermolysis conditions without purging any gases which have not been reported to date. In addition, the reported nanomaterial synthesis method is environmentally friendly, highly reproducible, template-free and easy to handle. It was found that the fabricated dual doped NS-VXC catalyst exhibited more highly active and durable to facilitate the ORR process than that of the precious Pt/C catalyst under alkaline media.

2 Experiments

2.1 Preparation of NS-VXC precursor

The NS-VXC precursor was prepared by modification of the synthesis method reported in the literature [18]. The carbon nanoparticles (VXC 72R) with an amount of 50 mg were added into a 50 ml flat bottom flask containing 25 ml ethanol. The admixture was ultrasonically dispersed for 30 min. At the same time, iron thiocyanate solution was prepared by dissolving 0.166 g $FeCl_3 \cdot 6H_2O$ and 0.29 g KSCN in 25 ml DI water. Once the carbon mixture was homogeneously dispersed, the iron thiocyanate solution was slowly transferred into the carbon slurry and the admixture was continually ultrasonicated for 60 min. After that, the bipyridine solution was prepared by dissolving 0.315 g 2,2-bipyridine in 10 ml ethanol and the obtained solution was poured into the carbon flask. The slurry flask was stirred for 20 h at room temperature. The solid product was collected by centrifuging the slurry at 13,000 rpm and finally drying at 80 °C overnight.

2.2 Synthesis of NS-VXC electrocatalyst

The dried NS-VXC precursor was finely ground using agate mortar and pestle. The fine powder was then transferred into a Hylok union cell and it was tightly sealed with another plug. A sample cell was put on a rectangular ceramic boat and it was delivered in the muffle furnace at the centre position. The heat treatment process was carried out at 700 °C for 1 h by maintaining the ramp rate at 10 °C min^{-1} under ambient conditions without purging any inert or ammonia gases. Once the calcination process was over, the sample was naturally cooled down to room temperature. The calcined sample was taken into a vial containing 1.0 M H_2SO_4 solution. The acid treatment process was programmed at 80 °C for 8 h to eliminate the unreacted species. The acid treated sample was collected by centrifuging the solution at 13,000 rpm and it was repeatedly washed with a large amount of DI water until the pH of the solution become neutral. The washed sample was then dried at 80 °C in oven overnight and the dried sample was denoted as NS-VXC.

2.3 Physical characterizations

A field-emission scanning electron microscope (FE-SEM) was conducted using the Hitachi S-4800II instrument. SEM images were taken under an accelerated voltage condition of 3 kV, the sample must be coated with osmium before SEM investigation. Brunauer Emmett and Teller (BET) technique was participated to determine the surface area and pore diameter size of the catalysts. In BET analysis, the sample was purified by degassing at 150 °C for 8 h. Powder X-ray diffraction (XRD, PANalytical-Empyrean) technique with Kα radiation and generation voltage of 40 kV was acquired to verify the crystal structure of the samples. Moreover, the elemental compositions of catalysts were analyzed using X-ray photoelectron spectroscopy (XPS, Thermo Fisher Scientific, ESCALAB250 XPS system, Theta Probe XPS system) with monochromated Al Kα source at 15 kV and 150 W. Binding energy values on the X-axis were calibrated using C1s from a carbon value taken as 284.6 eV).

2.4 Electrochemical ORR characterizations

The electrochemical ORR performance of catalysts was estimated using a rotating ring disk electrode (RRDE) technique (Pine instrument). A three-electrode system comprised of glassy carbon, Pt wire and Ag/AgCl which functioned as working, counter and reference electrodes, respectively, were employed. The reference Ag/AgCl potential was converted to the reversible hydrogen electrode (RHE) potential following the equation reported in the literature [19]. The catalyst ink with the composition of 5 mg catalyst, 250 µL of DI-water, 20 µL of Nafion (5%), and 500 µL of isopropyl alcohol was ultrasonically dispersed for 30 min to obtain a homogenous ink [20-22]. The homogeneous ink was then carefully coated onto a clean RRDE electrode and left it dry under room temperature. The content of NS-VXC catalyst loaded on the working electrode was maintained at 0.4 mg cm^{-2}. For comparison, the noble Pt/C catalyst (10% Johnson Matthey) was taken to prepare catalyst ink and it was coated on an RRDE working electrode with an amount of 30 µg cm^{-2}. The ORR catalytic activity was conducted under an O_2-saturated 0.1 M KOH solution at a scan rate of 10 mV s^{-1}. RRDE experiments were repeatedly investigated at less than 3 times.

3 Results and Discussion

3.1 Fabrication of NS-VXC electrocatalyst

The porous N and S codoped carbon product was fabricated through a single-step doping method. The NS-VXC complex was easily prepared by solution mixing the commercial carbon nanoparticle, KSCN and 2,2-bipyridine compounds as carbon, sulfur and nitrogen resources, respectively. The collected fine precursor powder was subjected to solid state thermolysis at 700 °C for 1 h under autogenic pressure conditions without flowing gas. The impurity contained in the as-prepared sample was chemically treated using 1.0 M H_2SO_4 and the purified product was centrifugally separated and finally dried in an oven overnight. X-ray diffraction analysis was utilized to evaluate the crystal structure of the acid treated NS-VXC sample and the XRD pattern result is presented in Figure 1a. There were two peaks located at 2θ of 24.3 and 43.8 degrees, corresponding to the typical single phase of graphitic C(002) and C(100) planes of the turbostratic carbon structure. The fact that the C(002) plane of the carbon is generally center at 26.57 degree [23]. The C(002) plane for the prepared NS-VXC sample appeared at 2θ lower than that of the standard C(002) plane of the turbostratic carbon, indicating an increase in the interlayer spacing due to an increase in the d-spacing which is associated with the incorporation of N and S atoms into the graphitic layers of the carbon framework [23].

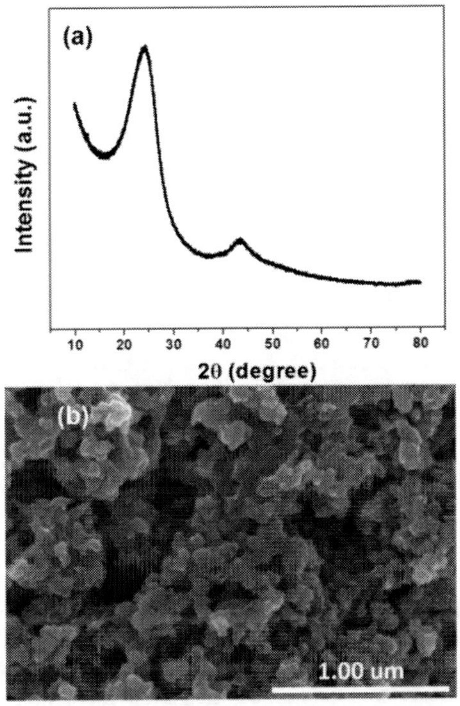

Figure 1 (a). X-ray diffraction pattern and (b) FE-SEM image of the acid treated NS-VXC sample

The morphology of the synthesized NS-VXC sample was subsequently determined using the FE-SEM technique and the SEM image of the acid treated NS-VXC sample is shown in Figure 1b. SEM image clearly reveals uniform nanosized particle connected chain-like structure which is a unique characteristic morphology of VXC 72R. The particle size was approximately in the range of 30-50 nm [21, 23]. In addition, the prepared NS-VXC sample exhibited noticeably high porosity which has benefit in the facile diffusion of reactant and product through the sample. It could be highlighted that thermal annealing of the NS-VXC precursor at 700 °C under autogenic pressure conditions resulted in the decomposition of sulfur and nitrogen precursors into individual atoms, whereas carbon is very stable, and it does not decompose under these

 https://doi.org/10.1051/e3sconf/202235501006

conditions. Atomized sulfur and nitrogen under solid state thermolysis conditions had high enough kinetic energy to randomly incorporate into carbon structure, maintaining the same morphology as the carbon host. This phenomenon was also observed in the morphology of doping heteroatom into graphene [12-13] and CNT materials [14-15].

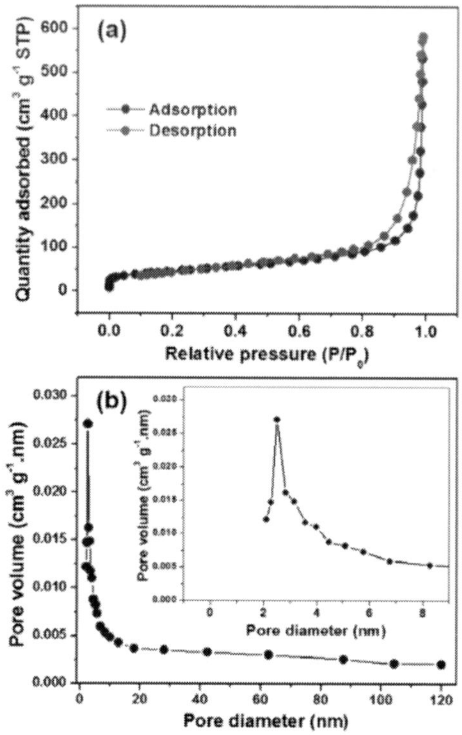

Figure 2 (a) Nitrogen adsorption/desorption isotherm and (b) the corresponding pore diameter size distribution of the acid treated NS-VXC sample

The porous structure and surface property of the acid treated NS-VXC sample was evaluated using the BET technique and the sorption isotherm and pore diameter size distribution results are depicted in Figure 2. The nitrogen sorption isotherm result of the purified NS-VXC sample displays the obvious hysteresis loop, suggesting that the sorption isotherm of the NS-VXC sample was of type IV [24]. According to the International Union of Pure and Applied Chemistry (IUPAC) classification, the formation of the hysteresis loop indicates the capillary condensation taking place in the mesopores [24]. Furthermore, the maximum pore volume of the synthesized NS-VXC sample was found at a pore diameter size of 3 nm (inset Figure 2b). Basically, pore diameter size smaller than 2 nm is classified as micropores, between 2 to 50 nm are known as mesopores and larger than 50 nm are called macropores [24]. Nitrogen adsorption/desorption isotherm and pore size distribution results suggested the calcined NS-VXC product was a mesoporous structure. The BET surface area of the prepared NS-VXC samples was calculated to be 232 m^2 g^{-1}. It has been reported that a catalyst with a mesoporous structure is capable of effectively facilitating the diffusion of the reactant in the ORR process [11, 13].

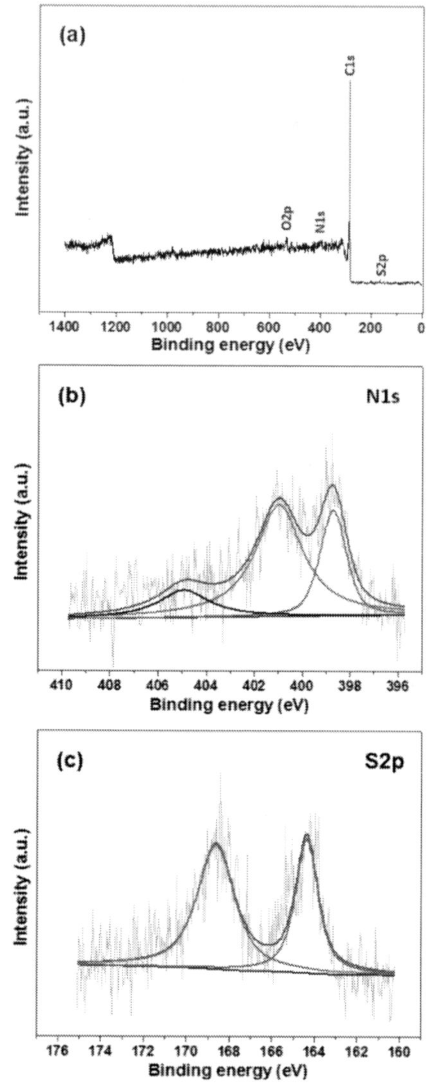

Figure 3 (a) XPS survey spectra, (b) deconvoluted XPS N1s spectra and (c) S2p XPS spectra of NS-VXC sample.

To verify the elemental compositions decorated in the prepared NS-VXC sample, the XPS experiment was acquired and XPS results are displayed in Figure 3. In XPS survey spectra (Figure 3a), elements C, N, S, and O were apparently observed. The presence of N and S elements in the purified NS-VXC sample indicate that N and S atoms were successfully doped into the carbon structure of the synthesized NS-VXC sample. The content of C, N, S and O elements presented in the NS-VXC sample was 93.7, 3.2, 1.1 and 2.0%, respectively. The fact that ferric ion was also added in the preparation of NS-VXC precursor. However, there was no information on Fe species in the XPS survey spectra (Figure 3a), indicating Fe species was not incorporated into carbon structure during the solid state thermolysis process and the remaining Fe compound was totally removed during the acid treatment process. Figure 3b shows the deconvoluted N1s XPS spectra of the prepared NS-VXC sample. The deconvoluted N1s XPS result provided 3 peaks at a binding energy of 398.7, 400.9 and 404.3 eV, which were corresponding to pyridinic-type, pyrrolic-type and graphitic-type nitrogen, respectively [23]. Figure 3c represents the deconvoluted S2p XPS spectra of the purified NS-VXC sample. The deconvoluted S2p XPS spectra generated 2 peaks at a binding energy of 164.3 and 168.4 eV [14]. The peak at a binding energy of 164.3 eV was indexed to a thiophene-like structure where a sulfur atom is embedded between two carbon atoms (C-S-C) [14]. On the other hand, the peak at a binding energy of 168.4 eV was associated with an oxidized sulfur (C-SO_X-C) species that has been proven to be ORR inactive [11]. It has been mentioned that the incorporation of a higher electronegativity N atom into a lower electronegativity carbon atom is attributed to relatively high positive charge density on the adjacent sp^2-bond carbon atoms. Especially, pyridinic-type nitrogen wherein nitrogen atoms present at the edge of the carbon framework bind to two carbon atoms majorly responsible for an ORR active site [11]. On the other hand, the S atom has an identical electronegativity value as the C atom. Doping a single S atom into the C framework resulted in the actively positive charge on the S atom mainly due to the mismatch of the outermost orbital of the S and C atoms [11]. Furthermore, simultaneous incorporation of both electron accepting N atom and S atom into carbon structure significantly contributed to the synergistic effects of asymmetrical spin and charge density which substantially increase the number of active C atoms, yielding superior ORR performance [11].

3.2 Electrochemical ORR performance of NS-VXC catalyst

Next, the purified NS-VXC sample was employed as electrocatalyst material to promote the ORR process in an alkaline solution. The homogeneous NS-VXC catalyst ink was firstly prepared by dispersing the calcined NS-VXC powder in the mixture solution of DI water, isopropyl alcohol and Nafion. A drop of the catalyst slurry was uniformly coated on a clean RRDE working electrode. The amount of NS-VXC catalyst on the RRDE working electrode was kept constant at 0.4 mg cm^{-2}. The catalyst coated RRDE working electrode was then assembled into an electrochemical cell equipped with a Pt wire as counter electrode and an Ag/AgCl as reference electrode under O_2 saturated 0.1 M KOH solution. Figure 4 presents ORR polarization plots using the linear swept voltammogram technique (LSV) for the dual-doped NS-VXC catalyst (Figure 4a) in comparison with LSV results obtained from the undoped commercial carbon nanoparticle (VXC) (Figure 4b) and the benchmark Pt/C (Figure 4c) catalysts at a rotation speed of 800 to 2,000 rpm and 10 mV s^{-1} scan rate under O_2 saturated 0.1 M KOH solution. It was found that the current density for the codoped NS-VXC (Figure 4a), pristine VXC (Figure 4b) and Pt/C (Figure 4c) catalysts noticeably enhanced when the rotation speed was raised from 800 to 2,000 rpm. This typical observation is principally attributed to the shortening flux-diffusion distance at high rotating speeds [23].

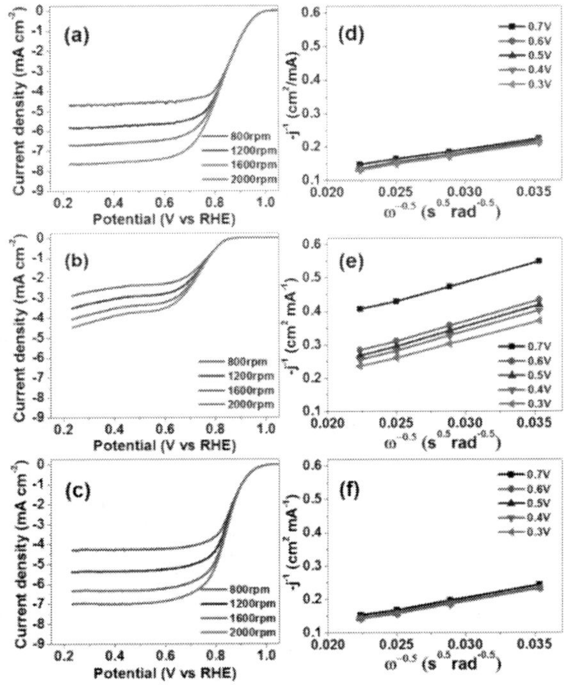

Figure 4 ORR polarization plots conducted by linear sweep voltammograms (LSV) for (a) NS-VXC, (b) VXC and (c) Pt/C catalysts in 0.1 M KOH at 10 mV s^{-1} at various electrode rotation speeds. Corresponding Koutecky-Levich (K-L) plots at different electrode potentials of (d) NS-VXC, (e) VXC and (f) Pt/C catalysts

To verify the ORR kinetic for the catalysts, inversion of current density (j^{-1}) and rotation speed ($\omega^{-0.5}$) at different electrode potentials was qualitatively plotted known as Koutecky-Levich (K-L) plot. The K-L plot results for the codoped NS-VXC (Figure 4d), undoped VXC (Figure 4e) and the commercial Pt/C (Figure 4f) catalysts over the potential range of 0.3 to 0.7 V were linear and parallel, implying the number of electron transfer facilitating ORR process in that potential range was same [23]. Particularly, the K-L plot results obtained from the dual doped NS-VXC (Figure 4d) and the benchmark Pt/C catalysts were almost identical over the potential range of 0.3 to 0.7 V, suggesting the ORR kinetic catalyzed by the dual-doped NS-VXC catalyst was definitely as efficient as promoted by the expensive Pt/C catalyst. The number of electron transfer values calculated from the slope of K-L plots for NS-VXC, VXC and Pt/C catalysts at a potential of 0.5 V was found to be 4.0, 3.1 and 4.0 respectively, suggesting that the fabricated NS-VXC catalyst promotes the ORR process with a directed-four electron transfer number which is as sufficient as the noble Pt/C catalyst.

In order to compare the ORR electrocatalytic activity as well as to understand the mechanism insight about the ORR process, LSV measurement using the RRDE technique was carried out for the codoped NS-VXC catalyst as compared to undoped VXC and the noble Pt/C catalysts at a rotation speed of 1,600 rpm in O_2-saturated 0.1 M KOH solution and the corresponding ring and disk current density is presented in Figure 5a. The content of the inexpensive NS-VXC and VXC catalysts loaded on the RRDE working electrode was about 0.4 mg cm^{-2} while the amount of expensive Pt/C catalyst coated on the working electrode was 30 μg cm^{-2}. The performance of the ORR electrocatalyst can be monitored by 3 major parameters: onset potential (E_{on}), half-wave potential ($E_{1/2}$) and liming current density (I_L) and these parameters can be directly extracted from the disk current density. The ORR onset potential (E_{on}) determines the capability of the catalyst to promote the ORR process. The larger number of E_{on} values indicates a highly efficient electrocatalytic material to accelerate the ORR process. The smaller number of the E_{on} value implies the inefficient ORR catalyst. The E_{on} value observed for the dual-doped NS-VXC, undoped VXC and the noble Pt/C catalysts was found to be 1.01, 0.88 and 1.02 V vs RHE, respectively. In comparison, the E_{on} value extracted from the dual-doped NS-VXC catalyst was 130 mV larger than that of the undoped VXC catalyst and eventually the fabricated NS-VXC catalyst exhibited comparable E_{on} value to that of the benchmark Pt/C catalyst. This means that the fabricated codoped NS-VXC catalyst was greatly effective to facilitate the ORR process similar to the noble Pt/C catalyst. In addition to the E_{on} value, half-wave potential ($E_{1/2}$) is a key potential directly determining the ORR reaction rate. The larger number of $E_{1/2}$ values is responsible for the faster ORR kinetic. The less value of $E_{1/2}$ suggests the sluggish ORR kinetic. In this study, the $E_{1/2}$ value was measured at a disk current density of -3 mA cm^{-2}. the $E_{1/2}$ value obtained for the codoped NS-VXC, undoped VXC and the benchmark Pt/C catalysts was about 0.86, 0.63 and 0.85 V vs RHE, respectively. The codoped NS-VXC catalyst exhibited 230 mV greater the $E_{1/2}$ value than that of the undoped VXC catalyst. Compared to the expensive Pt/C catalyst, the ORR performance of the inexpensive NS-VXC catalyst was 10 mV outperform. Obviously, simultaneous incorporation of both N and S atoms into carbon nanostructure resulted in substantially improving the ORR kinetic rate and eventually overcoming the ORR performance of the benchmark Pt/C catalyst. Besides, E_{on} and $E_{1/2}$, limiting current density (I_L), implies the diffusion rate of the reactant O_2 molecule and ORR product (OH^- compound) through the surface of the catalyst material. The greater I_L value indicates a more highly efficient catalyst to electrochemically reduce reactant O_2 molecule into the OH^- product in the ORR process. In other words, the small number of I_L means unpreferable catalyst material to promote the ORR process. The number of I_L values at potential 0.3 V vs RHE for the codoped NS-VXC, undoped VXC and the noble Pt/C catalysts was found to be -7.2, -3.8 and -6.3 mA cm^{-2}, respectively. The synthesized dual-doped NS-VXC catalyst generated the highest I_L value relative to undoped VXC and the expensive Pt/C catalysts, suggesting a more favourable ORR electrocatalyst in alkaline media.

The outstanding E_{on}, $E_{1/2}$ and I_L values for the prepared codoped NS-VXC catalyst significantly derived from the simultaneous incorporation of N and S atoms into the C structure. Embedding a single N atom into a carbon framework creates an active positive charge on the neighbouring sp^2-bond carbon atoms [11]. In particular, pyridinic-type N has been accepted as the main major ORR active site [11, 23]. On the other hand, the presence of the S atom in the carbon structure leads to an active positive charge on the S atom due to the mismatch of the outermost orbital of the S and C atom [11]. Interestingly, simultaneous codoping of N and S atoms into the carbon framework significantly resulted in the synergistic effects of asymmetrical spin and charge density which substantially improved ORR performance [11]. Not only an electronic effect, but the mesoporous structure and the high surface area of the prepared dual-doped NS-VXC catalyst were essentially responsible for the superior ORR as well.

Generally, the mechanism for the ORR process under alkaline media can be either direct pathway or indirect pathway processes. In direct pathway mechanism, oxygen molecule gains 4-electron transfer directly and electrochemically reduces to OH^- species as the product. In the indirect pathway route, the ORR process is divided into 2 steps. In the first step, the oxygen molecule receives two electron transfers, generating peroxide (HO_2^-) species as an intermediate compound. In the second step, the intermediate HO_2^- compound electrochemically receives another 2-electron transfer to complete the ORR process. Therefore, the amount of peroxide intermediate and the number of electrons transfers can be used to classify the ORR mechanism route. The direct pathway mechanism route requires a 4-electron transfer number with less content of peroxide species. However, a high amount of intermediate compound with low content of electron transfer number suggests an indirect pathway mechanism process. The content of the peroxide intermediate and electron transfer numbers can be calculated based on the ring and disk current density obtained from RRDE measurement at a rotation speed of 1,600 rpm in an O_2-saturated 0.1 M KOH solution. As shown in Figure 5a, the ring current density value of the dual doped NS-VXC and Pt/C catalysts was similar, while, the pristine VXC catalyst generated a much higher ring current density value than those of the codoped NS-VXC and Pt/C catalysts. This implied that the fabricated dual doped NS-VXC catalyst accelerated the ORR process with less amount of peroxide compound. The actual content of OH_2^- species and the number of electron transfers are displayed in Figure 5b. It was found that the amount of intermediate OH_2^- species (blue color) of the dual-doped NS-VXC catalyst was about 2% which is similar to the content of peroxide species derived from the precious Pt/C catalyst. On the other hand, the undoped VXC catalyst delivered a peroxide content of about 40%. The calculated electron transfer number (red color) for the dual doped NS-VXC, pristine VXC and precious Pt/C catalyst were 4.0, 3.2 and 4.0 respectively. The number of electron transfers obtained from RRDE measurement was consistent with the electron transfer number derived from the K-L plot. It is clearly mentioned that the non-precious N and S codoped carbon catalyst was able to electrochemically reduce oxygen via a direct 4-electron reduction pathway in an alkaline medium.

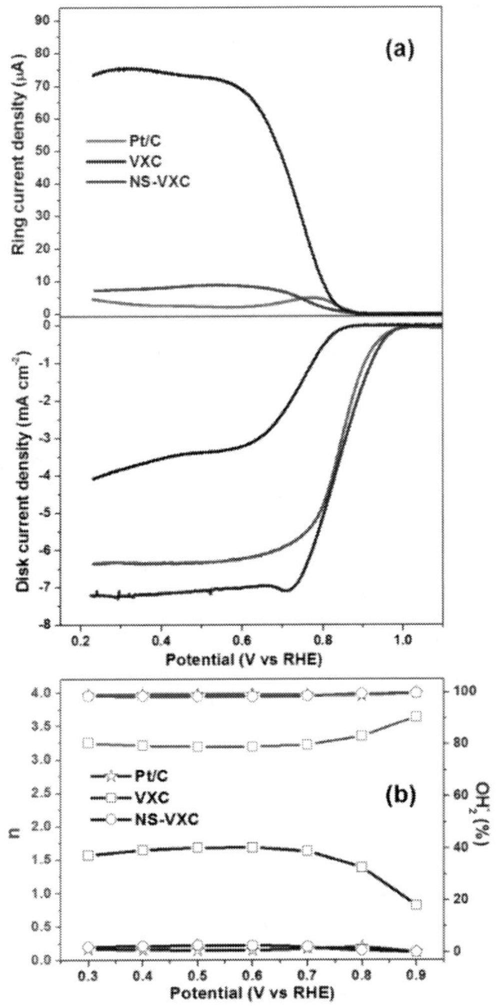

Figure 5 (a) LSVs measurement of NS-VXC, VXC and Pt/C (10% Pt/C, Johnson Matthew) catalysts obtained from rotating ring disk electrode measurement in O_2-saturated 0.1 M KOH at a sweep rate of 10 mV s^{-1}, 1,600 rpm and (b) Percentage of peroxide and the electron transfer number of catalyst at various electrode potential.

The durability is an important parameter indicating the performance degradation of the catalyst during ORR operation. Figure 6 compares the ORR performance degradation of the non-precious NS-VXC catalyst after 10,000 cyclic durability studies at a potential range of 0.6-1.0 V vs RHE with a scan rate of 50 mV s^{-1} in alkaline solution. It was found that the degradation of the number of $E_{1/2}$ values for the codoped NS-VXC catalyst was 1 mV (at a current density of -3 mA cm^{-2}) after 10,000 cycles of durability evaluation. The fabricated NS-VXC catalyst exhibited much durability to perform ORR in alkaline solution. It has been reported that the ORR performance of the noble Pt/C catalyst drastic decayed by 50 mV after 5,000 cycles of durability determination [23]. Pt nanoparticles supported on carbon easily aggregate each other becoming large particles [4]. The large particle size directly results in the obvious reduction of catalyst surface area, yielding less ORR active size and eventually low ORR activity [4]. Moreover, Pt dissolution and carbon oxidation during the accelerated durability test were also noticeably attributed to the degradation of the ORR performance of the noble Pt/C catalyst [5].

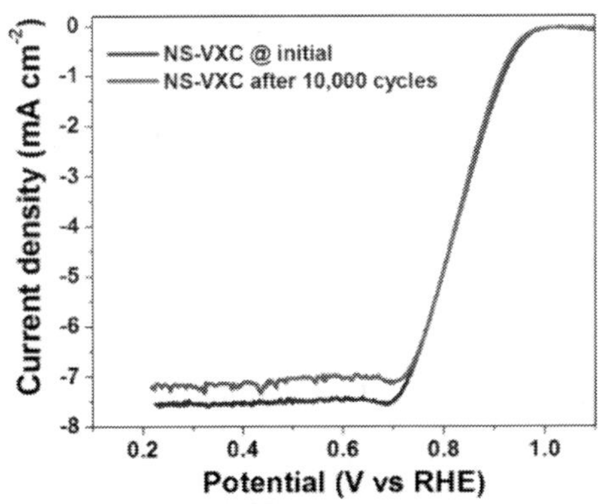

Figure 6 Polarization curves of fabricated NS-VXC catalyst during cycling durability test in O_2-saturated at 1,600 rpm (cycling test were carried out in potential range 0.6 - 1.0V vs RHE with 50 mV s^{-1} in 0.1 M KOH)

4 Conclusion

We succeeded in fabricating N and S codoped into porous carbon nanostructure as a high performance and durable metal free ORR catalyst using a simple and scalable method. The metal free catalyst was obtained by pyrolysis of the nitrogen, sulfur and carbon derived complex at 700 °C for 1 h under autogenic pressure conditions. The acid treated product exhibited a relatively high surface area with mesoporous structure as well as a large number of N and S dopants embedded into the carbon framework. The fabricated catalyst was subsequently electrochemically investigated the ORR performance using the RRDE technique in alkaline media. The N and S codoped carbon (NS-VXC) catalyst exhibited much more efficiency to facilitate the ORR process than the undoped carbon (VXC) and eventually the ORR performance of the dual doped NS-VXC catalyst was outperform the benchmark Pt/C catalyst. Furthermore, the fabricated NS-VXC catalyst demonstrated a highly durable ORR cycling test for 10,000 cycles with negligible performance loss.

Acknowledgements

This research was funded by King Mongkut's University of Technology North Bangkok, Contract no. KMUTNB-65-NEW-06.

References

[1] A. Kulkarni, S. Siahrostami, A. Patel, J. K. Nørskov, Understanding catalytic activity trends in the oxygen reduction reaction, Chemical Reviews, 118(5) (2018):2302-2312.

[2] M. Shao, Q. Chang, J.P. Dodelet, R. Chenitz, Recent advances in electrocatalysts for oxygen reduction reaction, Chemical reviews, 116(6) (2016):3594-3657.

[3] Y.J. Wang, N. Zhao, B. Fang, H. Li, X.T. Bi, H. Wang, Carbon-Supported Pt-Based Alloy Electrocatalysts for the Oxygen Reduction Reaction in Polymer Electrolyte Membrane Fuel Cells: Particle Size, Shape, and Composition Manipulation and Their Impact to Activity, Chemical reviews, 115(9) (2015):3433-3467.

[4] H. Yano, M. Kataoka, H. Yamashita, H. Uchida, M. Watanabe, Oxygen reduction activity of carbon-supported Pt-M (M = V, Ni, Cr, Co and Fe) alloys prepared by nanocapsule method, Langmuir, 23(11) (2007):6438-6445.

[5] H. Yano, M. Watanabe, A. Iiyama, H. Uchida, Particle-size effect of Pt cathode catalyst on durability in fuel cells, Nano Energy, 29 (2016):323-333.

[6] S. Cherevko, N. Kulyk, K. J.J. Mayrhofer, Durability of platinum-based fuel cell electrocatalysts: Dissolution of bulk and nanoscale platinum, Nano energy, 29 (2016) : 275-298.

[7] A.A. Gewirth, J.A. Varnell, A.M. DiAscro, Nonprecious metal catalyst for oxygen reduction in heterogeneous aqueous systems, Chemical reviews, 118(5) (2018):2313-2339.

[8] L. Dai, Y. Xue, L. Qu, H.J. Choi, J.B. Baek, Metal-free catalyst for oxygen reduction reaction, Chemical reviews, 115(11) (2015):4823-4892.

[9] K. Gong, F. Du, Z. Xia, M. Durstock, L. Dai, Nitrogen-doped carbon nanotube arrays with high electrocatalytic activity for oxygen reduction, science, 323(5915) (2009):760–764.

[10] Y. Guo, S. Yao, L. Gao, A. Chen, M. Jiao, H. Cui, Z. Zhou, Boosting bifunctional electrocatalyst activity in S and N co-doped carbon naosheets for high-efficientcy Zn-air batteries, Journal of Materials Chemistry A, 8(8) (2020):4836-4395.

[11] J. Liang, Y. Jiao, M. Jaroniec, S. Z. Qiao, Sulfur and nitrogen dual-doped mesoporous graphene electrocatalyst for oxygen reduction with synergistically enhanced performance, Angewandte Chemie International Edition, 51(46) (2012):11496-11500.

[12] J. Li, Y. Zhang, X. Zhang, J. Huang, J. Han, Z. Zhang, X. Han, P. Xu, B. Song, S, N dual-doped graphene-like carbon nanosheet as efficient oxygen reduction reaction electrocatalyst, ACS Applied Materials & Interfaces, 9(1) (2017):398-405.

[13] K. Qu, Y. Zheng, S. Dai, S. Z. Qiao, Graphene oxide-polydopamine derived N, S-codoped carbon nanosheets as superior bifunctional electrocatalyst for oxygen reduction and evolution, Nano Energy,19 (2016):373-381.

[14] Q. Shi, F. Peng, S. Liao, H. Wang, H. Yu, Z. Liu, B. Zhang, D. Su, Sulfur and nitrogen co-doped carbon nanotubes for enhancing electrochemical oxygen reduction activity in acidic and alkaline media, Journal of Materials Chemistry A, 1(47) (2013):14853-14857.

[15] C. Dominguez, F.J P. Alonzo, S.A.A. Thabaiti, S.N. Basahel, A.Y. Obaid, A.O. Alyoubi, J.L.G. Fuente, S. Rojas, Effect of N and S co-doping of tytymultiwalled carbon nanotubes for the oxygen reduction, Electrochimica Acta, 157 (2015):158-165.

[16] Z. Liu, H. Nie, Z. Yang, J. Zhang, Z. Jin, Y. Lu, Z. Xiao, S. Huang, Sulfur-nitrogen co-doped three-dimensional carbon foams with hierarchical pore structures as efficient metal-free electrocatalysts for oxygen reduction reactions, Nanoscale, 5(8) (2013):3283-3288.

[17] C. Hu, L. Dai, Multifunctional carbon-based metal-free electrocatalyst for simultaneous oxygen reduction, oxygen evolution, and hydrogen evolution, Advanced Materials, 29(9) (2017):160492-16050.

[18] P. Chen, T. Zhou, L. Xing, K. Xu, Y. Tong, H. Xie, L. Zhang, W. Yan, W. Chu, C. Wu, Y. Xie, Atomically dispersed iron-nitrogen species as electrocatalysts for bifunctional oxygen evolution and reduction reactions, Angew Angewandte Chemie, 129(2) (2017):610-614.

[19] T. S. J. Wang, C. Qiu, X. Ling, B. Tian, W. Chen, C. Su, B, N codoped and defect-rich nanocarbon material as a metal-free bifunctional electrocatalyst for oxygen reduction and evolution reactions, Advanced Science, 5(7) (2018):1800036-1800045.

[20] J. Sanetuntikul, K. Ketpang, S. Shanmugam, Hierarchical nanostructured Pt_8Ti-TiO_2/C as an efficient and durable anode catalyst for direct methanol fuel cells, ACS Catalysis, 5(12) (2015):7321-7327.

[21] K. Ketpang, A. Boonkitkoson, N. Pitipuech, C. Poompipatpong, J. Sanetuntikul, S. Shanmugam, Highly active and durable transition metal-coordinated nitrogen doped carbon electrocatalyst for oxygen reduction reaction in neutral media, E3S Web of Conferences, 14 (2019):01005.

[22] K. Ketpang, J. Prathum, P. Juprasat, W. Junla, K. Wichianwat, A. Saejio, C. Poompipatpong, N. Chanunpanich, Electrochemical oxygen reduction reaction performance of water hyacinth derived porous non-precious electrocatalyst in alkaline media, E3S Web of Conferences, 14 (2019):01004.

[23] J. Sanetuntikul, S. Shanmugam, High pressure pyrolyzed non-precious metal oxygen reduction catalyst for alkaline polymer electrolyte membrane fuel cells, Nanoscale, 7(17) (2015):7644-7650.

[24] K. S. W. Sing, Reporting physisorption data for gas/solid systems with special reference to the determination of surface area and porosity, Pure and applied chemistry, 54(11) (1982):2201-2218.

* Corresponding author: kriangsak.k@sciee.kmutnb.ac.th

Understanding the performance degradation and recovery of passive direct ethanol fuel cell

Panuwat Ekdharmasuit[1,*]

[1]Faculty of Science, Energy and Environment, King Mongkut's University of Technology North Bangkok, Rayong Campus, Rayong, 21120, Thailand

Abstract. In developing a fuel cell, one of the major issues that obstruct the commercialization of fuel cells is cell degradation. Meanwhile, the recovery process is an important factor to upgrade the performance and durability of fuel cell system. In this work, a passive direct ethanol fuel cell (DEFC) recovered from the unused aged cell was investigated. The hydrogen evolution method was applied for recovering the cell performance. Electrochemical tools including cell polarization, anode polarization, electrochemical impedance spectroscopy (EIS), ethanol crossover measurement, and chronoamperometry were conducted to examine the activation phenomenon. The polarization curve of the fresh membrane electrode assembly (MEA) was obviously superior to the aged one. The maximum power density was decreased from 1.10 to 0.06 $mW \cdot cm^{-2}$ reaching an approximate 95% decrement after keeping the cell for 1 year. The polarization curve of the MEA after conducting hydrogen evolution was better than that before. The maximum power density was enhanced from 0.06 to 0.07 $mW \cdot cm^{-2}$ presenting an approximate 16.67% increment after recovering the cell. The MEA after the recovery process could reduce ohmic resistance by 67.40% indicating the enhancement of ionic and electronic conductivity and could improve kinetic reaction at the electrode. In conclusion, the recovery process would be helpful for the unused aged cell to improve the performance shortly.

Keyword. Direct ethanol fuel cell (DEFC), Passive fuel cell, Cell recovery process, hydrogen evolution method, Degradation

1 Introduction

Direct alcohol fuel cells (DLFCs) are devices that convert the chemical energy directly to electricity via a redox reaction of liquid fuel and oxidant. The advantages of DLFCs are simplicity, efficiency, low emissions, absence of moving parts and conveniently fuelling without a reforming step [1,2]. These make the DLFCs attractive to be a promising alternative power source for conventional batteries for portable devices.

Among all the investigated DLFCs, direct methanol fuel cells (DMFCs) are one of the most attention research due to the outstanding properties of methanol, such as high energy density, good electrochemical activity, biodegradable and relatively cheap [3]. It is notable that, to commercialize the DMFCs, the study of cell degradation behaviors is known to be another important issue. As we know, degradation behavior in a fuel cell is a combination of permanent and temporary degradation. Permanent degradation originates from electrode materials that are irrecoverable, such as MEA structure degradation, catalyst particle agglomeration/dissolution, catalyst dissolution and crossover, membrane degradation, gas diffusion layers and bipolar plates degradation, etc. While, temporary degradation is based on operating conditions and some reversible materials changes, such as the cathode water flooding, CO_2 accumulation at the anode, Pt oxides formation, etc. and can be recovered either partially or fully.

As of now, researchers have explored the strategies to recover and boost up cell performance due to temporary degradation during the cell operation in a long-term test, which is called the performance recovery technique. Hydrogen evolution is one of the recovery techniques for upgrading DMFC performance. It was observed that, after hydrogen evolution, cell performance was improved due to the improved porosity and tortuosity of the catalyst layers [4] and the oxidative removal of methanol and other reaction intermediates at the catalyst and thereby reducing the catalyst surface oxides leading to a very stable voltage in a long-term test [5,6].

In comparison with DMFCs, the direct ethanol fuel cells (DEFCs) have been introduced to be another great potential for commercialization on account of their advantages of ethanol over other types of liquid fuel, such as high gravimetric energy density, non-toxicity, and availability from renewable sources [7]. Theoretically, protons, electrons and carbon dioxide are obtained at the anode from the ethanol oxidation reaction, while the water is produced at the cathode from the oxygen reduction reaction. In addition, the main products of ethanol oxidation could change to acetaldehyde, acetic acid,

* Corresponding author: panuwat.e@sciee.kmutnb.ac.th

https://doi.org/10.1051/e3sconf/202235501007

carbon dioxide, protons and electrons in case of low temperature operating condition. The reactions are shown in equations (1)-(5) [8].

Ethanol oxidation:

$$C_2H_5OH + 3H_2O \rightarrow 2CO_2 + 12\,H^+ + 12e^- \quad (E^0_{anode} = 0.084\ V) \qquad (1)$$

Oxygen reduction:

$$3O_2 + 12H^+ + 12e^- \rightarrow 6H_2O \quad (E^0_{cathode} = 1.229\ V) \qquad (2)$$

Overall reaction:

$$C_2H_5OH + 3O_2 \rightarrow 2CO_2 + 3H_2O \quad (E^0_{cell} = 1.145\ V) \qquad (3)$$

Partial ethanol oxidation:

$$C_2H_5OH \rightarrow CH_3CHO + 2H^+ + 2e^- \qquad (4)$$

or

$$C_2H_5OH + H_2O \rightarrow CH_3COOH + 4H^+ + 4e^- \qquad (5)$$

Unfortunately, the kinetic reaction and liquid crossover behavior in DEFC would be different features compared to that in DMFC because of the higher molecular weight of ethanol and the slower kinetic rate of ethanol reaction on the Pt catalyst utilized at the cathode. Hence, it was expected that the impact of the activation procedure and the degradation mechanism of a DEFC would be different from its impact on a DMFC.

Recently, the degradation and recovery techniques are investigated in the DEFC during cell operation. Li *et al.* [9] conducted a 520-h long-term durability test of the alkaline DEFC and observed that the cell performance was degraded mainly due to the reduction of the electrochemical-active surface area at the anode. Jablonski *et al.* [10] introduced the periodic short circuit in the pulsed mode of the DEFC electrodes, which could increase both the OCV and efficiency of a DEFC operation due to the oxidation of CO and ethanol to CO_2 and afterward free the catalyst surface at the anode. In addition, the operating concept of DEFC can be classified into an active system requiring power for auxiliary parts and a passive system that involves no energy-consuming auxiliary device for the supply of reactants [8]. Passive DEFC exhibits different mass and heat transport patterns in comparison with active DEFC. The research on passive DEFC is rare and lacks sufficient information with respect to the exact changes, which occur in the MEA recovery behavior. Moreover, in the case of aged MEA of a fuel cell, the performance loss occurs due to cell degradation of both permanent and temporary features. The recovery techniques would be important in realizing a cell performance improvement from the temporary degradation.

In this work, as an attempt at passive DEFC recovery from the unused aged cell, we applied the hydrogen evolution method for recovering the cell performance. Electrochemical tools including cell polarization, anode polarization, electrochemical impedance spectroscopy (EIS), ethanol crossover measurement, and chronoamperometry were conducted to investigate the cell degradation and activation phenomenon.

2 Materials and methods

2.1. Materials and cell fabrication

Nafion® 115 (Dupont) was used as a membrane in this work and prepared by three main steps consisting of cleaning in 3% H_2O_2 solution, boiling in 0.5 M H_2SO_4 solution, and washing in de-ionized water with the condition of 70°C for 1h at each process. The anode catalyst was Pt-Sn (3:1 a/o) on Vulcan XC72 carbon (Johnson Matthey), while the cathode was the commercial gas diffusion electrode (GDE, E-TEK Somerset NJ, USA) containing 0.5 mgPt·cm^{-2} on Vulcan XC72 carbon.

To prepare the anode catalyst ink solution, the anode catalyst powder, Nafion ionomer solution (5wt% in alcohol/water), and isopropyl alcohol were mixed with the aid of an ultrasonic bath for 1 h. The 5 cm^2 anode catalyst layer was prepared by painting the catalyst ink on the commercial gas diffusion layer (E-LAT®, E-TEK) and dried in an oven at 80°C for 1 h. The obtained anode catalyst layer had a metal loading of 2 mg·cm^{-2} with the Nafion ionomer loading of 20% of the dry catalyst layer. The anode and the cathode were fabricated onto each side of the Nafion membrane to construct the membrane electrode assemblys (MEAs) and then were inserted into the in-house fuel cell stack by compressing between two acrylic blocks as shown in Figure 1.

Figure 1. In-house fuel cell stack.

2.2. Cell performance test and electrochemical measurements

The performance and characterization results were obtained with a potentiostat/galvanostat apparatus (Metrohm Autolab, PGSTAT 302N, Utrecht, The Netherlands). The diagram of the apparatus set-up is shown in Figure 2.

Before cell testing, to obtain a steady-state cell behavior, a single cell was activated for 2 h by flowing ethanol solution at 1 M and 1 ml·min^{-1} at the anode and using ambient air, room temperature, atmospheric pressure, at the cathode. Thereafter, the single fresh cell performance was tested by storing an ethanol concentration of 1 M at the anode and contacting with ambient air, room temperature, and atmospheric pressure, at the cathode. Then, the cell was kept in the box for 1 year without any cell operations (unused aged cell). After that, the cell was performed to investigate the performance change. In the beginning, the ethanol/air activation process was applied for 2 h to obtain a steady-state cell behavior with 1 M ethanol concentration, 1 ml·min^{-1} ethanol flow rate, at the anode and ambient air at room temperature, atmospheric pressure, at the cathode. Before H_2 evolution, cell characterizations were analyzed in a passive mode with 1 M ethanol concentration at the anode and ambient air at room temperature, and atmospheric pressure, at the cathode, using potentiostat/galvanostat. Whole cell polarization was measured by a potentiodynamic polarization mode from open circuit voltage (OCV) down to 0.1 V at a scan rate of 10 mV·s^{-1} and the time interval was set as 10 seconds per point. The anode polarization measurement was performed in a potentiodynamic polarization mode at a scan rate of 10 mV·s^{-1} by storing ethanol solution at the anode side and supplying unhumidified hydrogen gas to the cathode at a flow rate of 100 mL·min^{-1} as a dynamic hydrogen electrode (DHE). The protons generated in the anode due to the ethanol oxidation reaction migrated to the cathode and reduced, thereby delivering the hydrogen gas that acted as DHE [11].

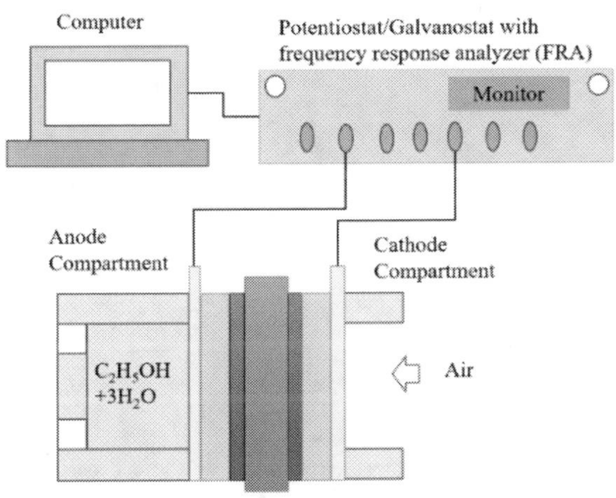

Figure 2. A schematic diagram of apparatus set-up.

EIS measurement was recorded in potentiostatic mode at 0.2 V and the measured frequency was ranged from 5 kHz to 0.05 Hz with 10 points per decade and the amplitude of 20 mV. In the case of the whole cell, impedance spectra were carried out by storing ethanol solution at the anode as a working electrode (WE) and opening to natural air at the cathode as reference (RE) and counter electrode (CE), while the anode impedance spectra were conducted by storing ethanol solution at the anode as a working electrode (WE) and hydrogen gas at the cathode at a flow rate of 100 mL·min^{-1} as a DHE. Further, ethanol crossover by the voltammetric method was conducted by taking purity nitrogen gas (99.98%) to the cathode at a flow rate of 100 mL·min^{-1} as WE and the ethanol solution of 1 M to the anode in passive mode as RE and CE. The applied voltage was varied from 0 to 1.2 V at a scan rate of 2 mV·s^{-1}. The reaction at the anode is hydrogen evolution, while the reaction at the cathode is the oxidation of ethanol crossed from the anode to the cathode [12]. The current discharging with time (chronoamperometry) at the voltage of 0.3 V for 20 min was conducted to investigate the performance behavior during steady state condition.

2.3. Recovery method

The hydrogen evolution technique at the cathode was conducted to recover the cell performance following the procedure elsewhere [6]. Ethanol solution with a concentration of 1.0 M at 1.0 ml·min^{-1} was fed to the anode, while N_2 gas was supplied to the cathode at 100 sccm. The anode was connected with the positive terminal of a and the cathode was connected with the negative terminal. This technique was controlled at a current density of 0.3 V for 10 min. The ethanol was oxidized at the anode to generate protons, electrons, and other products. Protons and electrons moved to the cathode and then they recombined to generate hydrogen. A schematic diagram explaining the hydrogen evolution process at the cathode is presented in Figure 3. Also, The hydrogen evolution at the anode was performed. It was the reverse of the previous procedure at the cathode. For comparison, cell characterizations were analyzed again after hydrogen evolution using the same procedure as conducted before activation.

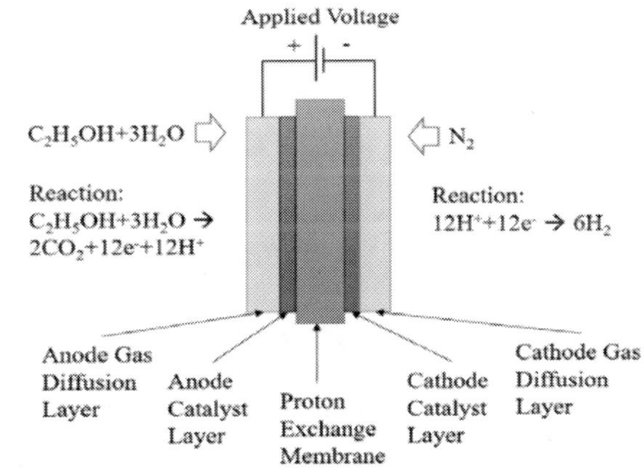

Figure 3. A schematic representation of the H_2 evolution process at the DEFC cathode.

3 Results and discussions

3.1. Cell performance degradation

At the beginning, single fresh cell performance was tested with an ethanol concentration of 1 M at the anode and fed

 https://doi.org/10.1051/e3sconf/202235501007

with ambient air, room temperature, and atmospheric pressure, at the cathode. After that, the cell was kept in the box for 1 year without any cell operations (unused aged cell). The performance of the unused aged cell was tested and compared to that of the fresh cell. Figure 4 shows the comparison of the polarization and power density curves of the fresh and aged MEA. The polarization curve of the fresh MEA was much better than that of the aged one. The maximum power density is decreased from 1.10 to 0.06 mW·cm^{-2} reaching an approximate 95% decrement after keeping the cell for 1 year, corresponding to the degradation rate of around 1.19 x 10^{-4} mW·cm^{-2}/h. The degradation rate in this case is different compared with that in aged DEFC in continuous operation of 500 h (0.03 mW·cm^{-2}/h) [9], in aged DMFC in continuous testing of 4044 h (6.18 x 10^{-3} mW·cm^{-2}/h) [13] and in aged passive DMFC in continuous operation of 3000 h (3.33 x 10^{-3} to 4.33 mW·cm^{-2}/h) [14].

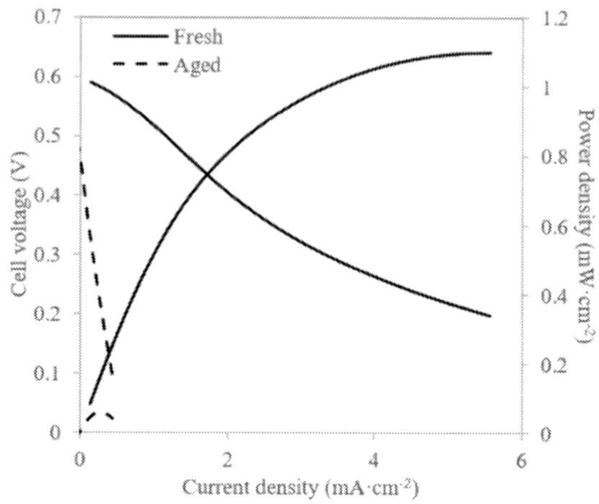

Figure 4. Influence of cell degradation on the cell performance curve.

To better understand the resistance mechanism, the comparison of Nyquist plots by EIS for the MEA before and after hydrogen evolution was carried out, shown in Figure 5. The EIS data were obtained by considering the values at the x-axis intercept, which represented the cell ohmic resistance. The intersection of the fresh MEA takes place at the lower resistance value in the real axis in a comparison with that of the aged MEA. The resistance decreases from 0.4 ohms for the fresh MEA to 1.9 ohms for the aged MEA. This demonstrates that the fresh MEA had much lower ohmic resistance caused by the lower both ionic and electronic conductivity. Besides, the size of the semicircle regards the kinetics of ethanol oxidation. It was found that the size of the fresh MEA is much smaller than that of the aged MEA due to the lower kinetic resistance of the fresh MEA.

3.2. Cell performance after the recovery process

Consequently, as an attempt to recover the unused aged cell, the hydrogen evolution method was devoted to both the anode and cathode for recovering the cell performance. Figure 6 shows the comparison of the polarization and power density curves of MEA before and after the recovery process by hydrogen evolution at both electrodes. The polarization curve of the MEA after conducting hydrogen treatment is better than that before. The maximum power density was enhanced from 0.06 to 0.07 mW·cm^{-2} presenting an approximate 16.67% increment after hydrogen evolution, similar to that reported in Ref. [6], which was improved by an approximate 9.5 to 14.3%.

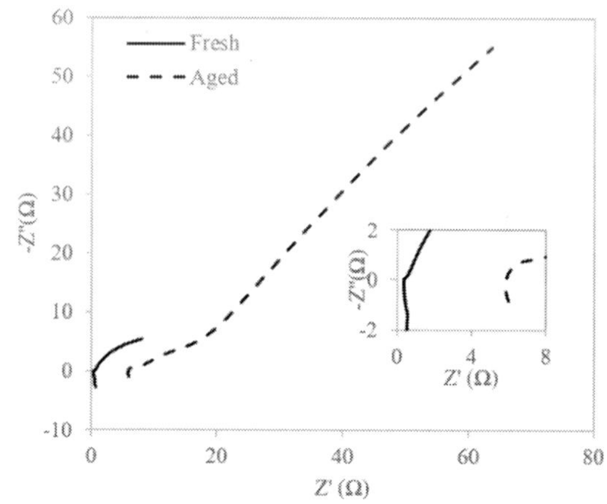

Figure 5. Influence of cell degradation on whole cell Nyquist plot.

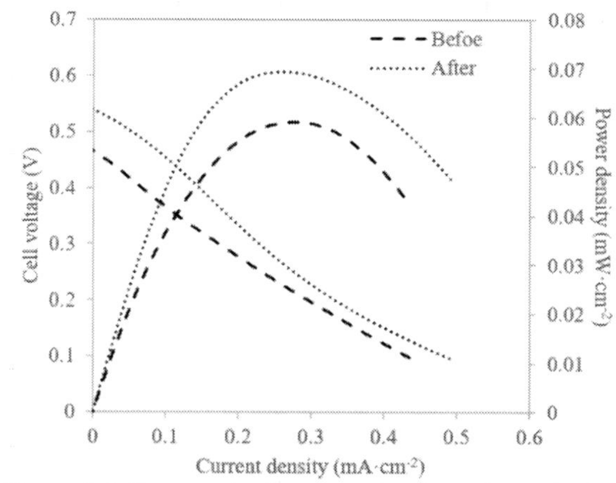

Figure 6. Influence of cell recovery on the cell performance curve.

Figure 7 shows the comparison of Nyquist plots for the MEA before and after the recovery process. The ohmic resistance value (the intersection value at high frequency) of the MEA before the recovery (approximately 5.89 ohms) was higher than that of the MEA after the cell recovery (approximately 1.92 ohms). The cell ohmic loss was reduced by 67.40% after applying the recovery process. This indicates that the MEA after the recovery could enhance cell ohmic resistance caused by the improvement of ionic and electronic conductivity. Moreover, it was observed that the size of the semicircle for the MEA after recovery tends to be smaller than that

 https://doi.org/10.1051/e3sconf/202235501007

of the MEA before the recovery due to the lower kinetic resistance of the MEA after the recovery process.

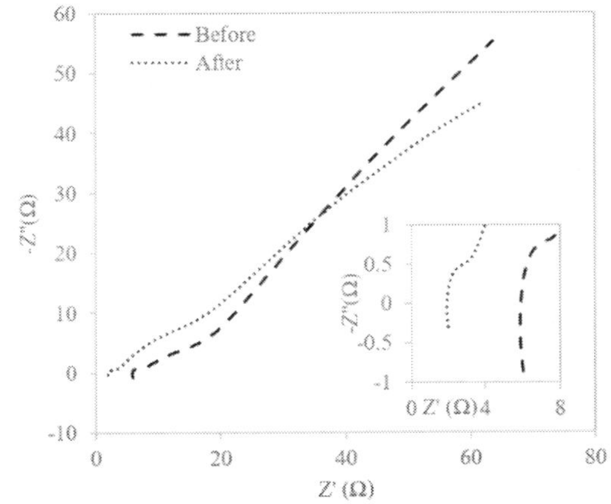

Figure 7. Influence of cell recovery on whole cell Nyquist plot.

Figure 8 shows the comparison of the anode potentials of MEA before and after the recovery process. The result shows that the current of MEA after the recovery process was higher than that of MEA before recovery throughout the voltage range. The current density at the high voltage region of around 0.6 V versus DHE was considered the anode-limiting current density. The MEA after recovery had the anode limiting current density of 1.13 mA·cm^{-2}, which was 10.78% higher than that of MEA before recovery (1.02 mA·cm^{-2}). This can explain that the recovery process could upgrade both the anode kinetic of ethanol oxidation and mass transfer resistance.

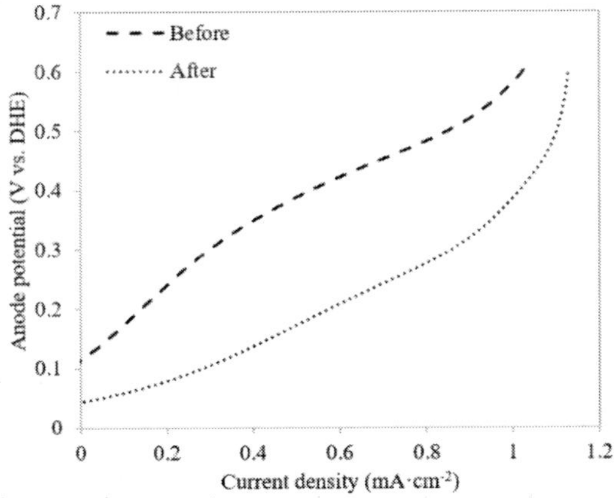

Figure 8. Influence of cell recovery on anode current density.

Figure 9 shows the comparison of Nyquist plots for the anode of MEA before and after the recovery process. The intersection of the MEA before the recovery takes place at the higher resistance value in the real axis in a comparison with that of the MEA after the recovery. The anode ohmic resistance value of the MEA before the recovery (5.27 ohms) is much higher than that of the MEA after the cell recovery (2.58 ohms). It is seen that the ohmic loss at the anode electrode could be reduced by 51.04% after conducting the recovery process. Apparently, it is found that the ohmic loss was mainly reduced at the anode leading to the improvement of cell performance. It should also be mentioned that the order of the resistance observed in the anode half cell is quite consistent with the cell resistance in the whole cell.

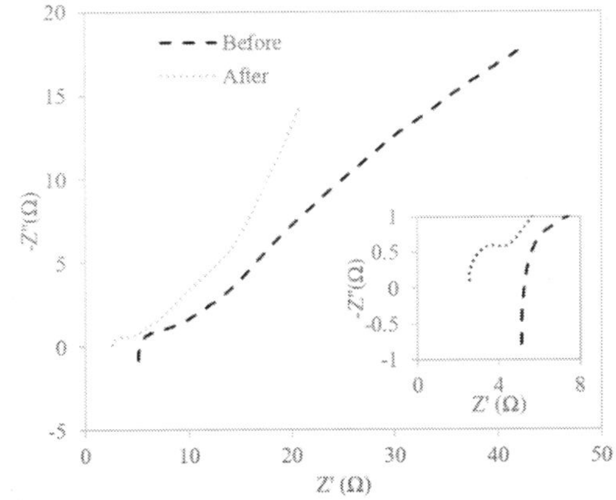

Figure 9. Influence of cell recovery on anode Nyquist plot.

In addition, the size of the semi-circle represents the charge transfer resistances of the cell. Therefore, the charge transfer resistances of the MEA after the recovery seems to be smaller than that of the MEA before the recovery. This result indicates that the recovery process could enhance the electrochemical reactions on the anode electrode.

Figure 10 shows the ethanol crossover current density versus the applied voltage of MEA before and after the recovery process. The current density was considered as the amounts of ethanol permeated from the anode to the cathode. The highest value of current density and a so-call peak crossover current density at the highest voltage was selected for the comparison. A peak crossover current density of MEA before the recovery process was 0.66 mA·cm^{-2}, which was lower than that of MEA after the recovery process at 1.44 mA·cm^{-2}. This means that the MEA after the recovery process increased the crossover of ethanol solution through the membrane. It might be explained that the membrane is completely saturated during the recovery process promoting ethanol diffusivity across the membrane.

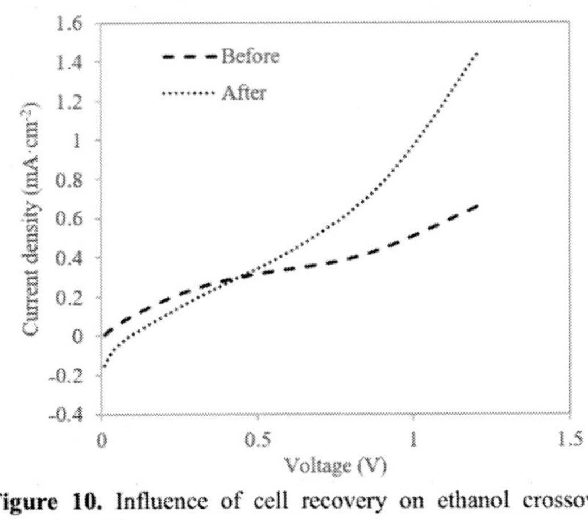

Figure 10. Influence of cell recovery on ethanol crossover current density.

Figure 11 shows the current discharging with time at the voltage of 0.3 V. The result shows that the current of MEA after the recovery process was higher than that of MEA before recovery for the entire time. Moreover, the signal of current discharging of MEA after the recovery process was obviously stable and uniform. It is believed that the MEA after the recovery process could effectively reduce the catalyst surface oxides and maintain a stable current density during cell operation [6].

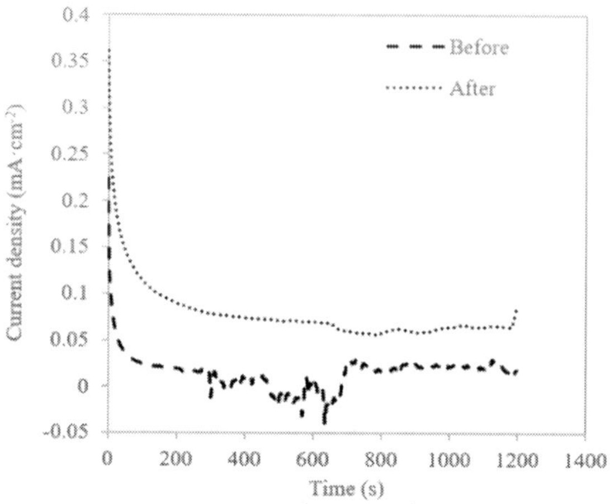

Figure 11. Influence of cell recovery on cell current density variation with time for 20 min.

4 Conclusions

A passive DEFC recovery from the unused aged cell was investigated. The hydrogen evolution method was performed to recover the cell performance. The polarization curve of the fresh MEA was obviously superior to the aged one. The maximum power density was decreased from 1.10 to 0.06 mW·cm^{-2} reaching an approximate 95% decrement after keeping the cell for 1 year. The loss of cell performance was mainly for both the ohmic and the kinetic resistance. Consequently, the cell was recovered by the hydrogen treatment process.

The performance of the MEA after conducting the recovery was improved. The maximum power density was enhanced from 0.06 to 0.07 mW·cm^{-2} presenting an approximate 16.67% increment after recovering the cell. The cell ohmic loss was reduced by 67.40% after applying the recovery process and the kinetic reaction was improved at the electrode. At the anode, the ohmic loss and kinetic activation loss could be reduced after conducting the recovery process. It can be seen that the ohmic loss was mainly reduced at the anode. Also, the crossover of ethanol solution through the membrane was improved after the activation due to the complete saturation in the membrane during the recovery process promoting ethanol diffusivity across the membrane.

Finally, the current discharging with time at the voltage indicated that the MEA after the recovery process could effectively reduce the catalyst surface oxides leading to a stable current density during cell operation. In conclusion, the recovery process would be helpful for the unused aged cell to improve the performance effectively and shortly.

Acknowledgements

This research was funded by King Mongkut's University of Technology North Bangkok [Contract no. KMUTNB-61-DRIVE-030] and the author would like to acknowledge the facility supported from Faculty of Science, Energy and Environment (SCIEE), King Mongkut's University of Technology North Bangkok, Rayong Campus.

References

1. J. Wang, Barriers of scaling-up fuel cells: Cost, durability and reliability, Energy, 80 (2015): 509-521
2. J.P. Pereira, D.S. Falcao, V.B. Oliveira, and A.M.F.R. Pinto, Performance of a passive direct ethanol fuel cell, Journal of Power Sources, 256 (2014): 14-19
3. C. Lamy, A. Lima, V. LeRhun, F. Delime, et al., Recent advances in the development of direct alcohol fuel cells (DAFC), Journal of Power Sources, 105 (2002): 283-296
4. Q. Ye, T.S. Zhao, Electrolytic Hydrogen Evolution in DMFCs Induced by Oxygen Interruptions and Its Effect on Cell Performance, Electrochemical and Solid-state Letters, 8(4) (2005): A211-A214
5. J. Prabhuram, T.S. Zhao, H. Yang, Methanol adsorbates on the DMFC cathode and their effect on the cell performance, Journal of Electroanalytical Chemistry, 578 (2005): 105-112
6. A. Mehmood, H.Y. Ha, Performance restoration of direct methanol fuel cells in long-term operation using a hydrogen evolution method, Applied Energy, 114 (2014): 164-171
7. L. An, T.S. Zhao, Y.S. Li, Carbon-neutral sustainable energy technology: Direct ethanol fuel cells, Renewable and Sustainable Energy Reviews, 50 (2015): 1462-1468
8. V.B. Oliveira, J.P. Pereira, A.M.F.R. Pinto, Modeling of passive direct ethanol fuel cells, Energy, 133 (2017): 652-665
9. Y.S. Li and T.S. Zhao, Understanding the performance degradation of anion-exchange membrane direct ethanol fuel cells, International Journal of Hydrogen Energy, 37, (2012): 4413-4421
10. A. Jablonski and A. Lewera, Improving the efficiency of a direct ethanol fuel cell by a periodic load change, Chinese Journal of Catalysis, 36, (2015): 496-501
11. J. Prabhuram, N.N. Krishnan, B. Choi, T.H. Lim, H.Y. Ha, S.K. Kim, Long-term durability test for direct methanol fuel cell made of hydrocarbon membrane, International Journal of Hydrogen Energy, 35 (2010): 6924-6933
12. S. Song, G. Wang, W. Zhou, X. Zhao, G. Sun, Q. Xin, et al., The effect of the MEA preparation procedure on both ethanol crossover and DEFC performance, Journal of Power Sources, 140 (2005): 103-110

13. F. Jing, R. Sun, S. Wang, H. Sun and G. Sun, Effect of the anode structure on the stability of direct methanol fuel cell, Energy Fuels, 34 (2020): 3850-3857
14. A.M. Zainoodin, S.K. Damarudin, M.S. Masdar, WR.W. Daud, A.B. Mohamad and J. Sahari, Investigation of MEA degradation in a passive direct methanol fuel cell under different modes of operation, Applied Energy, 135 (2014): 364-372

Lead (Pb^{2+}) removal in water using different forms of spent arabica coffee grounds at varying contact time

Jennifer Bandao-Antonio [1,*], *Giovanni* Malapit[1], *John Paul* Mendoza[1], and *Joerandel* Valdez[1]

[1]Department of Physical Sciences, University of the Philippines Baguio, Governor Pack Road Baguio City, 2600 Philippines

Abstract. The availability of clean water for drinking and domestic use is a basic human right, but it is oftentimes denied to some users due to lack of facilities that could clean water. The use of cheap and abundant biomaterials can help solve this problem. Spent coffee grounds is a biomaterial that can remove heavy metals, such as lead (Pb^{2+}) in water. In this study, the Ultraviolet Visible Spectrophotometry analysis resulted in the computation of the Total Removal Efficiency (R%) on lead (Pb^{2+}) in water using raw, crushed, carbonized, and chemically treated spent coffee grounds of the Arabica variety. The crushed and carbonized samples had the highest Total Removal Efficiency (R%) at a shorter coffee-lead contact time. This is due to the exposure of more surfaces as seen in the Scanning Electron Microscopy micrographs, where more open pores and deeper cages or crevices are exposed for metal adsorption. The Atomic % of elemental Carbon was also high for the crushed sample. FTIR analysis revealed that the presence of charged amines ($C{=}NH^+$) in the carbonized and chemically treated samples resulted to lead (Pb^{2+}) adsorption in a short time. Abundant and cheap waste biomaterials, such as spent Arabica coffee grounds can exceed the Total Removal Efficiency (R%) on lead (Pb^{2+}) of commercially available activated carbon.

Keyword. Biomaterials, Spent coffee grounds, Heavy metals, Adsorption

1 Introduction

The need for clean and potable water is also recognized by the United Nations General Assembly in 2010; water should be free of contaminants, readily available, and cheap [1]. Water is a very important commodity for everyone, but it is not oftentimes safe, accessible and cheap. Thus, there is a need to look for viable ways to increase water quality. One of the measures that can be done is the use of water filters but water filters are expensive for the low-average income earners because the cost is approximately ranging from Php 2,500.00 – Php 37,000.00 in the Philippines. Hence, there is a need to fabricate a low-cost filter from locally sourced materials so that everyone can be able to gain an access to safe and quality water. The province of Benguet in the Philippines is well known for producing tons of vegetables being supplied to the different parts of the country. Copper and gold mining is also done in some areas of the province. While this is economically sound, it is observed that these activities are causing water pollution. The vegetable terraces are often infested with pesticides and other chemical agents that leach into the soil and thereby go into the water system, such as springs and rivers. Chemical leaching from mines occurs and is directly affecting the water system. The springs and rivers are the water source that people in Benguet, Philippines use for drinking and other domestic needs. It is a basic human right to have an access to clean water but oftentimes people are not aware of this due to a lack of information. Water filters are also quite expensive and some users do not have the means to buy one. Locally available materials that could be used as low-cost water filter includes coffee grounds, which are oftentimes considered as biodegradable waste materials. In the study of Moelyaningrum *et al.*, Robusta coffee ground residues were able to adsorb lead in water [2].

In Baguio City, Philippines alone coffee is a favorite commodity that is served in homes, coffee shops, and restaurants as a beverage. Ground coffee beans are brewed, the aromatic drink is enjoyed by coffee drinkers and the spent grounds are thrown away as waste material. This widely abundant and cheap waste biomass can be studied for its capacity to remove pollutants from water such as Pb^{2+} (lead). Figure 1 shows a picture of a locally available raw Arabica coffee ground that can be used to adsorb lead (Pb^{2+}) in water.

Figure 1. A raw coffee ground

* Corresponding author: jbantonio4@up.edu.ph

Lead (Pb) is a p-block element in the periodic table found in group 14, period 6. It is a heavy metal used in industries for the production of batteries, ammunitions, solders, alloys, and sheathings. It is also found in the Polyvinyl chloride (PVC) pipes of water distribution system, which may possibly leach into tap water. Pb is also a by-product of copper, gold, or silver mining industries thus the possibility of leakage into our bodies of water. Human exposure to Pb includes neurological and behavioural effects, mortality due to cardiovascular disease, impaired renal function, hypertension, impaired fertility, adverse pregnancy outcomes, delayed sexual maturation, and impaired dental health [3]. The study of the use of cheap and low-cost material from abundant waste biomass could help in improving the water quality in each and everyone's tap.

Studies on the use of waste coffee grounds in removing pollutants from water have been performed by several researchers. Iron (Fe) ion removal in water was investigated by Mohamed and Lee. About 92.9 % of Fe ions were adsorbed by 0.10 g of spent coffee grounds at pH 4 in 60 minutes [4]. Untreated or raw coffee grounds alone can be able to remove lead in water as evidenced by the study by Medici *et al.* Using the Langmuir equation, the group found that the raw coffee grounds had a maximum adsorption of 2.46 mg/g [5]. A similar study on the efficiency of untreated coffee grounds by Azouaou *et al.*, but this time using a fixed-bed column found that the metal uptake capacity of the coffee grounds was 78.95 mg/g [6]. In the study of Seniunaite, *et al.*, they found out that coffee grounds were able to adsorb copper and lead in water and the cleaning efficiency is increased if the particle size is decreased to <200 μm [7]. Activated carbon produced from coffee grounds was effective in removing methyl orange in water via the adsorption process, which was investigated by Rattanapan *et al.* The adsorption process using the activated coffee grounds was found to be feasible and spontaneous [8]. The pH does also affect the adsorption potential of coffee grounds as a low-cost material used in the removal of heavy metals in water. Kyzas *et al.* found that at pH5, coffee ground as an adsorbent had its maximum performance in the adsorption of Copper from water at 76% [9]. In another study using coffee grounds and coffee husks from the Arabica variety, which are both effective in the removal of lead (Pb^{2+}) in water, pH was also seen to greatly affect the heavy metal biosorption process. At increasing pH, the repulsive forces between H^+ and metal ions decrease thus the sorption process is facilitated but at a very high pH, precipitation was observed as stated by Pariona-Palomino *et al.* in their discussion [10]. Malachite green, a synthetic dye was also effectively adsorbed by spent coffee grounds as studied by Mat *et al.* The contact time and percentage removal of dye both increased until equilibrium. It was also revealed that % the removal of dye increased from 77.35% to 94.47% as the dosage adsorbent increased [11]. Tanada *et al.* in 2005 investigated the possibility of removing lead ions in drinking water, findings implied that proteins and fats found in coffee grounds were able to cause greater adsorption of lead ions in water [12]. With the previous works done by other researchers, it was the aim of this study to investigate the use of spent Arabica coffee grounds in removing heavy metals, such as lead (Pb^{2+}) in water. The use of cheap and abundant resources, such as spent coffee grounds to remove heavy metals, such as lead (Pb^{2+}) could benefit people who cannot afford to buy expensive water filters. Thus, the parameters in the preparation of a spent coffee ground sample to remove lead (Pb^{2+}) in water, if proven to be efficient can contribute to the fabrication of an adsorbent for better water quality.

2 Experimental

2.1 Materials and Methodology

2.1.1 Preparation of coffee ground samples

Arabica variety of coffee grounds is available in the local markets of Baguio City, Philippines. Arabica spent coffee grounds used in this study were collected from the waste from a coffee maker. Four forms of Arabica coffee grounds were produced for testing and these are (1) raw, (2) crushed, (3) carbonized, and (4) chemically treated samples. The raw coffee grounds used in the study were boiled for 10 minutes with stirring on a hot plate. This is to remove impurities and water soluble substances. The coffee grounds were vacuum filtered and then oven-dried for 12 hours at 100°C. To produce the crushed sample, the previously boiled raw coffee grounds were crushed using a food processor at maximum speed for 5 minutes and sieved using ASTM (American Society for Testing and Materials) test sieve to produce a uniform sample of 850 μm size. The carbonized sample was also produced from the previously boiled raw coffee grounds. The sample was subjected to a furnace at 600°C for 4 hours. It was crushed using a food processor at maximum speed for 5 minutes and sieved using ASTM test sieve to produce a uniform sample of 212 μm size. The treated sample was taken from the carbonized sample and was soaked in 0.1 M HCl for 48 hours. The coffee grounds were vacuum filtered and oven-dried for 24 hours at 105°C. Figure 2 shows the flowchart of the preparation of coffee ground samples.

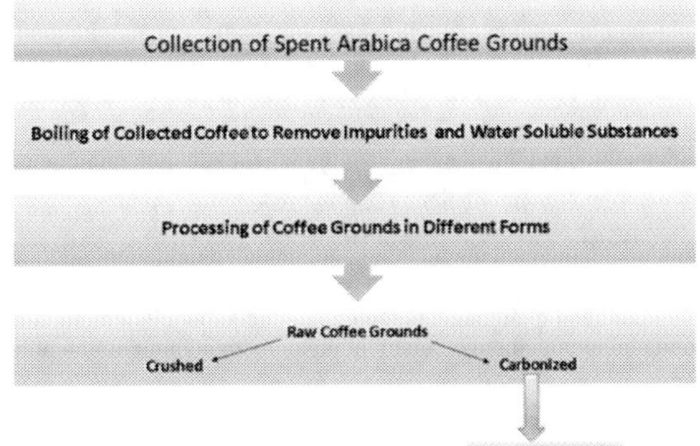

Figure 2. Flowchart for the preparation of coffee ground samples.

The four forms of Arabica coffee ground samples were placed in a sealed polyethylene container and stored in a desiccator. Figure 3 shows an actual set-up in the preparation of the sample where vacuum filtration and sieving were employed in the study.

Figure 3. An actual laboratory set-up for processing of the coffee grounds.

2.1.2 Determination of Total Removal Efficiency (R%) of coffee grounds at varying contact time

For Ultraviolet Visible Spectrophotometry (UV-Vis) analysis, 4.83 x 10^{-3} M Lead (II) Standard Solution (from $PbNO_3$) was prepared by dissolving 160.00 mg of $PbNO_3$ in enough water, pouring the solution into the volumetric flask and filling the flask up to the 100 mL mark. This concentration was chosen in consideration to the response of the UV-Vis spectrophotometer to the Pb solution. Higher concentrations might not give a reliable calibration curve. The cationic micelle cetyltrimethyl ammonium bromide (CTAB) solution of 0.3 M was prepared by dissolving 10.93 g in enough water, pouring the solution in the volumetric flask and filling it up to the 100 mL mark.

The photometric reagent 1,5-Diphenylthiocarbazone (dithizone) solution with a concentration of 1.95 x 10^{-4} M was prepared by dissolving 5.00 mg dithizone in enough ethanol, pouring the solution in the volumetric flask and filling it up to the 100 mL mark. The dithizone-lead solution was acidified using 4 x 10^{-3}M HCl. It was prepared by pouring enough water into a 100 mL volumetric flask, pipetting out 0.0333 mL concentration. HCl from the stock then into the flask, and filling the flask with water up to the 100 mL mark. The procedure was similar to the work reported by Khan *et al.* [13] with some modifications.

For the UV-Vis analysis of the sample, triplicates of each gram of coffee ground samples were placed on a 100 ml beaker, 50 ml of 0.000483 M $Pb(NO_3)_2$ solution was added. The samples were stirred for 1 hour, 2 hours, 3 hours contact time. The coffee-Pb sample solutions after 1 hour, 2 hours and 3 hours contact time were filtered into a cotton plugged funnel. A 25 ml of the filtrate was collected and transferred into a 100 ml volumetric flask containing 1.50 ml 0.000195M Dithizone solution, 4 ml 0.3M CTAB solution and 1 ml of 0.0004M HCl solution then diluted to mark with distilled water and shaken to mix the solution. The coffee-Pb was slightly acidic at pH 6.00. For comparison purposes, a similar procedure was also performed with a commercial activated carbon. The Pb^{2+} concentration in parts per million (ppm) was recorded before and after the batch adsorption experiment. The initial concentration was 197 ppm.

All of the reagents used in this experiment were purchased from Chemline Scientific Corporation in the Philippines and are of analytical reagent grade of known purity.

2.2 Characterization

The Shimadzu UV mini – 1240 UV-Visible (UV-Vis) spectrophotometer was used to determine the Total Removal Efficiency (R%) of coffee samples at various exposure times. The analytical wavelength for lead determination was found at 525 nm. The calibration curve has a linearity of 99.57 %.

Fourier Transform Infrared with Attenuated Total Reflectance (FTIR-ATR) Agilent Cary 630 FTIR spectrophotometer was used to determine the functional groups present in the coffee ground samples. The samples were scanned at a wavelength range of 650 to 4000 cm^{-1} at a resolution of 4 cm^{-1} and with sample and background scans of 32.

Scanning Electron Microscope with Energy Dispersive X-ray Analysis (SEM-EDS) JEOL JSM IT500HR/LA was used to determine the surface morphology and the atomic % of elements found in the different forms of coffee grounds. The samples were Platinum coated and processed using a landing voltage of 15.0 kV at 1,000 X magnification under high vacuum mode.

3 Results and discussion

3.1 Use of UV-Vis Spectrophotometry to obtain the concentration of Lead (Pb^{2+}) before and after treatment

Figure 4 is the summary of results from the UV-Vis analysis of Lead (Pb^{2+}) water samples considering the contact time and types of coffee samples employed in the study. The Total Removal Efficiency (R%) was computed from the amount of lead in water samples before and after contact. This formula was used :

$$R\% = (Co - Ce) / Co \times 100 \quad (1)$$

Where: R% is the total removal efficiency
Co is the initial Lead (Pb^{2+}) concentration
Ce is the Lead (Pb^{2+}) concentration at Equilibrium

The total removal efficiency (R%) was computed from an initial lead (Pb^{2+}) concentration of 197 ppm and an equilibrium concentration of 88 ppm and 89 ppm for both crushed and carbonized coffee ground samples at 1-hour exposure time. These were the parameters with the best adsorption capacity.

Considering the exposure time of commercial activated carbon and coffee ground samples with lead (Pb^{2+}), it can be seen in Figure 4 that the raw coffee ground has the optimum removal efficiency at 3 hours of contact time, which also means that it is a better adsorbent than the commercial activated carbon. As the coffee grounds become smaller, the surface area exposed to lead (Pb^{2+}) becomes bigger thus crushing to produce a uniform size of 850 μm resulting in a better lead (Pb^{2+}) removal efficiency. It can be seen that the optimum removal efficiency occurred at 1 hour contact time, which means that it has better adsorption capacity in a short period of time when compared to the raw coffee ground. This is also observed with the carbonized sample at 212 μm where 1 hour contact time resulted in a similar result to that of the crushed sample. For the chemically treated sample, the optimum removal efficiency was seen at 2 hours of contact time. For 1 hour and 3 hours contact time, the chemically treated sample was not that efficient in removing lead from the solution. This might be due to the soaking of the coffee grounds with 0.1 M HCl. The H^+ might have competed with the Pb^{2+} on the coffee ground surface. Gupta *et al.* mentioned in their review that desorption occurs at a low pH where the H^+ ions replace the metal ions on the surface of the adsorbent [14].

The results also imply that the most practical parameter is 1 hour contact time for both crushed and carbonized coffee ground samples. The shorter the time, the more energy and resources are conserved in removing Pb^{2+} in water. In general, the chemically activated coffee ground has a poor adsorption capacity. The HCl activating agent might not be suitable so there is a need to explore other modes of activation for the Arabica coffee ground variety. At 1 hour contact time, the crushed and carbonized coffee gave the best result in terms of Total Removal Efficiency (R%) at around 55%, which is also seen to be higher than that of the commercial activated carbon with Total Removal Efficiency (R%) of 28%.

The raw coffee ground has also a high Total Removal Efficiency (R%) at 58%. This means that when using raw coffee ground, the contact time between lead (Pb^{2+}) and water sample should be longer, but the 3 hours of contact time might not be practical in some other applications. It is recommended to explore crushing and carbonization as the Pb^{2+} removal efficiency was observed to be maximum at these parameters. When compared to commercially available activated carbon, the coffee ground samples have the potential to remove Pb^{2+} in water. In general, it can be gleaned from Figure 4 that the spent coffee ground is a promising low-cost and widely available material that can be able to remove pollutants such as lead (Pb^{2+}) in water.

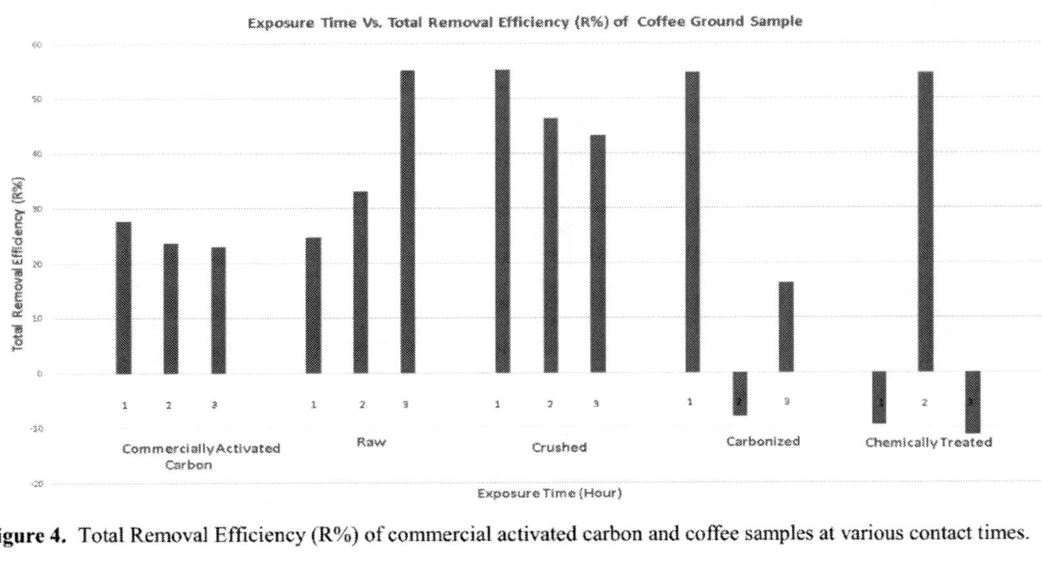

Figure 4. Total Removal Efficiency (R%) of commercial activated carbon and coffee samples at various contact times.

https://doi.org/10.1051/e3sconf/202235502001

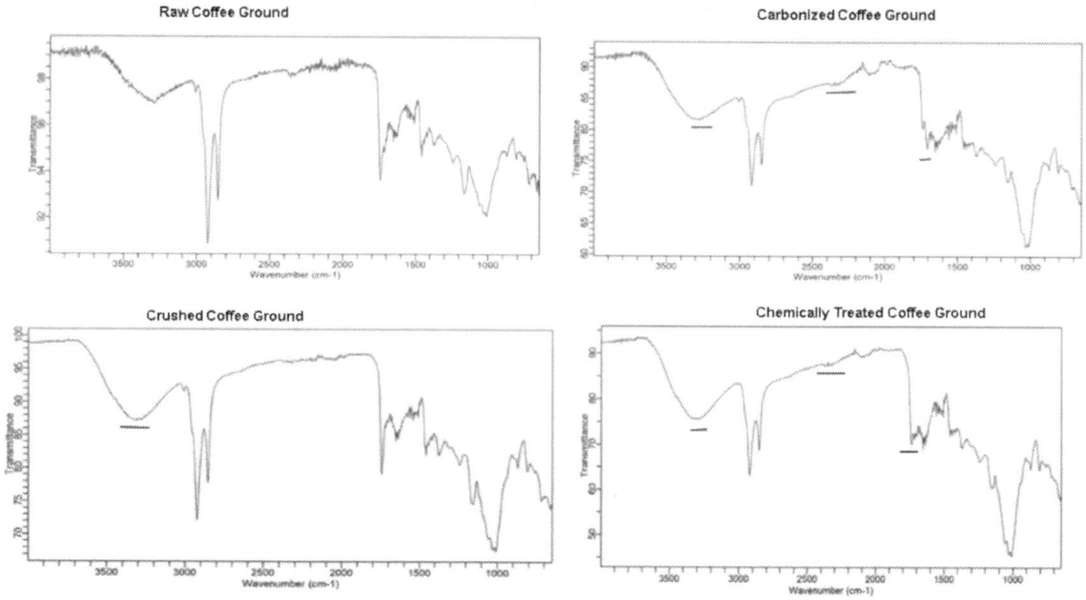

Figure 5. FTIR spectra of raw coffee ground, crushed coffee ground with broad and prominent O-H stretching, carbonized coffee ground and chemically treated coffee ground samples which both exhibit a broad and prominent O-H stretching, presence of charged amines, and shorter C=O peaks.

3.2 FTIR analysis

The FTIR spectra of the coffee samples are characteristics of a very good adsorbent with the following functional groups present: -OH stretching of alcohol at around 3,400 cm^{-1}, charged amines ($C=NH^+$) at around 2,400 cm^{-1}, C=O stretching at around 1,740 cm^{-1}, and –COOH stretching at around 1,450 cm^{-1}. The FTIR peaks were also identified by Mariana, et al. where coffee grounds used as bio sorbents can perform well in the adsorption process [15].

The FTIR spectra of the different coffee ground samples are shown in Figure 5. Comparing the O-H stretching due to the inter- and intra-molecular hydrogen bonding of phenolic compounds, such as alcohol, and carboxylic acid at around 3,200 to 3400 cm^{-1}, the samples exhibited a broad and prominent peak except for the raw coffee ground sample. This may explain why there is a need for more exposure time for the raw coffee ground sample to reach an optimum removal efficiency as previously discussed. The appearance of a peak from 2,200-2,400 cm^{-1} for charged amines ($C=NH^+$) in the carbonized and chemically treated samples are seen in the spectra. The absence of charged amines in the raw and crushed samples and its presence in the carbonized and chemically treated samples might have caused the faster adsorption of Pb^{2+} on the coffee ground surface as evidenced by the optimum removal efficiency at 1 hour and 2 hours of contact time. On the other hand, the carboxyl linkage, C=O derived from xanthine derivatives, such as caffeine at 1,700 – 1750 cm^{-1} became smaller and shorter for the carbonized and chemically treated samples. The better adsorption capacity of the coffee grounds was due to the presence of O-H bonds and charged amines that allowed forces of attraction between the coffee ground surface and lead (Pb^{2+}).

3.3 SEM-EDS analysis

Table 1 presents the Atomic % of the elements present in the coffee ground samples. For the chemically treated sample, the presence of Chlorine came from the soaking of the sample to 0.1 M HCl as an attempt to chemically activate the coffee ground. The good performance of the crushed sample in terms of Total Removal Efficiency (R%) at a shorter contact time may be due to the presence of more carbon atoms, 77.67 ± 0.61% among the prepared samples. According to Zhou and Wang, the high efficiency of bio-based material for metal ion adsorption is due to its high carbon and low ash content [16]. The processing parameters for carbonization and chemical treatment of the coffee grounds might be further investigated and improved to optimize their adsorption capacity. There is a possibility that the temperature used burned more of the carbon in the coffee ground sample.

Table 1. Atomic percentage (%) of elements present in the coffee ground samples and their corresponding uncertainties.

Element	Sample			
	Raw	Crushed	Carbonized	Chemically treated
C	66.13 ± 0.59	77.67 ± 0.61	69.45 ± 0.64	67.56 ± 0.54
O	33.87 ± 0.87	22.33 ± 0.78	30.55 ± 0.92	32.20 ± 0.76
Cl				0.24 ± 0.03

Figure 6 presents the SEM micrographs of the coffee ground samples. It can be seen that the raw sample has lesser pores or openings on the surface. This may explain the longer exposure time it needs to adsorb the Pb^{2+} on its surface. For the crushed sample, there is the presence of open pores and deeper crevices that looks like cages. This may explain the faster adsorption of Pb^{2+} on the coffee ground surface. For the carbonized and chemically treated samples, it can be seen that there is the presence of smaller surfaces with open pores and smaller cages. The open pores and cages increased the surface area of the coffee grounds that were able to trap and adsorb more lead (Pb^{2+}) from the aqueous solution.

Figure 6. SEM micrographs of coffee ground samples at 1000 X.

4 Conclusion

In this study, it can be implied that low-cost and abundant bio-materials can exceed the adsorption performance of commercially available activated carbon. Spent Arabica coffee grounds have a strong potential in removing metals such as lead (Pb^{2+}) in water when processing parameters are optimized. Raw coffee samples can be able to adsorb lead (Pb^{2+}) in water but there is a need for a longer contact time for it to be more efficient. Decreasing the particle size by crushing increases the surface area for lead (Pb^{2+}) adsorption and opens deeper crevices and cage-like structures for further interaction between the functional groups on the coffee ground surface. Thus, more lead (Pb^{2+}) is adsorbed at a shorter contact time. Carbonization and chemical treatment may even be better in removing metals such as lead (Pb^{2+}) in water if processing parameters such as carbonization temperature and activating reagents are further investigated. Despite the low Total Removal Efficiency (R%) in some samples, it can be seen that spent Arabica coffee grounds can be better than that of the commercial activated carbon. The Arabica coffee ground's Total Removal Efficiency (R%) on lead (Pb^{2+}) can be attributed to the presence of an increased surface area and the presence of functional groups such as –OH, $C=NH^+$, C=O, and –COOH which are characteristics of good adsorbent. Moreover, the crushing and carbonization of a spent coffee ground sample can be explored in further studies for its capacity to adsorb pollutants present in water.

Acknowledgements

The authors are thankful to the University of the Philippines Baguio-Cordillera Studies Center (CSC ITRG 2019-2020) for funding this research endeavor.

References

[1] World Health Organization. Guidelines for drinking-water quality: incorporating first addendum, Recommendations. 3rd ed. ISBN 9241546946, 1(2006)
[2] A.D. Moelyaningrum, K. Ngibad, L.P. Lestari, P. Sunu, D.A. Sari, E.P. Permana, D. Hertina, C.F.B. Hartanto, N. Kurniasih. The Robusta coffee grounds residues to adsorb the heavy metal Lead (Pb) in the water, Journal of Physics: Conference Series, IOP Publishing, 1114 (1) (2018): 012058
[3] World Health Organization. Lead in Drinking-water. https://www.who.int/water_sanitation_health/dwq/chemicals/lead, WHO (2011)
[4] M. Khairul Nizam, and Y. Lee Lai, Removal of Fe Ion from Polluted Water by Reusing Spent Coffee Grounds, Pertanika Journal of Science & Technology 27 (3) (2019): 1077-1090
[5] L. Roberto, M. Franco, P. María Silvina, Z. Antonio, Lead removal from water by adsorption on spent coffee grounds, Chemical Engineering Transactions, 47 (2016): 295-300
[6] A. Naima, S. Zahra, M. Hassiba, Removal of lead from aqueous solution onto untreated coffee grounds: a fixed-bed column study, Chemical Engineering Transactions 38 (2014): 151-156
[7] J. Seniūnaitė, R.Vaiškūnaitė, and V. Bolutienė. Coffee grounds as an adsorbent for copper and lead removal form aqueous solutions. The 9th International Conference "Environmental Engineering", Vilnius, Lithuania, (2014)
[8] S. Jurgita, V. Rasa, and B. Violeta,Coffee grounds as an adsorbent for copper and lead removal form aqueous solutions, The 9th International Conference "Environmental Engineering", (2014):1-6
[9] K. George Z., B. Dimitrios N., K. Margaritis, L. Nikolaos K., Copper removal from aqueous systems with coffee wastes as low-cost materials, E3S Web of Conferences, 1(2013):25004
[10] P. Jordy, V. Elvito, V. Armando, Removal of lead from aqueous solutions using different coffee by-products (*Coffea arabica* L.), South Sustainability 2(1) (2021): e031-e031
[11] M. Siti Syamimie Atirah, Z. Sharifah Zati Hanani Syed, A. R. Siti Kartini Enche, S. Khairunissa Syairah Ahmad, H. Noor Amirah Abdul, Z. Nor

 https://doi.org/10.1051/e3sconf/202235502001

Fauziah, Y. Nor Aida, R. Nor Munirah, I. Noor Hidayah, A. Adilah, M. S. Mohd Sharizan, Malachite green adsorption by spent coffee grounds, IOP Conference Series: Materials Science and Engineering. IOP Publishing, 318(1) (2018):012015
[12] T. Toshimitsu, K. Naohito, N. Takeo, A. Jyunichi, T. Seiki, Removal of lead ions in drinking water by coffee grounds as vegetable biomass, Journal of Colloid and interface Science 281(1) (2005): 56-61
[13] H. Khan, A. M. Jamaluddin, B. M. Iqbal, A rapid spectrophotometric method for the determination of trace level lead using 1, 5-diphenylthiocarbazone in aqueous micellar solutions, Analytical sciences 23(2) (2007): 193-199
[14] G. Vinod Kumar, N. Arunima, B. Brij, A. Shilpi, A critical analysis on the efficiency of activated carbons from low-cost precursors for heavy metals remediation, Critical Reviews in Environmental Science and Technology 45(6) (2015): 613-668
[15] M. Marwan, M. Farid, Yunardi, I. Tamer Ahmed, M. F. Hafdiansyah, Activation and characterization of waste coffee grounds as bio-sorbent, IOP conference series: materials science and engineering. IOP Publishing 334(1) (2018): 012029
[16] C. Zhou, W. Yixiang, Recent progress in the conversion of biomass wastes into functional materials for value-added applications, Science and Technology of Advanced Materials 21(1) (2020):787-804

* Corresponding author: jbantonio4@up.edu.ph

Carbon footprint analysis of the cultivated banana cultivation in Prachinburi Province, Thailand

Chalita Suwan[1,*], and *Thanutyot* Somjai[2]

[1] Department of Construction Design and Management, Faculty of Industrial Technology and Management, King Mongkut's University of Technology North Bangkok (Prachinburi Campus), Prachinburi, Thailand

[2] Department of Industrial Management, Faculty of Industrial Technology and Management, King Mongkut's University of Technology North Bangkok (Prachinburi Campus), Prachinburi, Thailand

Abstract. This study aimed to estimate the carbon footprint and analyze the environmental hotspots of the cultivated banana cultivation in a case study of Prachinburi Province in Thailand. In addition, the study proposed the mitigation of the carbon footprint. Primary data were obtained from eleven Prachinburi banana farmers. The banana farmers were interviewed using a questionnaire. Planting area, plant spacing, yield, land preparation, planting, fertilization, irrigation, and harvesting were among the data collected. The results revealed that the carbon footprint of cultivated banana cultivation was 117 g CO_2e/kg of fresh cultivated banana, with fertilization accounting for 83 percent of the total. To predominantly reduce greenhouse gas emissions, the usage of fertilizer should be reduced. Different parts of the plants, such as trimmed leaves, de-suckered shoots, and stems should be chopped into small pieces and spread over the soil where the bananas are planted to nourish the plants without supplemental fertilizer. However, the reduced usage of fertilizer for banana cultivation should be studied along with the productivity of the crop.

Keyword. Carbon footprint, Cultivated banana, Cultivation

1 Introduction

One of the most important challenges of our day is global warming. Excessive greenhouse gas emissions caused by human activities have resulted in many changes including shifting global weather patterns, causing a disturbance in nature's normal balance. Current and future emission rates will likely cause additional global warming, further instigating such changes with irreversible impacts [1].

Effective mitigation of greenhouse gas emissions has been encouraged and practiced globally as a drastic consequence of global warming. The attempts include the reduction of emissions along the product or service supply chain, which are the producers in the manufacturing and agricultural sectors, the providers in the service sector, and the consumers. It is necessary to provide relevant information about the product or service, such as its carbon footprint [2], to assist customers in making ecologically friendly decisions.

Thailand is known as an agricultural country. It has tropical weather because of its proximity to the equator. This weather, which is warm and humid, allows the cultivated banana, or *kluai nam wa*, as it is known in Thai, to be grown everywhere all year round. Therefore, the cultivated banana should be marketed as another potential economic crop supplied in the local marketplace and the export marketplace as either fresh fruit or processed products [3].

The cultivated banana is a cross between two species of wild bananas - *Musa acuminata* and *Musa balbisaina* and belongs to the ABB genome group [4]. The scientific name of the cultivated banana is *Musa ABB* cv. *Kluai 'Namwa'* or *Musa sapientum* Linn. (family Musaceae) [5].

The cultivated banana is a 3–4 meters tall herbaceous perennial that is big, leafy, monocotyledonous, and clump-forming. Underground rhizomes produce leafy shoots. What looks to be a trunk is actually a pseudostem made up of tubular sheaths that are closely furled. Leaf-blades are up to one meter long, oblong, green, waxy, and smooth. The inflorescence develops from the top of the pseudostem once the plant has reached maturity and consists of numerous bracts covering rows of flowers. Fruits grow from female flowers in a large hanging bed cluster known as a bunch. [6]. Banana trees have a lifespan of approximately six years. Every year, each stem produces only one crop of bananas. The pseudostem dies after fruiting. Suckers (ratoons or followers), which are small plants that sprout from the rhizome at the base of the parent plant, are how bananas reproduce and furnish the next crop. [7].

It has been reported that 100 grams of edible ripe cultivated banana have 143 kilocalories, 1.1 grams of protein, 0.2 grams of fat, 33.1 grams of carbohydrate, 2.3 grams of dietary fiber, 7 milligrams of calcium, 43 milligrams of phosphorus, 0.8 milligrams of iron, 5

* Corresponding author: chalita.s@fitm.kmutnb.ac.th

micrograms of vitamin A, and 11 milligrams of vitamin C [8]. Cultivated bananas can be used for a wide range of dietary uses due to their high nutritional value. The fruits can be eaten raw or cooked when they are ripe. They can be mashed and used as baby food or dried and turned into snacks, such as banana chips [6]. Unripe green fruit contains a significant amount of starch, making it a rich source of resistant starch [9]. Banana flowers are consumed as a vegetable. Furthermore, after the fruits have been eaten, the banana peels can be extracted to yield prebiotics that can be used in a variety of foods [10, 11].

For other purposes, especially in Thailand, banana leaves are used for food or dessert wrapping. Parts of the banana trunk are used for decoration in traditional or religious activities [12]. The residual parts of the banana can be used for feedstock or soil improvement.

As a research project under a research program entitled "Raising local plants through research and innovation to a new economic crop", it is necessary to add value to fresh fruit and processed products for the cultivated banana by being conscious of environmentally friendly production to maintain its competitiveness and to lead a low-carbon and environmentally friendly economy [3]. However, it does not appear that a study on the environmental impact of cultivated banana cultivation has been conducted in Thailand or possibly elsewhere. Therefore, this study aimed to analyze the carbon footprint of cultivated banana cultivation in a case study of Prachinburi Province in Thailand. The study also identified the possible hotspots and proposed the mitigation of greenhouse gas emissions.

2 Methodology

2.1 Life cycle assessment and carbon footprint

In this study, a life cycle assessment (LCA) was obtained as an evaluation tool. LCA compiles inputs and outputs and evaluates the potential environmental impacts of products or services throughout their life cycle from raw material extraction through manufacturing, transportation, usage, and disposal [13]. Therefore, LCA is a holistic assessment approach for products or services and has been proved to be a valuable tool for documenting the environmental considerations that need to be part of decision-making in order to achieve environmental sustainability [14].

Carbon footprint is the method to evaluate the potential global warming based on the LCA principle. It is expressed in terms of the amount of carbon dioxide (CO_2), with the addition of its equivalent (CO_2e) in other greenhouse gas emissions specified by the Kyoto Protocol including methane (CH_4), nitrous oxide (N_2O), hydrofluorocarbons (HFCs), perfluorocarbons (PFCs), and sulfur hexafluoride (SF_6). Carbon footprint evaluations provide valuable background information for strategic and operational planning to promote greenhouse gas emission reductions.

2.2 Study scope and system boundary

This study focused on the cultivation of the cultivated banana in Prachinburi, Thailand. The purpose of the study was to analyze the carbon footprint of cultivated banana cultivation, in a case study of Prachinburi Province, to identify possible hotspots and propose the mitigation of greenhouse gas emissions. The functional unit providing a reference to which the inputs and outputs are related was defined as "1 kg of the cultivated banana in a whole bunch form".

The system boundaries of this study are illustrated in Fig. 1. The system boundaries follow a cradle-to-gate approach including land preparation, planting, fertilization, irrigation, harvesting, as well as the production of resources and energy utilized during cultivation. Transportation was excluded from the system boundaries.

2.3 Calculation of cultivated banana cultivation carbon footprint

In this study, primary data were obtained from eleven Prachinburi banana farmers. A questionnaire was prepared and the banana farmers were interviewed. Information collected included: planting area, plant spacing, yield, land preparation, planting, fertilization, irrigation, and harvesting. Practices vary among farmers. Amounts of resource and energy consumption related to the system boundaries were directly collected from the farmers interviewed in Prachinburi Province, Thailand. Greenhouse gas emissions from all cultivation activities were calculated using the multiplication amount of resources, or energy and emission factors obtained from the national database and other literature (Table 1) as shown in Eq. 1 and expressed as the amount of the carbon dioxide equivalent (CO_2e).

$$CO_2e = activity\ data \times emission\ factor \quad (1)$$

Table 1. Greenhouse gas emission factors.

Activity	Unit	Emission factor (kg CO_2e/unit)	Sources
Resource and energy production			
- Diesel	kg	0.3522	[15]
- Lime	kg	1.0676	
- Chicken manure	kg	0.1097	
- Electricity	kWh	0.5986	
Cultivation			
- Combustion of diesel for land preparation	L	2.745	[16]
- Chicken manure application	kg	0.3157	

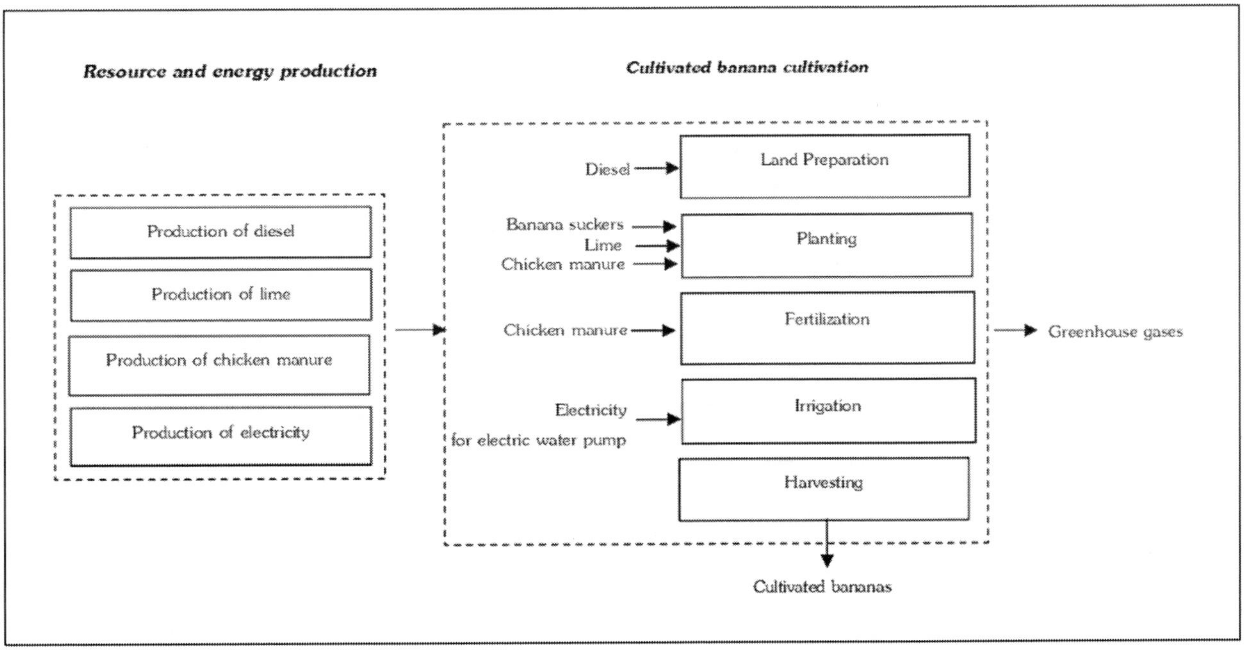

Figure 1. The system boundaries of the study.

3 Results and discussion

3.1 Cultivated banana cultivation in Prachinburi

The following is information on cultivated banana cultivation in Prachinburi.

3.1.1 Land preparation

For the preparation of the land, tractors were employed for tillage. From the information obtained from the field test in Prachinburi, the amounts of diesel consumed for primary tillage and secondary tillage were 5.5 L and 4.5 L per rai (1 rai = 0.16 ha), respectively [17].

3.1.2 Planting

Bananas are perennials that generate succeeding generations of crops. The first cycle after planting is called the plant crop. The ratoon is the sucker (also called the follower) succeeding the harvested plant.

Prachinburi farmers used banana suckers obtained from their orchards or from their neighbors to establish new banana orchards.

According to the recommendation, the planting pit should be around 50 cm × 50 cm × 50 cm with 4 m × 4 m spacing (100 plants per rai). As in the backyard garden and conventional practice, the cultivated bananas in Prachinburi averaged 50 parent plants per rai. For this activity, 3 kg of chicken manure and 2 kg of lime (to reduce soil acidity) were applied per planting pit. Human labor was used in every step of the planting process.

3.1.3 Fertilization

According to the interviews, no farmers in Prachinburi used chemical fertilizers, pesticides, or herbicides on their banana orchards. Farmers applied an average of 7.50 kg of chicken or cow manure per 1 pit every year to nourish their plants. The manure was applied during the vegetative stage and before the flowering stage. The inventory was analyzed based on the preference for chicken manure.

3.1.4 Irrigation

Supplemental irrigation has been necessary for the dry season. In general, an electric pump is used to pump water from a river, pond, or other water sources for banana irrigation. According to the questionnaire, the size of the electric pump, frequency, and interval of irrigation varied among the farmers. From background calculations, the average electricity supplied for the electric water pump was 0.0041 kWh/week for 1 planting pit of cultivated banana. Meteorological data showed the amount of precipitation, which can determine the dry period (precipitation less than 60 mm) of Prachinburi, which lasted for five months, from November to March [18]. For the inventory analysis of the study, it was determined that 22 weeks/year of supplemental irrigation was required.

3.1.5 Harvesting

Around 12 months after planting, bananas are ready to harvest. For this study, it was assumed that in one planting period, bananas can be harvested from the plant crops (the parent plant) and their ratoons for 4 years. Within 4 years,

 https://doi.org/10.1051/e3sconf/202235502002

one plant crop produced 11 ratoons. With the average of 50 parent plants/rai in Prachinburi, there were 600 plants per rai including the parent plants and their ratoons.

A single crop is produced by each banana stem. As a result, 600 banana plants yielded 600 banana bunches within one planting period. In Prachinburi, the average yield was 12 kg per bunch, 8 hands per bunch, and 15 bananas each hand. The bananas were harvested by hand and carried to local markets.

3.2 Life cycle inventory of the cultivated banana cultivation in Prachinburi

Information on the cultivation practices obtained from Prachinburi farmers was analyzed. Table 2 presents the inventory data from the cultivated banana cultivation in Prachinburi based on 1 rai of the banana orchard in one planting period (4 years). Suckers, diesel, lime, chicken manure, electricity, and banana yield were among the items in the inventory of banana cultivation.

Table 2. Life cycle inventory analysis of banana cultivation.

Inventory	Unit	Quantity (per rai)
Input		
Suckers	Plants	50
Diesel	L	10
Lime	kg	100
Chicken manure	kg	1,650
Electricity	kWh	18.04
Output		
Banana yield (a whole bunch)	kg	7,200

3.3 Carbon footprint analysis of the cultivated banana cultivation in Prachinburi

Greenhouse gas emissions from cultivated banana cultivation were calculated and illustrated in Table 3. The cradle-to-gate carbon footprint of the cultivated banana cultivation was 117 g CO_2e per 1 kg of the whole bunch of the cultivated bananas. Fig. 2 provides a relative breakdown of the carbon footprint by source. Fertilization either in terms of fertilizer production or N_2O emissions, along with fertilizer application, contributed significantly to greenhouse gas emissions. Fertilization was responsible for 83% of greenhouse gas emissions.

Table 3. Greenhouse gas emissions from cultivated banana cultivation in Prachinburi Province.

Activity	Greenhouse gas emissions	
	(g CO_2e/kg banana)	Percent
Resource and energy production		
- Diesel	0.416	0.35
- Lime	14.828	12.62
- Chicken manure	25.140	21.40
- Electricity	1.500	1.28
Banana cultivation		
- Combustion of diesel during land preparation	3.241	2.76
- Chicken manure application for plant nourishment	73.248	61.59
Total	117.472	100

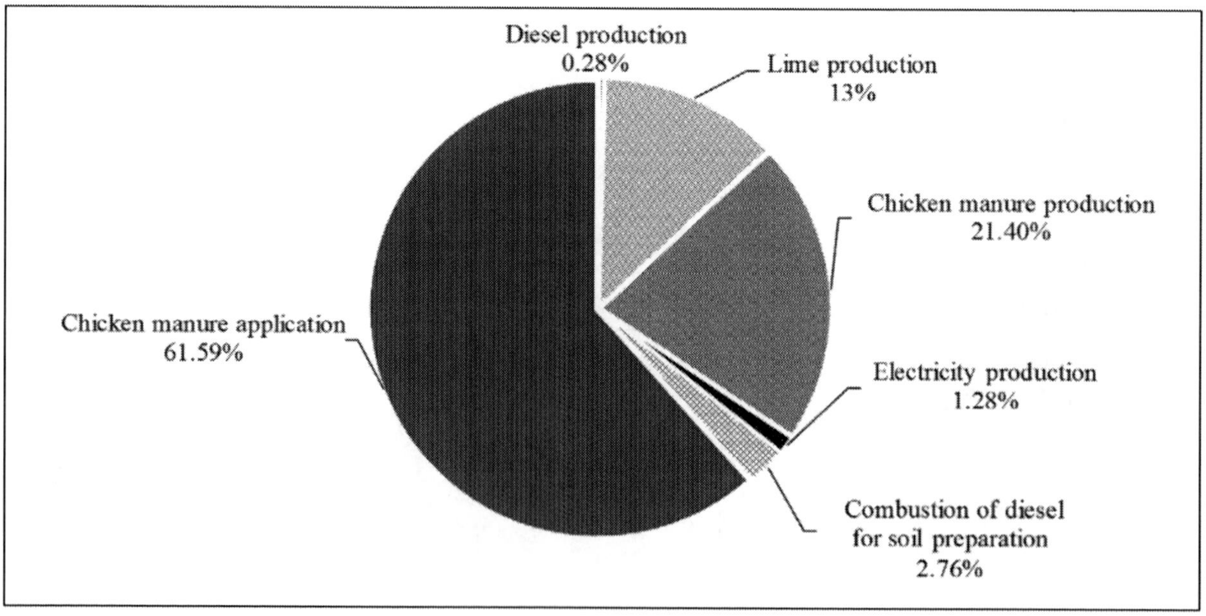

Figure 2. Carbon footprint breakdown of cultivated banana cultivation in Prachinburi Province.

As fertilization was dominant in contributing to greenhouse gases despite organic fertilizer usage, the mitigation of the carbon footprint should be achieved by reducing the use of fertilizer. One article [19] suggested that different parts of bananas, such as trimmed leaves, de-suckered shoots, and stems cut after harvesting, should be chopped into small pieces and spread over the soil where bananas are planted to increase soil nourishment without any supplemental fertilizer. The best way to replenish the nutrients that plants require is to bring pieces of the same plant to nourish it. However, the reduced use of fertilizer in banana cultivation should be investigated alongside the crop's productivity.

3.4 Discussion

3.4.1 Carbon footprint uncertainty

Several factors may cause the uncertainty of the carbon footprint analysis. Gathering all the essential data proved difficult, due to the variety of cultivation practices. The planting technique, the fertilization application, the irrigation, and other factors varied among farmers.

This study also used the lower end and the upper end of the inventory data to provide a minimum value and a maximum value of the possible carbon footprint for this case study. The carbon footprint varied from 28 to 246 g CO_2e/kg of banana. The findings were in line with the prior assessment that fertilizer was responsible for the majority of greenhouse gas emissions from banana production.

3.4.2 Carbon footprint comparison of the cultivated banana with other banana cultivars

It does not appear that there has been a study on the environmental impact of cultivated banana cultivation in Thailand or perhaps in other countries. For Cavendish bananas, only a few carbon footprint analyses have been published, and the results have varied depending on data and the methodology, such as defined system boundaries. Carbon footprints of a Cavendish banana supply chain ranged from 324 to 1,124 g CO_2e/kg of banana. Focusing on the primary stage, the greenhouse gases generated on farms, cradle to retail, were between 16% to 20% [20]. Therefore, the amount of carbon footprint from the cultivation was 52 to 225 g CO_2e/kg of banana. Another study analyzed the carbon footprint of Cavendish banana cultivation in Costa Rica. The result revealed a carbon footprint of 220 g CO_2e/kg of banana [21]. The carbon footprints of Ecuadorian bananas grown on conventional farms and organic farms were 302 g and 249 g CO_2e/kg of banana, respectively [22].

These amounts of carbon footprint were compared to the 117 g CO_2e/kg banana obtained from our study. Our outcome was lower because of the use of manure instead of chemical fertilizer, no energy-intensive activity for land preparation and planting, as well as no use of any pesticide or herbicide. Furthermore, the system boundaries of our study followed a cradle-to-gate approach, excluding packing bananas in carton boxes. The carton boxes are the most important factor that produces a high carbon footprint [20]. However, fertilization played a significant contribution to greenhouse gas emissions for banana cultivation in all studies.

3.4.3 Carbon footprint comparison of the cultivated banana with the other crops in Prachinburi Province

There were studies of carbon footprint analysis of crops in Prachinburi, which covered sugarcane [23], cassava [24], and Napier Pakchong 1 grass [17]. The carbon footprints of these crops as well as the cultivated banana were calculated per 1 hectare (ha), and were 3,463, 1,537, 1,727 and 1,320 kg CO_2e/ha, respectively. Sugarcane contributed the highest carbon footprint as compared to other crops. The majority of the carbon footprint of sugarcane was from greenhouse gas emitted from pre-harvest biomass burning. For other crops, the carbon footprint was affected by fertilizer application in terms of type and quantity.

4 Conclusion

This study aimed to estimate the carbon footprint and to analyze the environmental hotspots of the cultivated banana cultivation in a case study of Prachinburi Province in Thailand. In addition, the study proposed the mitigation of the carbon footprint. Primary data were obtained from eleven Prachinburi banana farmers. The banana farmers were interviewed using a questionnaire. Planting area, plant spacing, yield, land preparation, planting, fertilization, irrigation, and harvesting were among the data collected.

According to the results, the carbon footprint of cultivated banana cultivation was 117 g CO_2e/kg of fresh cultivated banana, with fertilization accounting for 83 percent of the total. To predominantly reduce greenhouse gas emissions, the consumption of fertilizer should be reduced. However, the reduced usage of fertilization for banana cultivation should be further studied along with the productivity of the crop.

Due to the variety of cultivation practices that cause the uncertainty of the carbon footprint analysis, more banana orchard samples should be considered. In addition, the orchards should be classified by size to make the analysis and the conclusion comprehensible.

Carbon footprint assessment of plant-based products is often a part of efforts to shift to more sustainable practices, both at the farm or cultivation stage and throughout the supply chain [25]. Further research along the supply chain is highly recommended for the cultivated banana in Thailand as fresh fruit or processed products, in order to raise local plants as a new economic crop.

Acknowledgments

This work was supported by (I) King Mongkut's University of Technology North Bangkok (KMUTNB), and (II) the National Science, Research and Innovation Fund (NSRF), Contract Number: KMUTNB-BasicR-64-38-3.

References

[1] Ministry of Natural Resources and Environment, Thailand (2021). Mid-century, Long-term Low Greenhouse Gas Emission Development Strategy, Thailand, Submitted under the Paris Agreement, https://unfccc.int/sites/default/files/resource/Thailand_LTS1

[2] Carbon Footprint International. https://carbon footprintinternational.com/thailand/

[3] Manager Online (2016). Beneficial Banana Planting Strategy, https://mgronline.com/smes/detail/9630000126929

[4] B. Silayoi, C. Babprasert, and K.Suwitavas (2017), List of Banana Cultivars. http://www3.rdi.ku.ac.th/ exhibition/Techno_ku60/res-16/index16.html

[6] Faculty of Pharmacy, Silapakorn University (2017). Medicinal Plant Information. https://pharmacy. su.ac.th/herbmed/herb/text/herb_detail.php?herbID=16

[7] Pha Khao Lao (2021), Banana. https://www.phakhaolao.la/en/kb/0000362

[8] Svarsth Life. How many year banana tree live? https://www.svasthlife.com/organic-products/how-many-years-banana-tree-live.html

[9] Bureau of Nutrition, Department of Health, Ministry of Public Health (2018), Food Composition of Thai Food. https://thaifcd.anamai.moph.go.th/nss

[10] P. Jaiturong, N. Laosirisathian, B. Sirithunyalug, S. Eitssayeam, S. Sirilun, W. Chaiyana and J. Sirithunyalug. Physicochemical and prebiotic properties of resistant starch from Musa sapientum Linn., ABB group, cv. Kluai Namwa Luang, Helyon, 6(2020):e05789

[11] S. K. Syahpura, Zulfahmi and A. Syofiana. Analysis of Fructuktooligosaccharide Isolution Banana Peel (Musa paradisiaca var. sapientum (L.) as Prebiotic in the Ration (In Vitro), International Journal of Applied Agricultural Research, 14 No.2 (2019):87-92

[12] H. F. Zahid, C. S. Ranadheera, Z. Fang and S. Ajlouni. Utilization of Mango, Apple and Banana Fruit Peels as Prebiotics and Functional Ingredients. Agriculture, 11(2021)

[13] R. Termpitipong. Banana by-products in Thailand Exploring its feasibility as bioplastics feedstock for food packaging, Master Thesis, Faculty of Engineering LTH, Lund University (2020)

[14] ISO 14044, Environmental management - Life Cycle Assessment - Principles and Framework

[15] C. Suwan and S.H. Gheewala, Application of LCA to support solid waste management policy in Phuket,Waste Management, 10(2012):222 – 238

[16] Thailand greenhouse gas management organization, Greenhouse gas Emission Factors. http://thaicarbonlabel.tgo.or.th/admin/uploadfiles/ emission/ts_f2e7bb377d

[17] Thailand greenhouse gas management organization, Greenhouse gas Emission factors for organization carbon footprint.http://cfologov.com/search/search

[18] T. Somjai and C. Suwan. Carbon Footprint of Napier Pakchong 1 Grass Plantation in Prachinburi Province, In Proceedings of the Research, Invention, and Innovation Congress (RI2C) (December 11-13, 2019). Arnoma Grand Hotel, Bangkok: Thailand. (2019):1109-1116

[19] Meteorological Department (2016) Agricultural Meteorology to know for Prachinburi. http://www.arcims.tmd.go.th/DailyDATA/Eastern.html

[20] Rakbankerd (2016). The Miracle of Microorganisms VS. Banana Shoots. https://www.rakbankerd.com /agriculture/

[21] Food and Agriculture Organization of the United Nations. Carbon footprint of banana supply chain. https://www.fao.org/world-banana-forum/projects/good-practices/carbon-footprint/en/

[22] Erik Svanes and Anna K. S. Aronsso. Carbon footprint of a Cavendish banana supply chain. Life Cycle Assessment, 18(2013):1450-1464

[23] L. Roibás, A. Elbehri and A. Hospido. Evaluating the sustainability of Ecuadorian bananas: Carbon footprint. Water usage and wealth distribution along the supply chain, Sustainable Product and Consumption, 2(2015):3-16

[24] C. Suwan and T. Somjai. Carbon Footprint of Sugarcane Cultivaiton in Sakaeo and Prachinburi Provinces, RMUTI Journal, Special Issue 1, The 6th International Conference on Science, Technology and Innovation for Sustainable Well-Being (STISWB VI), 28-30 August 2014, Siem Reap, Kingdom of Cambodia, (2015):469-478

[25] T. Somjai and C. Suwan. Carbon Footprint Analysis of Cassava Cultivation in Sakaeo and Prachinburi Provinces, RMUTI Journal, Special Issue 1, The 6th International Conference on Science, Technology and Innovation for Sustainable Well-Being (STISWB VI), 28-30 August 2014, Siem Reap, Kingdom of Cambodia, (2015):464-469

[26] L. Roibás, A. Elbehri and A. Hospido. Carbon footprint along the Ecuadorian banana supply chain, methodological improvements and calculation tool. Cleaner Production. 12(2016):2441-2451

The impact of light intensity and wavelength on the performance of algal-bacterial culture treating domestic wastewater

Keerthi Katam[1]*, *Rishika* Ananthula[2], *Sushmitha* Anumala[2], *Malinee* Sriariyanun[3], *and Debraj* Bhattacharyya[2]

[1]Department of Civil Engineering, École Centrale School of Engineering, Mahindra University, Telangana, 500043, India
[2]Department of Civil Engineering, Indian Institute of Technology Hyderabad, Kandi, Telangana, 502285, India
[3]Biorefinery and Process Automation Engineering Center, Chemical and Process Engineering, The Sirindhorn International Thai-German Graduate School of Engineering, King Mongkut's University of Technology North Bangkok, Bangkok, 10800, Thailand

Abstract. Light is the main energy source for microalgae, and the intensity and wavelength of light influence cell metabolism and biomass composition, which, in turn, affects wastewater treatment. The objective of this study is to examine how different light intensities and light wavelengths affect the growth of mixed algal-bacterial culture while treating sewage. Three different light intensities (100, 200, and 300 µmol/m2-s) of four different light wavelengths (blue, red, white, and yellow) were selected for this study. The dissolved organic carbon (DOC), dissolved nitrogen (DN), and dissolved phosphorus (DP) in influent and effluent samples were measured, along with chlorophyll content in the biomass. The highest chlorophyll concentration of 3.5 mg/L was observed at 100 µmol/m2-s intensity of red light. The concentration of chlorophyll decreased as light intensity increased, with exception of white light. The highest DOC removal of 84% was observed at 300 µmol/m2-s intensity of blue light whereas the highest DN (51%) and DP (80%) removal was observed with a red light intensity of 100 µmol/m2-s. Overall, blue light with an intensity of 300 µmol/m2-s and red light with an intensity of 100 µmol/m2-s were found to be the most efficient at removing carbon and nutrients. The results suggested that the color and intensity of light influence algal-bacterial growth and wastewater treatment efficiency.

Keyword. Artificial light, Carbon removal, Chlorophyll, Nutrient removal

1 Introduction

The application of algae in wastewater treatment has been recognized as a sustainable approach for simultaneous treatment and useful biomass production in the last couple of decades [1,2]. Despite the known benefits, algae have only been employed in wastewater treatment as a polishing unit in algal ponds. One of the most challenging aspects of using microalgae for commercial applications is determining the best conditions for large-scale biomass production. The design of the cultivation system has a significant effect on microalgal growth in wastewater treatment. The large-scale algal reactor configurations mainly include open suspended-growth bioreactors, closed suspended-growth photobioreactors, and biofilm or attached growth photobioreactors. Open algal systems use solar energy to grow and have a low energy cost but it has drawbacks in regulating operating parameters, such as light intensity or temperature, thereby, impacting efficient algal cultivation [3–6]. Due to a lack of natural light on rainy days and too much light in the summer, open algal wastewater treatment systems are inefficient in nutrient conversion. Light is one of the important factors for microalgal photosynthesis, both light wavelength and intensity have effect on algal growth and lipid accumulation.

Microalgae require light to synthesize essential molecules, including Adenosine triphosphate and Nicotinamide Adenine Dinucleotide Phosphate Hydrogen [7]. These reactions take place in photosystems in microalgae, and light energy is absorbed by chlorophyll pigments and carotenoids that make up a photosystem antenna complex. The light energy is absorbed and delivered to the reaction center, where it is transformed into chemical energy for photosynthesis [8]. Chlorophylls are magnesium-tetrapyrrole molecules that are crucial to the charge-separation processes that take place in reaction centers and in the light-harvesting complexes, which capture light energy [9,10]. The light reaction and the dark reaction are the two processes of photosynthesis. The two photoreactions in the light reaction are photosystem I and photosystem II, which utilize photon energy associated with two different wavelengths (700 nm and 680 nm) [11]. Chlorophyll-a is the major pigment in algal cells, while chlorophylls b, c, d, and f are minor pigments. The primary electron donor in the reaction centers of photosystems I and II is chlorophyll-a [9]. The Calvin cycle, also known as the dark reaction, is the component of the light reaction in which the energy acquired is used to produce glucose.

* Corresponding author: keerthi.katam@mahindrauniversity.edu.in

Microalgae use light with wavelengths ranging from 400 to 700 nm for photosynthesis. Depending on the species, microalgae absorb different wavelengths. In green microalgae, for example, chlorophylls, a primary pigment that absorbs light in the 450–475 nm and 630–675 nm wavelength ranges, and carotenoids, an accessory pigment that absorbs light in the 400–550 nm wavelength range [12,13]. Red light (600–700 nm) and blue light (400–500 nm) stimulate microalgae; however, their growth rates and lipid content fluctuate with light intensity [14]. The growth is accelerated by increasing light intensity up to a point, which varies based on the microalgal species [15]. Photo-inhibition can be caused by high light intensity beyond the saturation point [16].

Therefore, closed systems with artificial lights are preferred for optimizing algae growing conditions for biofuel production and nitrogen removal in wastewater treatment facilities because operating variables, such as light conditions and intensities can be controlled [17–20]. In a closed suspended or attached cultivation system, artificial lights are used to provide constant light intensity for the algal culture. Artificial light, especially light-emitting diodes (LEDs), can be an attractive way for effective algal cultivation. LEDs are now the most efficient artificial light sources, with a photosynthetically active radiation (PAR) efficiency of 80-100 percent. LEDs provide color combinations with wavelengths that correspond to blue, green, yellow, orange, red, and other light, as well as a high intensity that closely resembles the spectrum of sunlight [15,21,22]. The advantages include low power consumption, a wide range of intensity and a small size. Considering the high costs of fluorescent lights, LED light sources are a more cost-effective choice because of their lower operating costs, longer lifespan, and spectrum benefits.

The light intensity ranging between 26 - 400 $\mu mol/m^2$-s is the ideal light intensity for microalgae growth. The activation of lipid synthesis is triggered by an increase in light intensity. For maximum lipid productivity, microalgal species and strains require flux between 60 and 700 $\mu mol/m^2$-s [14]. The effect of light intensity and photoperiod on the algal activity is species-dependent; the optimal light intensity for photosynthesis is 200 $\mu mol/m^2$-s (~ 8700 lux). Some algal species like *Selenastrum minutum* and *Scenedesmus obliquus* can grow at light intensities of 400–500 µmol/m2-s [7,23]. In real wastewater treatment plants, most of the algal species interact with the bacterial culture present in the wastewater. In mixed algal and bacterial culture systems, ammonia oxidizing bacteria (AOB) and nitrite-oxidizing bacteria (NOB) are more sensitive to light than algae [24,25]. The exposure to 75 $\mu mol/m^2$-s under 12:12 h light/dark conditions inhibited both AOB and NOB [24]. The photo-inhibition of AOB and NOB can affect N removal during wastewater treatment in algal-bacterial systems, especially under high light conditions. Although several studies have examined the effect of distinct wavelengths of light (blue, green, red, and white) and their effect on algae growth, lipid accumulation and oxygen production [15,26–28], limited research has been done on understanding the effects of light on algal-bacterial co-culture systems [29,30]. The objective of this study is to examine the response of algal-bacterial consortium to different LED light wavelengths and intensities in terms of chlorophyll and the removal of carbon and nutrients from wastewater.

2 Materials and methods2.1 Materials and Methodology

2.1 Collection of seed culture and wastewater

The algal culture was collected from Mehboob Sagar lake, Hyderabad, Telangana, India. The obtained culture was centrifuged to remove the debris and was cultivated in the algal medium for about 15 days. The culture was examined under a microscope after 15 days and it was observed that *Chlorella sp.*, *Scenedesmus sp.*, and pennate diatoms were the dominant species in the biomass. *Chlorella sp.* and *Scenedesmus sp.*, are green algal species.

The aerobic bacterial culture was collected from a decentralized domestic wastewater treatment plant in Hyderabad, Telangana, India. The obtained biomass was kept in continuous aeration until further use. The wastewater was collected on a daily basis from the student dorms (about 3000 students) and dining hall disposal point at the Indian Institute of Technology Hyderabad, Telangana, India. The sterilization of wastewater was not performed in order to use real wastewater for the study.

The characteristics of the domestic wastewater are; pH of about 7.0 ± 0.4 The dissolved organic carbon (DOC), dissolved nitrogen (DN), and dissolved phosphorus (DP) concentrations were 65.8 ± 8.09, 25.7 ± 5.2, and 4.3 ± 0.6 mg/L respectively. The carbon and nutrient values were observed to be on the lower side of typical domestic wastewater [31]. The C: N: P of the wastewater is 15:6:1 which is less than the optimal C: N: P molar ratio (56:9:1) suitable for algal growth [32]. The optimum N/P value for significant nutrient removal from domestic wastewater ranges from 5:1 to 30:1 [33,34].

2.2 Experimental design

Four different light wavelengths were used in this experiment as an illumination source. These were blue (450 – 460 nm), red (640 – 670 nm), yellow (570 – 590 nm), and white (400 – 700 nm). Three different levels of intensities (100, 200, and 300 µmol/m2-s) for each wavelength were tested. Artificial lighting using LED light strips was provided. The transparent glass bottles with a wide opening of 500 mL volume capacity were used as test reactors. LED light strips were installed around the outer perimeter of the reactor. To avoid the intrusion of external light, the test reactors were covered with aluminium foil and the opening was closed with cotton. In total, twelve reactors were set up as shown in Figure 1 (a). The experimental design and naming for each test reactor have been given in Table 1. The three light intensities of 100, 200, and 300 $\mu mol/m^2$-s were named as I1, I2, and I3, respectively. The test reactors that were illuminated to blue wavelengths with intensities of I1, I2, and I3 were given the names B1, B2, and B3, respectively. Similarly, the red wavelengths I1, I2, and I3 were given the names R1, R2, and R3, respectively. Yellow wavelengths with intensity I1, I2 and I3 are represented by Y1, Y2 and Y3, whereas white wavelengths with intensities I1, I2 and I3 are represented by W1, W2 and W3 respectively. Figure 1 (b) shows the light intensity of 300 μmol/m^2-s for all wavelengths in empty test reactors. The reactors were filled with an equal proportion of algae and bacteria at the beginning of the experiment. The initial cell density in all of the test reactors was the same (volatile suspended solids = 1500 mg/L). The test reactors were kept on a magnetic stirrer and were continuously mixed at 100 rpm for 7 days with a 12/12 h light-dark cycle. The experiment started with the light phase and then continued with the dark and light cycles. 2-mL mixed culture samples were collected from the test reactors every four hours for the first 24 hours and then for every 12 hours for the next six days. The supernatant was collected after centrifuging the samples for 20 minutes at 5,000 rpm. The chlorophyll concentration of the separated biomass was determined, and the supernatant was filtered through 0.45-m membrane filter paper and examined for dissolved carbon and nutrient concentrations.

Figure 1. Experimental set-up

Table 1. Experiment design of the study

Light color	Light intensity (μmol/m^2-s)			Temperature (°C)	
	I1 (100)	I2 (200)	I3 (300)	Light period	Dark period
Blue	B1	B2	B3	34 – 35	26 – 28
Red	R1	R2	R3	38 – 39	26 – 28
Yellow	Y1	Y2	Y3	36 – 37	26 – 28
White	W1	W2	W3	34 – 35	26 – 28

2.3 Analytical methods

The filtered supernatant samples were analyzed for total dissolved organic carbon (DOC), dissolved phosphorus (DP), and total dissolved nitrogen (DN). Total dissolved organic carbon (DOC) and nitrogen (DN) were measured using a TOC-L analyzer (Make: Shimadzu). The DP was measured using method 4500 P-C described in the Standard Methods for the Examination of Water and Wastewater [35]. The 10200 H method in Standard Methods for the Examination of Water and Wastewater was used to determine the chlorophyll concentration in the obtained biomass sample [35].

The removal efficiency was computed by deducting the final concentration from the initial concentration and dividing the result by the initial concentration. The obtained decimal number was multiplied by 100 to convert it to a percentage. The correlation matrix for five parameters – intensity, chlorophyll (Chl), DOC removal (%), DN removal (%), and DP removal (%) for different wavelengths was plotted using Excel Statistical toolpak.

3 Results and discussion

3.1 Chlorophyll concentration

Chlorophyll is an important pigment that converts light energy into chemical energy during the photosynthetic process by producing organic molecules. Chlorophylls effectively absorb light in the red and blue spectral regions. Each chlorophyll also has its unique absorption spectrum. Algal cells have chlorophyll-a as the primary pigment and other chlorophylls (b, c, d, and f) as minor pigments. The seed culture used in this study was a mixed algal culture with a, b, and c pigments. The total chlorophyll concentration (Chl_a + Chl_b + Chl_c) for various wavelengths at different light intensities is shown in Figure 2. It was observed that the chlorophyll concentration reached its peak value within 12 hours of the experiment and then started decreasing.

At light intensity I1, the maximum chlorophyll concentration, 3.5 mg/L was observed in red light spectra. The next highest chlorophyll concentration was observed in B1 and there was not much difference in W1 and Y1. The order of chlorophyll content at I1 light intensity is red > blue > white > yellow. At light intensity I2, it was observed that the chlorophyll concentration was higher in white and yellow wavelengths. A similar pattern was observed at I3 where the highest chlorophyll concentration was observed in yellow and white wavelengths and lower concentration was observed at B3 and R3. The sequence of chlorophyll at I2 and I3 is White = Yellow > Blue > Red.

The effect of light intensity for blue, red, white and yellow wavelengths is shown in Figure 3. The chlorophyll concentrations have been shown to be affected differently by varying the light intensity for different wavelengths. In blue and red-light illumination reactors, the low-light intensity (100 μmol/m2-s) test reactors showed higher chlorophyll concentrations compared to the high-light conditions (B1> B2> B3; R1> R2> R3). Whereas, high- light intensity test reactors had higher chlorophyll concentrations than low light intensity test reactors in both yellow and white light wavelength reactors. The maximum chlorophyll concentration was observed at red light wavelength with low intensity. The next maximum concentration was observed in yellow wavelength. Red and yellow light, each has longer wavelengths, can be used to avoid photoinhibition. Shorter wavelengths of blue light can be more energetic and efficient for photosynthesis, while causing photoinhibition at higher intensities.

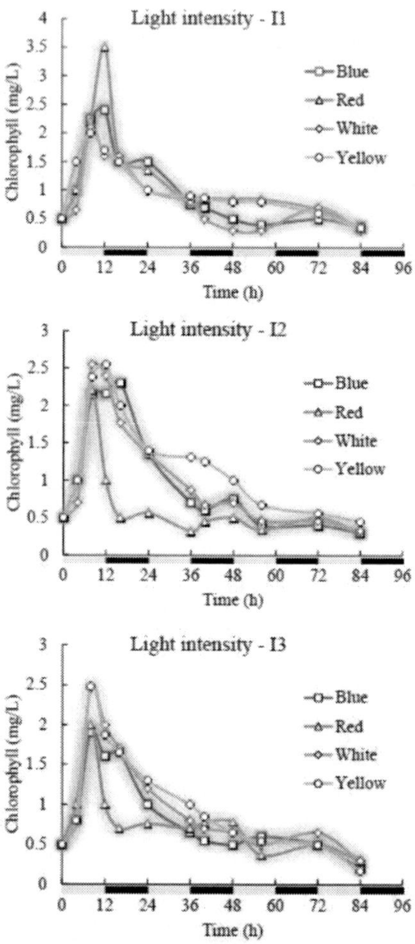

Figure 2. Total chlorophyll concentration for various wavelengths at different light intensities

The biomass production in R2 and R3 is lower than in other wavelengths due to the extensively absorbed red light, which reduces light utilization efficiency in algae [36]. The average chlorophyll concentrations were detected in white light because the emission spectrum band of white light completely covers that of red light, blue light and other growth-inefficient light wavelengths.

3.2 Carbon and nutrient removal

The average dissolved carbon removal at different wavelengths and intensities is shown in Figure 4. At I1 light intensity, the DOC removal at different light colors decreased in the order red > yellow > white > blue. The DOC removal decreased in order of blue> red > white > yellow at I2 and I3 light intensity. The highest DOC removal of 84% was observed in blue illumination test reactors (B2 and B3). The DOC removal increased with the increase in the intensity of blue light. The removal was more or less constant in red and white wavelengths whereas the removal efficiency decreased with an increase in intensity for yellow wavelength. The lowest

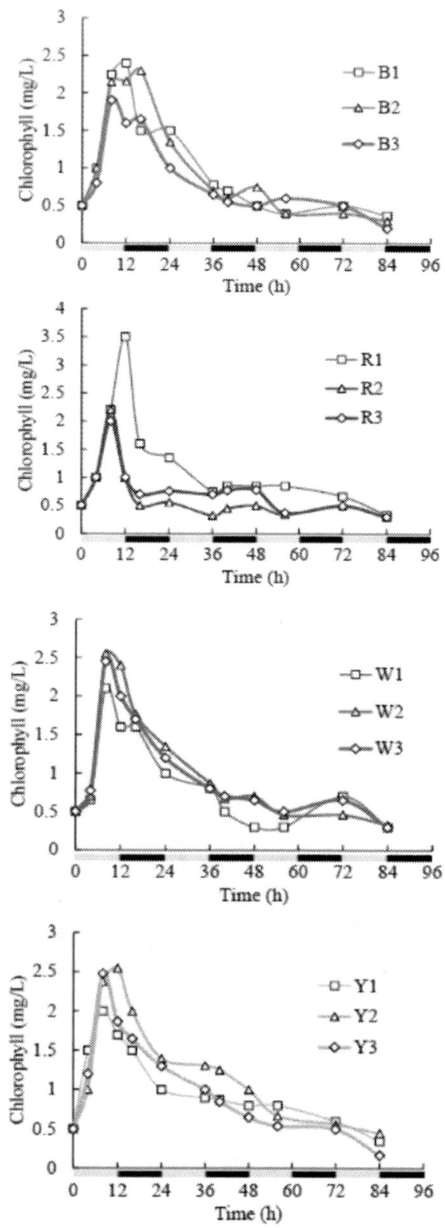

Figure 3. Effect of intensity on chlorophyll concentration

DOC removal of 44% was observed in B1. Overall, the order of DOC removal is B3 > B2 > R1 > Y1 ≥ R2 ≥ R3 > W1 > W2 > W3 > Y2 >Y3 > B1 at different intensities. The chlorophyll concentration and the organic carbon removal had not shown any relation indicating that a major portion of carbon was removed by the bacterial culture.

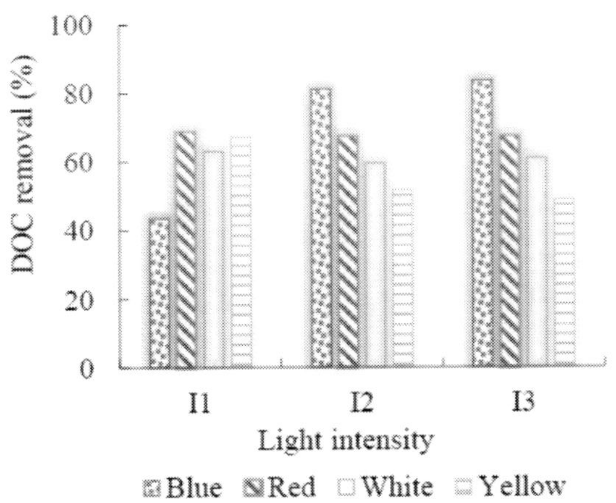

Figure 4. Dissolved carbon removal at different wavelengths and intensities

The average dissolved nitrogen removal at different wavelengths and intensities is shown in Figure 5. At I1 light intensity, the DN removal at different light colors decreased in the order red > blue > white > yellow. The DN removal decreased in order of yellow > red > white > blue at I2 light intensity. At I3 light intensity, the DN removal is in the order yellow > white > red > blue. The highest DN removal of 52% was observed in the red light test reactor (R1). The lowest was observed in B2 with 32%. The overall DN removal at different intensities is R1 > Y3> W3 > R3 > Y2 > B3 > R2 > B1 > W1 > W2 > Y1 > B2.

The average dissolved phosphorous removal at different wavelengths and intensities is shown in Figure 6. At I1 light intensity, the DP removal at different light colors decreased in the order red > yellow > white > blue. The DP removal decreased in order of yellow > red > white > blue at I2 light intensity. At I3 light intensity, the DP removal is in the order yellow > red > white > blue. The highest DP removal of 81% was observed in yellow wavelength test reactors (Y1, Y2, Y3) and white wavelength W2. The lowest was observed in B1 with 64% removal. The overall DP removal is Y3> Y2> Y1> W3> R1> R2> R3> W2> B3> B2> W1> B1. The low C/N/P ratio might be one of the reasons for low nutrient removal in test reactors. The higher chlorophyll concentration in red light and higher nutrient removal in R1 indicates that the algae had played role in removing nutrients from the wastewater.

During their cell and life cycles, algae undergo morphologic changes that are influenced by light quality and nutrition availability. In general, microalgae consume more nitrates when exposed to blue light than when exposed to red light. At this wavelength, nitrate and phosphate reductases are activated, leading to increased absorption of these nutrients [37,38]. However, in this study red and yellow wavelength has shown higher nutrient removal.

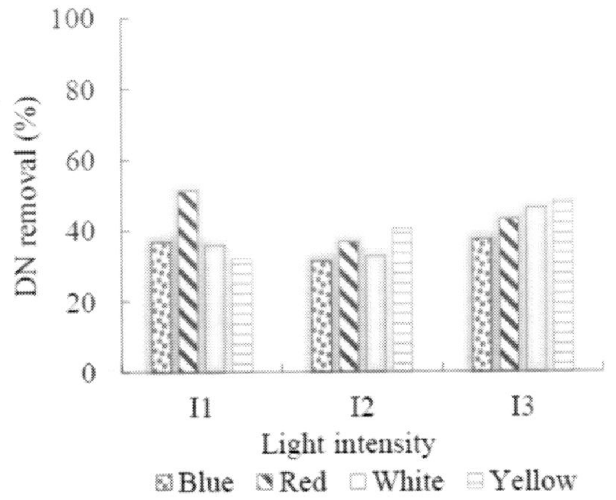

Figure 5. Dissolved phosphorus removal at different wavelengths and intensities

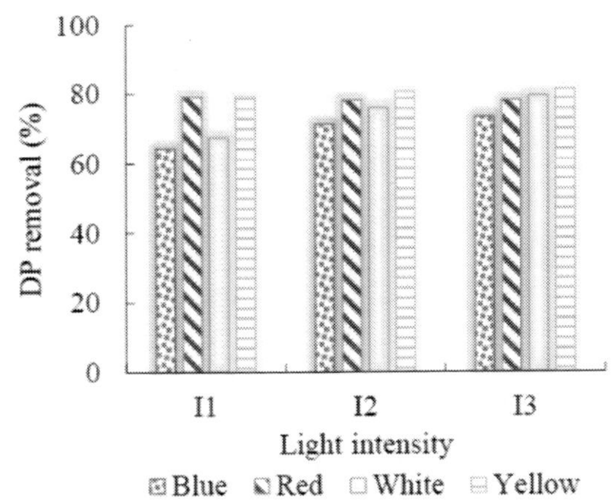

Figure 6. Dissolved phosphorus removal at different wavelengths and intensities

The correlation matrix for five parameters – intensity, chlorophyll (Chl), DOC removal (%), DN removal (%), and DP removal (%) for different wavelengths are given in Table 2. It was observed that each wavelength responded differently with intensity. In blue wavelength, it was observed that the intensity and chlorophyll concentration are negatively related, whereas it is strongly correlated to the carbon and nutrient removal. The chlorophyll concentration has a negative correlation with carbon and nutrient removal. In red wavelength reactors, it was observed that the intensity has a negative correlation with all the other parameters - algal growth, carbon and nutrient removal. The chlorophyll concentration has a strong correlation with carbon and nutrient removal. In white wavelength, it was observed that the intensity has a strong correlation with chlorophyll concentration and nutrient removal whereas negative relation with carbon removal. In yellow wavelength, the intensity has shown a positive correlation with chlorophyll and nutrient removal.

Table 2. Correlation coefficient matrix for different light wavelengths

BLUE	*Intensity*	*Chl*	*DOC (%)*	*DN (%)*	*DP (%)*
Intensity	1				
Chl	-1.00	1			
DOC (%)	0.89	-0.88	1		
DN (%)	0.10	-0.12	-0.36	1	
DP (%)	0.96	-0.95	0.99	-0.20	1
RED	*Intensity*	*Chl*	*DOC (%)*	*DN (%)*	*DP (%)*
Intensity	1				
Chl	-0.92	1			
DOC (%)	-0.87	0.99	1		
DN (%)	-0.56	0.84	0.89	1	
DP (%)	-0.87	0.99	1.00	0.90	1
WHITE	*Intensity*	*Chl*	*DOC (%)*	*DN (%)*	*DP (%)*
Intensity	1				
Chl	0.84	1			
DOC (%)	-0.60	-0.94	1		
DN (%)	0.74	0.25	0.11	1	
DP (%)	0.97	0.94	-0.76	0.56	1
YELLOW	*Intensity*	*Chl*	*DOC (%)*	*DN (%)*	*DP (%)*
Intensity	1				
Chl	0.77	1			
DOC (%)	-0.93	-0.95	1		
DN (%)	1.00	0.77	-0.93	1	
DP (%)	0.97	0.91	-0.99	0.97	1

In blue, white and yellow light wavelengths the chlorophyll concentration has shown a negative correlation to carbon removal. The high carbon removal in B2 and B3, a negative correlation of light intensity to chlorophyll concentration, and negative relation of chlorophyll to carbon removal indicated that bacteria have played a major role in eliminating the carbon from wastewater. Although there was a positive correlation between light intensity and chlorophyll in white and yellow wavelengths, the chlorophyll has shown negative relation to carbon removal indicating no role of algae in carbon removal. There might be a release of extracellular compounds by the biomass which resulted in relatively low carbon removal in yellow and white wavelengths. The bacterial consortium mineralizes organic carbon mostly by heterotrophic metabolism, which is a common mechanism in wastewater treatment plants. Photoautotrophic and mixotrophic processes are present in microalgae [39]. In their photoautotrophic metabolism, microalgae use light and CO_2 (or inorganic carbon) as energy and carbon sources. Microalgae have a mixotrophic metabolism in the presence of both light and organic carbon. Organic carbon is utilized in the algal-bacterial system via mixotrophic or heterotrophic metabolism [40]. Several algae species, including *Chlorella* and *Scenedesmus*, can use organic carbon molecules as a carbon source when exposed to sunlight [41]. Extracellular organic carbon compounds like polysaccharides, nucleic acids, proteins, lipids and other dissolved organic substances are released by autotrophic microalgae and lead to an increase in organic carbon concentration [42]. As per literature, algae will release 7-50 percent of the absorbed carbon back into the environment [43]. Under normal conditions, certain algal species can synthesize these organic compounds; others can produce them in response to low nutrient stress or other stressful conditions (e.g., unfavorable light, pH, or temperature), and others can produce them after the decomposition of algal cells [44].

Chlorophyll concentration has a negative correlation with nutrient removal in the blue wavelength, but a positive correlation in the red, white, and yellow wavelengths. As the intensity of red light increased, nutrient removal and chlorophyll concentration both dropped. Chlorophyll concertation and nutrient removal increased as light intensities in the white and yellow wavelengths increased. This suggests that microalgae have a significant role in wastewater nutrient removal. The nutrient removal by microalgae may be attributed to the assimilation process of the microalgae, precipitation of the insoluble nutrient, release in the form of gas due to aeration or stir, and bio-sorption by algal cells [45,46]. In general, microalgae can assimilate nitrogen in various forms including ammonium, nitrate, nitrite, and urea although ammonium is the preferred form for nitrogen assimilation [47].

In red and white wavelengths, carbon and nitrogen removal are positively related, but in blue and yellow wavelengths, they are negatively related. The carbon and phosphorus removal are positively related in blue and red. In microalgae, the utilization of nutrients and carbon and nutrient are interrelated. Microalgae use the tricarboxylic acid cycle to generate energy from the carbon that has been stored. [48]. The resultant energy is utilized to absorb inorganic nitrogen (ammonium) to produce the glutamine and glutamate amino acids, which need energy in the form of ATP and NADP and carbon skeletons in the form of 2-oxoglutarate and oxaloacetate [49,50].

The nitrogen removal has shown a positive correlation to phosphorus removal in red, white and yellow wavelengths. Microalgae metabolize nitrogen predominantly into proteins that are necessary for the synthesis of ribosomes and ribosomal RNA. The primary function of phosphorus absorption is the storage of the synthesized ribosomal RNA. Therefore, adequate quantity of nitrogen is required for the uptake of phosphate in order to prevent inhibition of cell's capacity to synthesize proteins [51]. Phosphorus uptake by microalgae is generally influenced by several additional elements, including light intensity, temperature, pH, algae physiology, phosphate concentration, and chemical form of accessible phosphate [52]. Compared to other species of algae, green algae require significantly more N and P. Algal growth stoichiometry states that in order to eliminate 1 mg of phosphorus, 5.4–7.2 mg of nitrogen are required [53,54]. This implies that P removal from municipal wastewater is significantly influenced by the nitrogen to phosphorous ratio, which also influences biomass growth and nutrient removal [52].

In addition to the wastewater characteristics, other factors that influence the algal wastewater treatment include the photobioreactor design, light intensity, and light period. This study demonstrates that compared to white and yellow light, blue and red light wavelengths are better suited for mixed algal-bacterial systems. Since only monochromatic light was used in this experiment, it was necessary to examine the effect of different light modes and dichromatic light illumination on mixed algal-bacterial culture. It is also crucial to study how the light affects the destruction of micropollutants along with carbon and nutrients.

Due to the strict pollution controls recommended by circular economy principles before wastewater is discharged and reused in the environment, existing wastewater treatment plants are currently having difficulties. Many technologies, including pyrolysis, anaerobic digestion, and hydrothermal carbonization, are now being researched and used to improve energy recovery from residual biosolids and eliminate excessive solids discharge [55,56]. A sustainable wastewater treatment and disposal waste management solution is being constrained by excessive solid waste landfilling, ineffective recycling, high energy consumption, and greenhouse gas emissions by existing treatment technologies. These drawbacks can be addressed by using algae-bacterial systems. Carbon fixation, nutrient uptake, and the consequent generation of value-added biomass, which may be used as a substrate for the production of biodiesel and bioethanol, are the potential environmental and economic advantages of integrating microalgae in wastewater treatment systems. However, since using artificial lighting might raise treatment costs, it is also necessary to do a cost-analysis for the photobioreactor when examining how light affects algal-bacterial systems.

4 Conclusion

Algal growth was regulated by the wavelength and intensity of light, which in turn influenced the overall effectiveness of the algal-bacterial system in wastewater treatment. The concentration of chlorophyll has a significant impact on nutrient removal. The R1 wavelength of low-intensity red light resulted in a better nutrient removal efficiency. Higher carbon removal was achieved using the blue wavelength with I3 intensity. Overall, the wavelength of blue light at I3 can be considered the best wavelength for removing carbon and nutrients. At I1, the red wavelength can be considered the second-best wavelength. In comparison to white and yellow light wavelengths, blue and red light wavelengths are more suited for algal-bacterial systems, according to this study.

Acknowledgements

This research was funded by the Ministry of Education of India through IMPRINT 2 program.

References

[1] W.J. Oswald, The high-rate pond in waste disposal, Developments in Industrial Microbiology, 4 (1963): 112–119

[2] W.J. Oswald, H.B. Gotaas, C.G. Golueke, W.R. Kellen, E.F. Gloyna, Algae in Waste Treatment, Sewage and Industrial Wastes, 29(4) (1957): 437–457

[3] K. Kumar, S.K. Mishra, A. Shrivastav, M.S. Park, J.W. Yang, Recent trends in the mass cultivation of algae in raceway ponds, Renewable and Sustainable Energy Reviews, 51 (2015): 875–885

[4] R. Craggs, J. Park, D. Sutherland, S. Heubeck, Economic construction and operation of hectare-scale wastewater treatment enhanced pond systems, Journal of Applied Phycology 27, 5 (2015): 1913–1922

[5] R.M. Handler, C.E. Canter, T.N. Kalnes, F.S. Lupton, O. Kholiqov, D.R. Shonnard, P. Blowers, Evaluation of environmental impacts from microalgae cultivation in open-air raceway ponds: Analysis of the prior literature and investigation of wide variance in predicted impacts, Algal Research, 1(1) (2012): 83–92

[6] S. Dahmani, D. Zerrouki, L. Ramanna, I. Rawat, F. Bux, Cultivation of Chlorella pyrenoidosa in outdoor open raceway pond using domestic wastewater as medium in arid desert region, Bioresource Technology, 219 (2016): 749–752

[7] S.P. Singh, P. Singh, Effect of temperature and light on the growth of algae species: A review, Renewable and Sustainable Energy Reviews, 50 (2015): 431–444

[8] R. Wimalasekera, Effect of Light Intensity on Photosynthesis, Photosynthesis, Productivity, and Environmental Stress, (2019): 65–73

[9] H. Lokstein, G. Renger, J.P. Götze, Photosynthetic Light-Harvesting (Antenna) Complexes—Structures and Functions, Molecules, 26(11) (2021): 3378

[10] H.G. Koh, N.K. Kang, S. Jeon, S.E. Shin, B.R. Jeong, Y.K. Chang, Heterologous synthesis of chlorophyll b in Nannochloropsis salina enhances growth and lipid production by increasing photosynthetic efficiency, Biotechnology for Biofuels, 12(1) (2019): 1–15

[11] A. Pavlou, J. Jacques, N. Ahmadova, F. Mamedov, S. Styring, The wavelength of the incident light determines the primary charge separation pathway in Photosystem II, Scientific Reports, 8(1) (2018): 2837

[12] D. Diaz-MacAdoo, M.T. Mata, C. Riquelme, Influence of Irradiance and Wavelength on the Antioxidant Activity and Carotenoids Accumulation in Muriellopsis sp. Isolated from the Antofagasta Coastal Desert, Molecules, 27(8) (2022): 2412

[13] S. Baer, M. Heining, P. Schwerna, R. Buchholz, H. Hübner, Optimization of spectral light quality for growth and product formation in different microalgae using a continuous photobioreactor, Algal Research, 14 (2016): 109–115

[14] Y. Maltsev, K. Maltseva, M. Kulikovskiy, S. Maltseva, Influence of Light Conditions on Microalgae Growth and Content of Lipids, Carotenoids, and Fatty Acid Composition, Biology, 10(10) (2021): 1060
[15] M.N. Metsoviti, G. Papapolymerou, I.T. Karapanagiotidis, N. Katsoulas, Effect of Light Intensity and Quality on Growth Rate and Composition of Chlorella vulgaris, Plants, 9(1) (2019): 31
[16] A. Difusa, J. Talukdar, M.C. Kalita, K. Mohanty, V. V. Goud, Effect of light intensity and pH condition on the growth, biomass and lipid content of microalgae Scenedesmus species, Biofuels, 6(1-2) (2015): 37–44
[17] D.M. Arias, E. Uggetti, M.J. García-Galán, J. García, Cultivation and selection of cyanobacteria in a closed photobioreactor used for secondary effluent and digestate treatment, Science of The Total Environment, 588 (2017): 157–167
[18] S.K. Prajapati, P. Kumar, A. Malik, V.K. Vijay, Bioconversion of algae to methane and subsequent utilization of digestate for algae cultivation: A closed loop bioenergy generation process, Bioresource Technology, 158 (2014): 174–180
[19] M.R. Tredici, N. Bassi, M. Prussi, N. Biondi, L. Rodolfi, G. Chini Zittelli, G. Sampietro, Energy balance of algal biomass production in a 1-ha "Green Wall Panel" plant: How to produce algal biomass in a closed reactor achieving a high Net Energy Ratio, Applied Energy, 154 (2015): 1103–1111
[20] T. Li, L.N. Liu, C.D. Jiang, Y.J. Liu, L. Shi, Effects of mutual shading on the regulation of photosynthesis in field-grown sorghum, Journal of Photochemistry and Photobiology B: Biology, 137 (2014): 31–38
[21] E. Darko, P. Heydarizadeh, B. Schoefs, M.R. Sabzalian, Photosynthesis under artificial light: the shift in primary and secondary metabolism, Philosophical Transactions of the Royal Society B: Biological Sciences, 369(1640) (2014): 20130243
[22] M. Glemser, M. Heining, J. Schmidt, A. Becker, D. Garbe, R. Buchholz, T. Brück, Application of light-emitting diodes (LEDs) in cultivation of phototrophic microalgae: current state and perspectives, Applied Microbiology and Biotechnology, 100(3) (2016): 1077–1088
[23] R. Bouterfas, M. Belkoura, A. Dauta, Light and temperature effects on the growth rate of three freshwater algae isolated from a eutrophic lake, Hydrobiologia, 489(1) (2002): 207–217
[24] T. Yoshioka, Y. Saijo, Photoinhibition and recovery of NH4+-oxidizing bacteria and NO-2-oxidizing bacteria, The Journal of General and Applied Microbiology, 30(3) (1984): 151–166
[25] C. Vergara, R. Muñoz, J.L. Campos, M. Seeger, D. Jeison, Influence of light intensity on bacterial nitrifying activity in algal-bacterial photobioreactors and its implications for microalgae-based wastewater treatment, International Biodeterioration and Biodegradation, 114 (2016): 116–121
[26] A. Khalili, G.D. Najafpour, G. Amini, F. Samkhaniyani, Influence of nutrients and LED light intensities on biomass production of microalgae Chlorella vulgaris, Biotechnology and Bioprocess Engineering, 20(2) (2015): 284–290
[27] C.E. Sharp, S. Urschel, X. Dong, A.L. Brady, G.F. Slater, M. Strous, Robust, high-productivity phototrophic carbon capture at high pH and alkalinity using natural microbial communities, Biotechnology for Biofuels, 10(1) (2017): 84
[28] B. Brzychczyk, T. Hebda, N. Pedryc, The Influence of Artificial Lighting Systems on the Cultivation of Algae: The Example of Chlorella vulgaris, Energies, 13(22) (2020): 5994
[29] D. Kang, K. Kim, Y. Jang, H. Moon, D. Ju, D. Jahng, Nutrient removal and community structure of wastewater-borne algal-bacterial consortia grown in raw wastewater with various wavelengths of light, International Biodeterioration & Biodegradation, 126 (2018): 10–20
[30] D. Kang, K.T. Kim, T.Y. Heo, G. Kwon, C. Lim, J. Park, Inhibition of Photosynthetic Activity in Wastewater-Borne Microalgal–Bacterial Consortia under Various Light Conditions, Sustainability, 11(10) (2019): 2951
[31] G. Tchobanoglous, F.L. Burton, H.D. Stensel, Wastewater Engineering: Treatment and Reuse. Metcalf & Eddy/Aecom (2014)
[32] S.M. Phang, O. Kim-Chong, Algal biomass production in digested palm oil mill effluent, Biological Wastes, 25(3) (1988): 177–191
[33] L. Xin, H. Hong-ying, G. Ke, S. Ying-xue, Effects of different nitrogen and phosphorus concentrations on the growth, nutrient uptake, and lipid accumulation of a freshwater microalga Scenedesmus sp., Bioresource Technology, 101(14) (2010): 5494–5500
[34] H.J. Choi, S.M. Lee, Effect of the N/P ratio on biomass productivity and nutrient removal from municipal wastewater, Bioprocess and Biosystems Engineering, 38(4) (2015): 761–766
[35] APHA/AWWA/WEF, Standard Methods for the Examination of Water and Wastewater, Standard Methods, (2012): 541
[36] T. de Mooij, G. de Vries, C. Latsos, R.H. Wijffels, M. Janssen, Impact of light color on photobioreactor productivity, Algal Research, 15 (2016): 32–42
[37] T.H. Kim, Y. Lee, S.H. Han, S.J. Hwang, The effects of wavelength and wavelength mixing ratios on microalgae growth and nitrogen, phosphorus removal using Scenedesmus sp. for wastewater treatment, Bioresource Technology, 130 (2013): 75–80
[38] P.S.C. Schulze, H.G.C. Pereira, T.F.C. Santos, L. Schueler, R. Guerra, L.A. Barreira, J.A. Perales, J.C.S. Varela, Effect of light quality supplied by light emitting diodes (LEDs) on growth and biochemical profiles of Nannochloropsis oculata and Tetraselmis chuii, Algal Research, 16 (2016): 387–398
[39] C.Y. Chen, K.L. Yeh, R. Aisyah, D.J. Lee, J.S. Chang, Cultivation, photobioreactor design and

harvesting of microalgae for biodiesel production: A critical review, Bioresource Technology, 102(1) (2011): 71–81

[40] A. Abeliovich, D. Weisman, Role of heterotrophic nutrition in growth of the alga Scenedesmus obliquus in high-rate oxidation ponds, Applied and Environmental Microbiology, 35(1) (1978): 32–37

[41] A.W. Mayo, T. Noike, Effect of glucose loading on the growth behavior of Chlorella vulgaris and heterotrophic bacteria in mixed culture, Water Research, 28(5) (1994): 1001–1008

[42] D. Vandamme, I. Foubert, I. Fraeye, K. Muylaert, Influence of organic matter generated by chlorella vulgaris on five different modes of flocculation, Bioresource Technology, 124 (2012): 508–511

[43] S. Babel, S. Takizawa, H. Ozaki, Factors affecting seasonal variation of membrane filtration resistance caused by Chlorella algae, Water Research, 36(5) (2002): 1193–1202

[44] L.O. Villacorte, Y. Ekowati, T.R. Neu, J.M. Kleijn, H. Winters, G. Amy, J.C. Schippers, M.D. Kennedy, Characterisation of algal organic matter produced by bloom-forming marine and freshwater algae, Water Research, 73 (2015): 216–230

[45] A.M. Silva-Benavides, G. Torzillo, Nitrogen and phosphorus removal through laboratory batch cultures of microalga Chlorella vulgaris and cyanobacterium Planktothrix isothrix grown as monoalgal and as co-cultures, Journal of Applied Phycology, 24(2) (2012): 267–276

[46] C. Yan, Y. Zhao, Z. Zheng, X. Luo, Effects of various LED light wavelengths and light intensity supply strategies on synthetic high-strength wastewater purification by Chlorella vulgaris, Biodegradation, 24(5) (2013): 721–732

[47] L. Wang, M. Min, Y. Li, P. Chen, Y. Chen, Y. Liu, Y. Wang, R. Ruan, Cultivation of green algae Chlorella sp. in different wastewaters from municipal wastewater treatment plant, Applied Biochemistry and Biotechnology, 162(4) (2010): 1174–1186

[48] X. Johnson, J. Alric, Central carbon metabolism and electron transport in chlamydomonas reinhardtii: Metabolic constraints for carbon partitioning between oil and starch, Eukaryotic Cell, 12(6) (2013): 776–793

[49] H.C. Huppe, D.H. Turpin, Integration of Carbon and Nitrogen Metabolism in Plant and Algal Cells, Annual Review of Plant Physiology and Plant Molecular Biology, 45(1) (1994): 577–607

[50] N. Liu, F. Li, F. Ge, N. Tao, Q. Zhou, M. Wong, Mechanisms of ammonium assimilation by Chlorella vulgaris F1068: Isotope fractionation and proteomic approaches, Bioresource Technology, 190 (2015): 307–314

[51] A. Beuckels, E. Smolders, K. Muylaert, Nitrogen availability influences phosphorus removal in microalgae-based wastewater treatment, Water Research, 77 (2015): 98–106

[52] P.L. Gupta, H.J. Choi, S.M. Lee, Enhanced nutrient removal from municipal wastewater assisted by mixotrophic microalgal cultivation using glycerol, Environmental Science and Pollution Research, 23(10) (2016): 10114–10123

[53] F.P. Healey, Inorganic nutrient uptake and deficiency in algae, Critical Reviews in Microbiology, 3(1) (1973): 69–113

[54] J. Yang, Y. Gou, F. Fang, J. Guo, H. Ma, X. Wei, B. Shahmoradi, Impacts of sludge retention time on the performance of an algal-bacterial bioreactor, Chemical Engineering Journal, 343 (2018): 37–43

[55] Y. Sun, Z. Chen, G. Wu, Q. Wu, F. Zhang, Z. Niu, H.Y. Hu, Characteristics of water quality of municipal wastewater treatment plants in China: implications for resources utilization and management, Journal of Cleaner Production, 131 (2016): 1–9

[56] H. Al-Jabri, P. Das, S. Khan, M. Thaher, M. Abdulquadir, Treatment of Wastewaters by Microalgae and the Potential Applications of the Produced Biomass—A Review, Water 2021, 13(1) (2020): 27

* Corresponding author: keerthi.katam@mahindrauniversity.edu.in

Identification of bacterial species from healthy wood of *Aquilaria crassna* using colony PCR

Vanarat Phakeenuya[1], *Bowonlak* Phetsri[1], *Thidarat* Khunsuksaen[1], *Weeradej* Meeinkuirt[2], *Phuri* Kalnaowakul[3], *Wawat* Rodiahwati[4] and *Theerawut* Phusantisampan[1,5*]

[1]Department of Biotechnology, Faculty of Applied Science, King Mongkut's University of Technology North Bangkok, Bangkok 10800, Thailand
[2]Water and Soil Environmental Research Unit, Mahidol University, Nakhonsawan Campus, Nakhonsawan 60130, Thailand
[3]Aircraft Maintenance Engineering, Department of Mechanical and Industrial Engineering, Rajamangala University of Technology Krungthep, Bangkok 10120, Thailand
[4]Department of Chemistry, Faculty of Science, University of New England, Australia
[5]Biorefinery and Process Automation Engineering Center, Department of Chemical and Process Engineering, The Sirindhorn Thai-German International Graduate School of Engineering, King Mongkut's University of Technology North Bangkok, Thailand

Abstract. Agarwood is one of the most expensive, sacred and valuable woods used in the biotechnology industry. This natural raw material is in high demand as a commodity and can be used to manufacture a variety of products. Isolating a microbe from its environment is crucial because several bacterial strains can produce novel compounds for biotechnological applications. Exploration of bacterial strains, such as bacterial endophytes, has been associated with bioactive compounds of the agarwood plant. The purpose of this study was to isolate and identify bacteria from Agarwood trees in the Kaeng Hang Maeo Sub-district of Chanthaburi Province, Thailand by using molecular biology techniques. A colony PCR technique without prior DNA extraction for rapid and simple detection of bacteria was performed. After the DNA sequence analysis, the results showed that most populations in the normal layer are belonged to *Pantoea dispersa* (99%). Through analysis of phylogenetic tree by neighbor-joining method, the sequences of the *16s rRNA* gene were analyzed to understand evolutionary relationships of bacteria found in agarwood. In conclusion, good amplification of the *16s rRNA* sequence (1.5 kb) was detected with the specificity of the target sequences of the *16s RNA*. In addition, the isolated bacterial strains are the most common species in agarwood, however, there has been no previous research on isolating these bacterial strains in *Aquilaria crassna*.

Keywords. *Aquilaria crassna*, Agarwood, Bacterial endophytes, Colony PCR, *16s rRNA*

1 Introduction

Aquilaria crassna is an agarwood producing tree of the family Thymalaeaceae [1,2]. This plant is found primarily in China, Bangladesh, India, and Thailand [3]. When aquilaria trees are wounded or stressed, their wood turns from white to yellow, brown, black, and eventually greasy black and spreads out. Furthermore, the resinous substances accumulated in the wood surrounding the wound can be secreted by the wood. It must, however, be left for several years until completely turning to black or brown color [4,5]. Many countries, including Egypt, China, Greece, and India, used agarwood essential oil or fragrant resins for medicinal therapy, incense in religious rites, perfumes, and ornamental materials [6,7].

Many bacteria are ubiquitous; that is, they are present everywhere. They are also highly adaptable to a variety of environmental conditions [8]. A previous report presented endophytic bacteria as playing a crucial role in the production of pharmaceutical compounds present in plants [9]. The wood was also used to decorate boxes, belts and clothing articles and its usage has recently expanded to include new products such as detergent and shampoo [10]. There is also evidence that it has tonic and therapeutic properties [11,12]. According to recent research, endophytic bacteria have been discovered to be abundant among the resident microflora of the healthy inner tissues of numerous plant species [13]. Because the function of these bacteria is unknown, the isolation and identification of bacterial strains in these groups may aid in understanding the relationship between microbes and plants, particularly aquilaria species that produce agarwood, as well as the discovery of novel bacterial species.

In 2005, molecular approaches based on DNA gene markers were widely used to identify the genetic variation of plant species in agarwood producing tree species [14,15]. In the present study, we presented the results of the isolation and identification of major bacterial strains associated with the non-resinous part of the agarwood plant. This study provided the results of the isolation and identification of major bacterial strains associated with the non-resinous part of the agarwood plant. Furthermore, this study was the first investigation to identify bacteria strains that could be employed as an alternative bio-

* Corresponding author: phusantisampan.t@gmail.com

resource for the future production of bioactive compounds.

2 Materials and methods

2.1 Bacterial isolation from healthy wood sample

The wood samples (*Aquilaria crassna* Pierre ex Lec) were collected in the Kaeng Hang Maeo Sub-district of Chanthaburi Province (Figure 1). They were pulverized into fine powder. The powder (0.3 g) was soaked in 10 ml distilled water before being incubated at 37°C for 48 h. One milliliter of suitable dilutions of the suspension was used to inoculate a set of three Petri dishes, each containing 15mL Nutrient Agar (NA) medium (HIMEDIA®, India). Plates were then incubated at 37°C for 24 h.

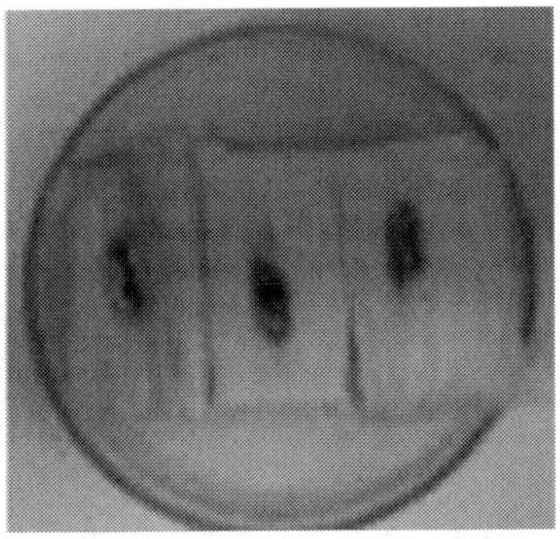

Fig. 1. Characteristics of ground wood for bacterial isolation

2.2 Colony PCR

The PCR reaction mixtures were carried out in a volume of 25 µl with 2.5U of Pfu DNA polymerase (Biotechrabbit, Germany), 2.5 µl of 10X reaction buffer, 200µM of each dNTP, 1.5mM $MgCl_2$, 25 pmole of each primer and were adjusted with distilled water. Primer forward (5'AGA GTT TGA TCM TGG CTC AG3') and reverse (5'TAC GGY TAC CTT GTT ACG ACT T3') to 16S rRNA PCR amplification were used for bacterial identification. A single colony of cream whitish- colored colonies was selected as the DNA template for PCR. The PCR reaction was carried out using a DNA Engine thermal cycler (BIORAD) with an initial denaturation step of 95°C for 5 min, followed by 40 cycles of denaturation at 95°C for 30 sec, annealing at 52°C for 30 sec, extension at 72°C for 30 sec, and an additional extension step at 72°C for 10 min. This process used distilled water as a negative control to check for contamination. The PCR reaction components and conditions are shown in Table 1 and Table 2.

Table 1. PCR reaction components and the volumes for PCR amplification

PCR component	Volume (µl)
Distilled water	19.0
10X PCR buffer	2.5
1.5 mM $MgCl_2$	1.5
200 µM dNTPs	0.5
2.5 pmol/ µl forward primer	0.5
2.5 pmol/ µl reverse primer	0.5
2.5 U/ µl Pfu DNA polymerase	0.5
Total volume	25.0

Table 2. Condition for colony PCR amplification

Step	Temperature (°C)	Time	Cycles
Initial denaturation	95	3 min	1
Denaturataion	95	30 sec	
Annealing	50	30 sec	35
Extension	72	30 sec	
Final extension	72	5 min	1

2.3 Detection of PCR products and sequencing

The presence of PCR products was investigated using 2% agarose gel electrophoresis. PCR yields were estimated by comparing them to a low DNA mass ladder (Promega, USA). The post PCR purification step was then carried out using a QIAquick PCR purification Kit (QIAGEN, USA). The purified PCR product was sequenced with the ABI PRISM™ BigDye Terminator Cycle Sequencing Kit. The amplification was carried out in a DNA Engine thermal cycler (BIORAD) under the following conditions: 25 cycles of 96°C for 10 sec, 55°C for 5 sec and 70°C for 1 min. The DNA was recovered by ethanol precipitation after amplification and the dye-labeled DNA was analyzed using an Applied Biosystems 377 sequencer (Perkin-Elmer, Norwalk, CT, USA).

 https://doi.org/10.1051/e3sconf/202235502004

2.4 Data analysis using Bioinformatics tools

After obtaining the nucleotide sequences of the *16s rRNA* gene, they were checked by using the nucleotide blast database (https://blast.ncbi.nlm.nih.gov/Blast.cgi). All alignment illustrations were created using the ClustalW program [16] and the GeneDoc software was used to display the output [17]. MEGA 7 was used to create the phylogenetic trees using the Neighbor-Joining method and Bootstrapping analysis [18]. The data has been bootstrapped 1,000 times.

3 Results and discussion

3.1 Characteristics of colonies obtained by isolating bacteria from wood

The selected isolates underwent presumptive phenotypic identification based on bacterial morphology after gram staining, colony morphology and the ability to grow on NA media and culture conditions. The results showed that the isolated microorganism displayed cream whitish-colored colonies, round, raised, smooth on each plate shown in Figure 2. From the gram staining, it was found gram-negative rod (data not shown). This pattern is similar to that seen in previous research [9, 13].

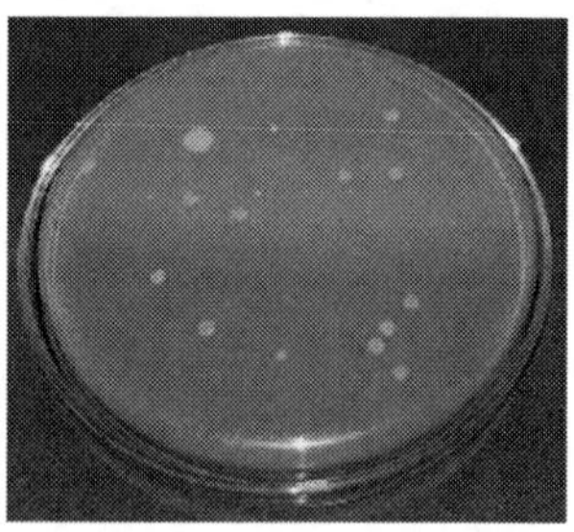

Fig. 2. The colony morphology of bacteria in wood cultures grew on NA media

3.2 Colony PCR for detection of the *16s rRNA* gene

In this study, colony PCR was used as the universal primers. The PCR products were about 1,500 base pairs long, which was the expected size. As shown in Figure 3, the universal primers produced the amplified fragment of the expected size on the gel when compared with the 100 bp DNA marker. The *16S rRNA* gene is universal in bacteria commonly due to its conservation across bacteria and is popularly used to establish the evolution of bacteria through the approximately 1,500 base pair (bp) locus of this gene [19]. According to the results, *16S rRNA* sequence analysis found isolates that were more accurate than traditional methods [20]. In addition, the *16S rRNA* gene sequences were also generated between groups of known bacteria and the novel isolates of related bacteria to study the evolutionary relationships of living organisms [21].

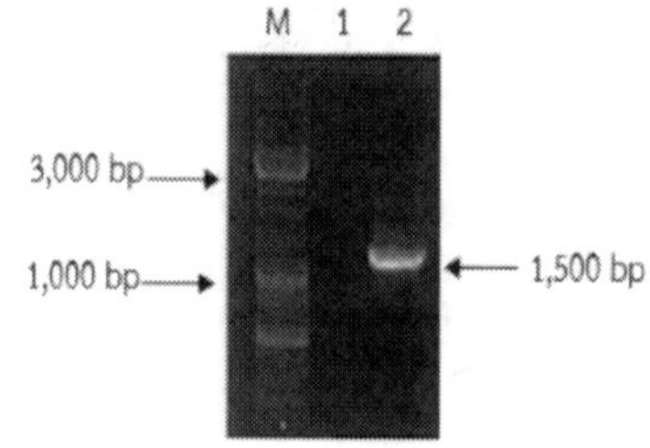

Fig. 3. Agarose gel electrophoresis of direct colony PCR using the universal primers: Lane M 100bp ladder, Lane 1 negative control, Lane 2 indicates PCR product of *Pantoea dispersa*

3.3 Sequence analysis of the *16S rRNA* gene and Phylogenetic analysis

After obtaining the sequences from this study, they were compared with the bacterial species recorded in the NCBI database (Figure 4). A single query sequence of query ID 31109 found their best match with the species belonging to the Enterobacteriaceae family. It was found that *Pantoea dispersa* had the highest identity at 99%. In addition, the sequences of all bacterial strains were aligned with species recorded in the database (No.2-No.12). From the result of multiple sequence alignment, the polymorphic nucleotides were examined using the Clustal W program [16] and the GeneDoc program [17]. It is composed of *Pantoea cypripedii* (NR_041973.1), *Pantoea cypripedii* strain LMG 2657 (NR_118394.1), *Enterobacter kobei* (LT5478211), *Kluyvera cryocrescens* (NR_114108.1), *Pantoea stewartii* (NR_104928.1), *Erwinia billingiae* (NR_118431.1), *Erwinia billingiae* strain Billing E63 (NR_104932.1), *Pantoea coffeiphila* (KJ427829.1), *Pantoea dispersa* strain DSM 30073 (NR_116797.1), *Pantoea dispersa* (AB907780.1) and *Aedes albopictus* (DQ397921.1). The results of multiple sequence alignment were presented in Figure 5.

Fig. 4. Graphical summary output of BLAST showing the homology coverage between query and the hits and description section in the BLAST report showing one-line summaries of sequences producing significant alignments

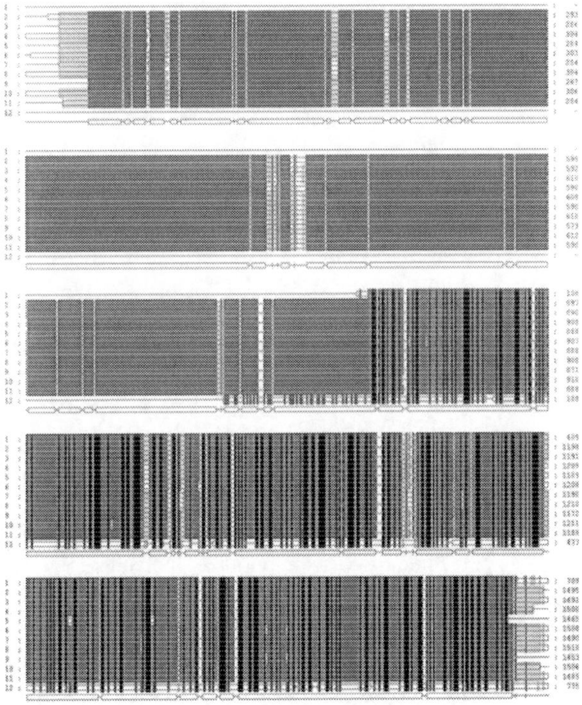

Fig. 5. Multiple sequence alignment of the *16s rRNA* gene from 13 species including 1) sequences from this study. 2) *Pantoea cypripedii* (NR_041973. 1) , 3) *Pantoea cypripedii* strain LMG 2657 (NR_118394. 1) , 4) *Enterobacter kobei* (LT547821. 1) , 5) *Kluyvera cryocrescens* (NR_114108. 1) , 6) *Pantoea stewartii* (NR_104928.1), 7) *Erwinia billingiae* (NR_118431.1), 8) *Erwinia billingiae* strain Billing E63 (NR_104932.1), 9) *Pantoea coffeiphila* (KJ427829.1), 10) *Pantoea dispersa* strain DSM 30073 (NR_116797.1), 11) *Pantoea dispersa* (AB907780. 1), 12) *Aedes albopictus* (DQ397921. 1). Black represents 100% identical nucleotide sequences and white has no DNA sequence similarity

Morphological identification becomes difficult because of the invisibility of some features followed by not well-developed specimens, which can even make the identification impossible. Therefore, molecular biology studies are necessary to support the morphological studies of bacteria isolated from natural sources. Phylogenetic trees were constructed based on the multiple sequence alignment of the 16s rRNA using neighbor-joining (NJ). In the phylogenetic trees, the laboratory sequences are checked according to the morphologically identified characteristic, while other species name is given for the sequences downloaded from NCBI GenBank. The evolutionary data was inferred using the Neighbor-Joining method. The percentage of replicating trees in which the associated taxa clustered together in the bootstrap test (1000 replicates) was analyzed to predict tree characteristics [18]. The phylogenetic tree (Figure 6) was established with the outgroup of *Aedes albopictus* (accession number DQ397921.1). Based on the sequence comparison and phylogenic tree construction, the bacterial isolate from the healthy wood of *Aquilaria crassna* presents 99% homology with *Pantoea dispersa*, a known Gram negative bacteria of the family Enterobacteriaceae. Genus *Pantoea* is isolated from Enterobacteriaceae in 1989 [22]. The genus *Pantoea* is classified into 20 different species named *Pantoea eucalyptii*, *Pantoea agglomerans*, *Pantoea vagans*, *Pantoea conspicua*, *Pantoea deleyi*, *Pantoea anthophila*, *Pantoea brenneri*, *Pantoea ananatis*, *Pantoea allii*, *Pantoea stewartii*, *Pantoea cypripedii*, *Pantoea calida*, *Pantoea gavinae*, *Pantoea dispersa*, *Pantoea septica*, *Pantoea wallisii*, *Pantoea eucrina*, *Pantoea rodasii*, *Pantoea rwandensis*, and *Pectobacterium carotovorum*. *Pantoea dispersa* is commonly isolated from plants, as evidenced by the study by *Erwinia* [22,23]. The genus *Pantoea* is a gram-negative, flagellated, nonencapsulated, nonspore, and capsule-forming ubiquitous *Brevibacterium cruentum* [24,25]. This genus is primarily associated to plant pathogens, with a few historical human diseases [25,26]. It was previously reported that this microorganism is useful as a biocontrol agent for black rot in sweet potatoes [27]. In addition, several members of the genus have been potentially found to enhance growth and yield, control plant pathogenic microbes, and increase abiotic tolerance in rice plants [28]. To our knowledge, this is the first report describing the isolation of bacterial species from healthy wood of *Aquilaria crassna* with rapid detection method.

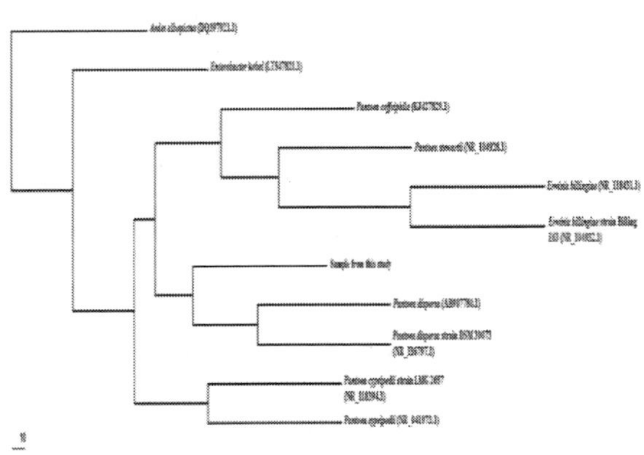

Fig. 6. Neighbor-joining phylogenetic tree based on the *16S rRNA* showing the relative positions of the *Pantoea* isolates and other *Pantoea* type strains. Evolutionary analyses were conducted in the MEGA7 program and bootstrap values (based on 1,000 replicates). The bar represents nucleotide substitution rate unit of 10.

4 Conclusion

The shape of the colony and the content in agarwood samples indicated cream whitish-colored colonies and gram-negative rods based on observation and isolation. According to the NCBI database, *Pantoea dispersa* showed the highest identity at 99%, and was regarded as a candidate bacterial endophyte. Furthermore, we found that species-specific colony PCR can be used to identify *P.*

dispersa without the need for DNA extraction from a natural sample. In comparison to conventional methods, it produces results in less time, requires less effort and resources, and is less expensive, rendering it suitable for biotechnological application.

5 Acknowledgements

This research was funded by Thailand Science Research and Innovation Fund, and King Mongkut's University of Technology North Bangkok with Contract no. KMUTNB-BasicR-64-41. We also would like to thank Mr. Prayoon Tiangyuttitom and Ms. Oratip Tiangyuttitom from the Krisanarak Small and Medium Community Enterprise for providing the samples of *Aquilaria crassna* in this study.

References

[1] I.A. Gibson, The role of fungi in the origin of oleoresin deposits (agaru) in the wood of *Aquilaria agallocha* Roxb, Bano Biggyan Patrika, 6, (1977): 16–26

[2] L.E. Hai, C.Y. Shyun, A.M. Yusoff, Early survival and growth in field trials of *Aquilaria malaccensis* (karas) and *Azadirachta excelsa* (sentang), Journal of Tropical Forest Science, 11, 4 (1999): 852–854

[3] S. Kamonwannasit, N. Nantapong, P. Kumkrai, P. Luecha, S. Kupittayanant, N. Chudapongse, Antibacterial activity of *Aquilaria crassna* leaf extract against *Staphylococcus epidermidis* by disruption of cell wall, Annals of Clinical Microbiology and Antimicrobials, 12, 20 (2013): 12-20

[4] D.T. Ahmaed, M. Mohammed, A.M. Masaad, S. N. Tajuddin, Investigation of agarwood compounds in *Aquilaria malaccensis* & *Aquilaria Rostrata* chipwood by using solid phase microextraction, Biomedical Journal of Scientific & Technical Research, 1, 6 (2017): 1609-1616

[5] M. Gao, X. Han, Y. Sun, H. Chen, Y. Yang, Y. Liu, H. Meng, Z. Gao, Y. Xu, Z. Zhang and, J. Han, Overview of sesquiterpenes and chromones of agarwood originating from four main species of the genus *Aquilaria*, The Royal Society of Chemistry, 9, 8 (2019): 4113–4130

[6] H. Chen, Y. Yang, J. Xue, J. Wei, ZZhang, Z., and, H. Chen, Comparison of compositions and antimicrobial activities of essential oils from chemically stimulated agarwood, wild agarwood and healthy *Aquilaria sinensis* (Lour.) Gilg trees, Molecules, 16, 6 (2011): 4884-4896

[7] A. LÓpez-Sampson, T. Page, History of use and trade of agarwood, Economic botany, 72, 1 (2018):107-129

[8] L.A. Ogilvie, P.R. Hirsch, Microbial ecology theory: current perspectives. Caister Academic Press, Norfolk, (2012): 71-85

[9] S. J. Bhore, J. Preveen, K. I. Kandasamy, Isolation and identification of bacterial endophytes from pharmaceutical agarwood-producing *Aquilaria* species, Pharmacognosy Research, 5, 2 (2013): 134–137

[10] K. Chakrabarty, A. Kumar, V. Menon, Trade in Agarwood, TRAFFIC India and WWF- India, New Delhi, (1994): 51pp

[11] B. I. Henry, A dictionary of the economic products of the Malay Penninsula. Vol. I. Governments of Malaysia and Singapore, Ministry of Agriculture and Co- operative (1966), HD9016 M3B9

[12] H. Okugawa, R. Ueda, K. Matsumoto, K. Kawanishi, A. Kato, Effects of agarwood extracts on the central nervous system in mice, Planta Medica, 59, 1 (1993):32-36

[13] J.J. Tian, X.X. Gao, W.M. Zhang, L. Wang, L.H. Qu, Molecular identification of endophytic fungi from *Aquilaria sinensis* and artificial agarwood induced by pinholes- infusion technique, African journal of biotechnology, 12, 21 (2013): 3115-3131

[14] M. Ito, K.I. Okimoto, T. Yagura, G. Honda, F. Kiuchi, Y. Shimada, Induction of sesquiterpenoid production by methyl jasmonate in *Aquilaria sinensis* cell suspension culture, Journal of Essential Oil Research, 17, 2 (2005): 175–180

[15] H. Chhipa, N. Kaushik, Fungal and Bacterial diversity isolated from *Aquilaria malaccensis* tree and soil, induces agarospirol formation within 3 months after artificial infection, Frontiers in Microbiology, 8, (2017): 1286

[16] J.D. Thompson, D.G. Higgins, T.J. Gibson, CLUSTAL W: improving the sensitivity of progressive multiple sequence alignment through sequence weighting, position-specific gap penalties and weight matrix choice, Nucleic Acids Research, 22, 22 (1994):4673–4680

[17] K. B. Nicholas, H. B. Nicholas Jr, D. W. Deerfield, GeneDoc: analysis and visualization of genetic variation, EMBNEW News, 4, (1997): 1-4

[18] S. Kumar, G. Stecher, K. Tamura, MEGA7: Molecular Evolutionary Genetics Analysis Version 7. 0 for Bigger Datasets, Molecular Biology and Evolution, 33, 7 (2016):1870–1874

[19] A. Cheng, C.Y. Liu, H.Y. Tsai, M.S. Hsu, C.J. Yang, Y.T. Huang, C.H. Liao, P.R. Hsueh, Bacteremia caused by *Pantoea agglomerans* at a medical center in Taiwan, 2000-2010, Journal of Microbiology, Immunology and Infection, 46, 3 (2013):187–194

[20] L. Benga, W.P. Benten, E. Engelhardt, K. Köhrer, C. Gougoula, M. Sager, 16S ribosomal DNA sequence-based identification of bacteria in laboratory rodents: a practical approach in laboratory animal bacteriology diagnostics, Laboratory Animals, 48, 4 (2014):305-312

[21] P. P. Bosshard, R. Zbinden, S. Abels, B. Böddinghaus, M. Altwegg, E.C. Böttger, 16S

rRNA gene sequencing versus the API 20 NE system and the VITEK 2 ID- GNB card for identification of nonfermenting Gram-negative bacteria in the clinical laboratory, Journal of Clinical Microbiology, 44, 4 (2006):1359-1366

[22] A. M. Walterson, J. Stavrinides, Pantoea: insights into a highly versatile and diverse genus within the Enterobacteriaceae, FEMS Microbiology Reviews, 39, 6 (2015): 968–984

[23] C. Brady, I. Cleenwerck, S. Venter, M. Vancanneyt, J. Swings, T. Coutinho, Phylogeny and identification of Pantoea species associated with plants, humans and the natural environment based on multilocus sequence analysis (MLSA) , Systematic and Applied Microbiology, 31, (2008): 447-460

[24] H. Schmid, S. Schubert, C. Weber, J.R. Bogner, Isolation of a *Pantoea dispersa*-like strain from a 71- year- old woman with acute myeloid leukemia and multiple myeloma, Infection, 31, 1 (2003): 66-67

[25] N. Asai, Y. Koizumi, A. Yamada, D. Sakanashi, H. Watanabe, H. Kato, A. Shiota, M. Hagihara, H. Suematsu, Y. Yamagishi, H. Mikamo, *Pantoea dispersa* bacteremia in an immunocompetent patient: a case report and review of the literature, Journal of Medical Case Reports, 13, 1 (2019): 33

[26] Y. Yang, H. Hu, C. Zhou, W. Zhang, Y. Yu, Q. Liu, T. Lu, Q. Zhang, Characteristics and accurate identification of *Pantoea dispersa* with a case of spontaneous rupture of hepatocellular carcinoma in China: A case report, Medicine (Baltimore), 14, (2022): e28541

[27] L. Jiang, J.C. Jeong, J.S. Lee, J.M. Park, J.W. Yang, M.H. Lee, S.H. Choi, C.Y. Kim, D.H. Kim, S.W. Kim, J. Lee. Potential of *Pantoea dispersa* as an effective biocontrol agent for black rot in sweet potato Scientific Reports, 9, (2019): 16354

[28] F. Doni, N. S. M. Suhaimi, B. Irawan, Z. Mohohamed, M. S. Mispan, Associations of Pantoea with Rice Plants: As Friends or Foes?, Agriculture, 11, 12 (2021): 1278

A comparative study of pH and temperature on rheological behaviour between Polyacrylamide (PAM) and its modified PAM

Jin Kwei Koh[1], *Chin Wei* Lai[1,*], *Mohd Rafie* Johan[1], *Sanjay Mavinkere* Rangappa[2], *Suchart* Siengchin[2]

[1]Nanotechnology & Catalysis Research Centre (NANOCAT), Level 3, Block A, Institute for Advanced Studies, University of Malaya (UM), 50603, Kuala Lumpur, Malaysia.
[2]Natural Composite Research Group Lab, Department of Materials and Production Engineering, The Sirindhorn International Thai-German Graduate School of Engineering (TGGS), King Mongkut's University of Technology North Bangkok (KMUTNB), Bangkok, Thailand.

Abstract. Drilling waste is a problem that affects the environment, society, and health. However, the rheological additive in drilling fluid is the source of the generation of drilling waste. Hence, a suitable rheological additive became a concern in bored pile construction. Conventional bentonite has been replaced by the usage of polymer in drilling operations, this is due to the operational, environmental, and economic aspects. Unlike bentonite, polyacrylamide (PAM) has a high molecular weight, good gelling behaviour, and is eco-friendly. However, there are limitations of PAM in terms of thermal stability and pH sensitivity. Therefore, the conventional polymer-based drilling fluid requires the hybridisation of functionalised material. In this paper, PAM with its enhancement is needed for water-based drilling through modification using silica (SiO_2) and sodium dodecyl sulphate (SDS). The paper presented a comparative analysis of rheological behaviour between polyacrylamide (PAM) and modified PAM. This research involved the influence of temperature (ambient to 80 °C) and pH (9 and 10). Results revealed that modified PAM performed better at 40 °C and pH 10. Furthermore, PAM and modified PAM had a better rheological performance at pH 10. Future studies can demonstrate the modified PAM as the drilling fluid in the bored pile construction to investigate the frictional resistance of the drilling fluid.

Keyword. Polyacrylamide, functionalised, rheological, temperature, pH

1 Introduction

Drilling waste is the second-largest waste. This issue has become a concern in drilling, exploration, and offshore industry because the waste can affect the soil quality, health, water, and entire ecosystem without proper management[1], [2]. Hence, the drilling fluid is a critical component that can generate waste after excavation and exploration. Generally, drilling fluid is composed of rheological additives, water, and clay. Water-based drillings can use with various rheological additives, such as bentonite, polymer, and others [3]. Currently, the polymer is the alternative to bentonite in drilling technology due to operational, environmental, and economic aspects [4]. In addition, polyacrylamide (PAM) is commonly utilised in bored pile and diaphragm construction because it has a high molecular weight, good gelation time, inexpensive, and eco-friendly industrial usage. In contrast, the consumption of bentonite requires higher amounts of dosage, which generate more waste than PAM [5].

Nonetheless, the poor thermal stability and pH sensitivity are the limitations of PAM usage in drilling. Additionally, the polymer can degrade at a specific temperature. For example, the thermal stability of PAM can endure a temperature below 175 °C, according to Europe *et al.* [6]. However, Xie *et al.* [7] reported a detrimental effect on drilling performance when the incline of temperature with the excessive hydrolysis in the polymer. In this study, the temperature of drilling fluid was investigated from ambient conditions to 80 °C due to safety concerns on the laboratory scale and the boiling point of water acting as the heating base of fluid.

Besides, the drilling fluid is always alkaline, with a pH of 8 to 10.5. Gamal *et al.* [8] studied the bentonite-based drilling fluid that kept better rheological performance at pH 9-10. However, more acidic or basic conditions of drilling fluid can contribute to the cost of facilities in drilling, such as pumping power and equipment corrosiveness. Hence, a drilling fluid with a pH of 9 to 10 can perform well in the drilling operation. Additionally, a low dosage of nanoparticles in polymer composite influenced the rheological behaviour and pH of modified drilling marginally [9], [10]. Inversely, the pH of the drilling fluid was more acidic after graphene oxide was involved in the drilling fluid, as investigated by Kusrini *et al.* [11]. Therefore, this study investigated the rheological behaviour between PAM and modified PAM at pH 9 and 10 because the water-based drilling fluid is usually favourable at pH 9 and 10.

Due to the poor thermal stability and chemical stability of PAM, the modification of PAM is popular with the hybridisation with additives, such as functionalised materials, organic modifiers, cross-linking

* Corresponding author: cwlai@um.edu.my.

https://doi.org/10.1051/e3sconf/202235502005

agents, and others. Among all the functionalised materials, silica is the most simple and economical material as it is easy to access and less harmful to the environment. Hence, this study enhanced PAM by using sodium dodecyl sulphate (SDS) and silica (SiO_2) as the modifiers. Furthermore, several researchers investigated the synergistic effect of SDS and SiO_2 involved in PAM-based drilling applications, contributing the better viscosity, surface tension, and others [10], [12]–[15]. The paper studied the rheological behaviour with the effects of temperature and pH between bare PAM and modified PAM. Further, chemical and surface characterisations were studied to investigate the surface properties between PAM and modified PAM.

2 Experimental

2.1 Materials

Polyacrylamide was provided by Synergy Lite Sdn Bhd. Sodium dodecyl Sulphate was purchased from Merck. Silica was purchased from Sigma-Aldrich. Soda Ash was purchased from Solvay S.A.

2.2 Synthesis of PAM and modified PAM

Bare PAM fluid was mixed in 1000 ppm. Besides, 0.5 wt % SiO_2 and 0.2 wt % SDS were mixed with 1000 ppm PAM as a modified PAM fluid. These were mixed well at a relative speed with Joanlab overhead stirrer.

2.3 Characterisation

PAM and modified PAM were characterised using Fourier Transform Infrared Spectroscopy (FTIR) and tensiometer.

2.4 Rheological testing

PAM and modified PAM were characterised using HST-6ST 6-speed rotational viscometer, Qingdao Heng Taida. This test was further analysed with the effect of pH (9, 10) and the influence of temperature (ambient, 40, 60, and 80 °C) on both polymers. The procedure of testing is followed, according to American Petroleum Institute (API).

3 Results and discussion

3.1. Appearance of fluid

As depicted in **Fig. 1**, the appearance of PAM is transparent. However, the colour shifted from transparent to milky white after modification, indicating the addition of SiO_2 and SDS in the PAM solution. Additionally, the top layer of modified PAM formed with white foam, which is attributed to SDS dispersion in SiO_2/PAM solution. This also indicates that SDS increases the surface activity of SiO_2 in drilling fluid [16].

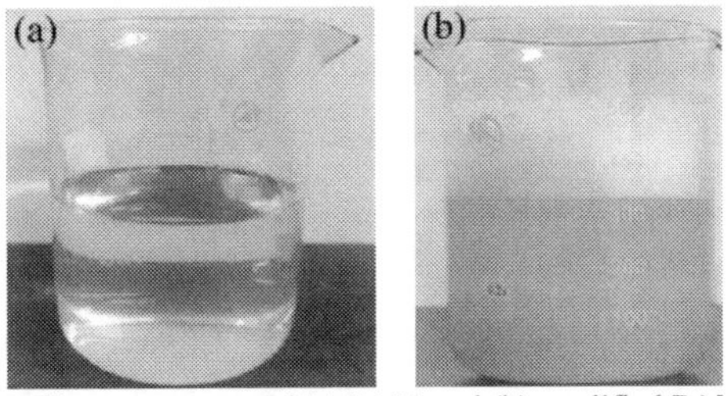

Fig. 1. The appearance of (a) PAM and (b) modified PAM doped with SiO_2 and SDS.

3.2 FTIR analysis

The FTIR spectra of bare PAM and the modified PAM are illustrated in **Fig. 2**. A spectrum of PAM exhibited the stretching vibration of the carbonyl group at the peak of 1650 cm^{-1} and the stretching vibration of the amide group at the peak of 3424 cm^{-1}. The stretching vibration of the C-H bonds is shown in two consecutive peaks at 2905 cm^{-1} and 2974 cm^{-1}. The ring structure vibration of the methyl C-H bond is observed at the peak of 1449 cm^{-1}, while the peak at 1378 cm^{-1} is C-N stretching vibration [17], [18].

Due to their low concentration of SiO_2 and SDS applied in surface modification, the functionality of both modifiers is poorly identified in the spectra of modified PAM. However, a prominent peak in the range 3200 – 3400 cm^{-1} is the silanol -OH vibration in all modified PAM spectra. Furthermore, the bending vibration of H-O-H is indicated at a sharp peak at 1630 cm^{-1} in the modified PAM spectra. Both peaks in modified PAM spectra exhibited a stronger linking between the -OH groups and the modified SiO_2 functionalised particles [17]. Moreover, the peak observed at 3424 cm^{-1} and 1650 cm^{-1} in PAM are shifted towards the lower peaks at 3280 cm^{-1} and 1630 cm^{-1} in modified PAM, showing stable dispersion of SiO_2 in PAM.

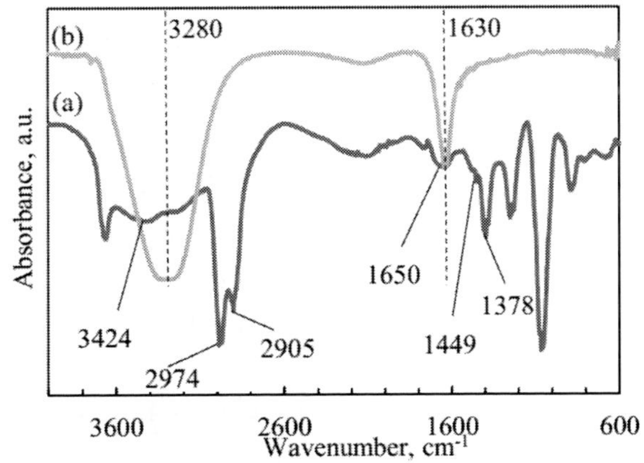

Fig. 2. FTIR spectra of (a) PAM and (b) modified PAM doped with SiO_2 and SDS.

3.3 Surface characteristic

Table 1 shows the surface tension and surface contact angle of PAM and modified PAM. Modified PAM

 https://doi.org/10.1051/e3sconf/202235502005

showed less surface tension mean than PAM, indicating the modified PAM can enhance the rheological properties of drilling. The lesser surface tension can reduce the rheological risk, such as pipe sticking, wellbore issue, and others [19], [20].

Meanwhile, the surface contact angle of modified PAM was lower than that of PAM, demonstrating the modified PAM is more hydrophilic than PAM. Based on the wettability theory, both have hydrophilic properties as their contact angles were below 90°.

Table 1. The surface characteristic of PAM and modified PAM.

Surface characteristic	PAM	Modified PAM
Surface tension mean, mN/m	43.83	30.11
Surface contact angle mean, °	42.19	24.58

3.4. Rheological test

3.4.1 Effect of temperature on rheological properties between PAM and modified PAM

As depicted in **Fig. 3**, overall showed the fluctuation in apparent viscosity (AV), plastic viscosity (PV), and yield point (YP) between PAM and modified PAM. In ambient conditions, the modified PAM had an incline of YP by 2.2 % but AV and PV reduced by 17.7 % and 25.8 %, respectively. After heating at 40 °C, modified PAM showed a reduction of AV, PV, and YP by 20.9 %, 15.1 %, and 30.2 %, respectively. After heating at 60 and 80 °C, the modified polymer demonstrated the incline of AV and PV with the decline of YP. When compared to the gel strength of PAM, modified PAM demonstrated lower over all the temperatures in **Fig. 4**. In conclusion, modified PAM can withstand temperatures up to 40 °C because modified PAM exhibited better rheological performance, especially the reduction of PV in this study.

Generally, the heating of the polymer can enhance the flocculation and adsorption capacity. This situation can affect the rheological performance of the drilling fluid. Additionally, the endothermic process also enhances the flocculation rate. Flocculation can change the chemical conformity and functionality of the polymer [21], [22]. Therefore, high PV in drilling fluid due to flocculation is undesired, which can influence the wellbore performance [23], [24]. This paper showed that the inclination of PV is affected by higher heating temperatures. The large gel-like floc formation that occurs at high temperatures potentially contributes to the higher viscosity with high pumping energy during drilling. Therefore, aggregation and flocculation can cause poor rheological performance at higher temperatures [24]–[26].

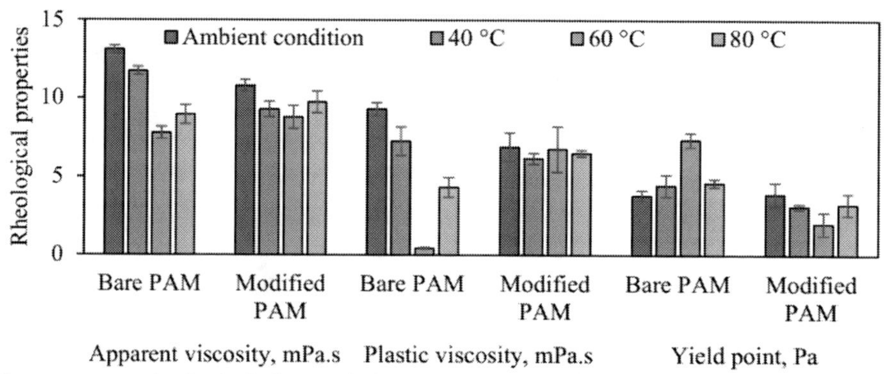

Fig. 3. Effect of temperature on the rheological properties between PAM and modified PAM doped with SiO_2 and SDS.

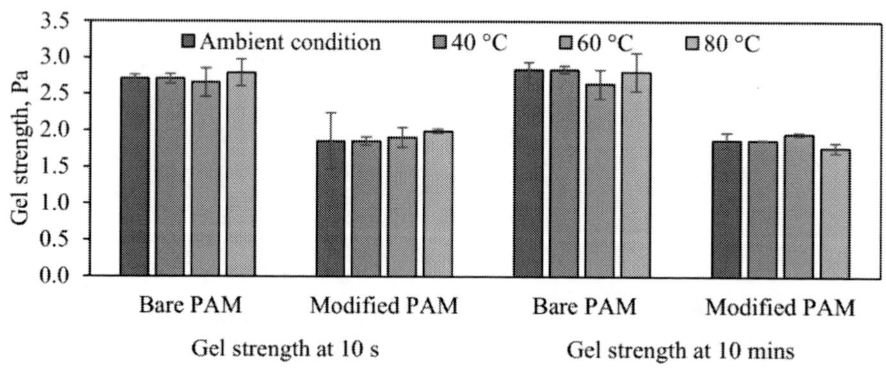

Fig. 4. Effect of temperature on the gel strength between PAM and modified PAM doped with SiO_2 and SDS

3.4.2 Effect of pH on rheological properties between PAM and modified PAM

The effect of pH on the rheological properties between PAM and modified PAM is compared in **Fig. 5** and **Fig. 6**. At pH 9, the modified PAM reduced AV and YP by 11.6% and 79.8%, respectively. However, modified PAM showed an incline of AV by 6.3% and a decline of YP by 48.1%, respectively, at pH 10. For PV, modified PAM inclined by 48%, and 89.4%, at pH 9 and 10, respectively. Although the rheological performance of modified PAM was not ideal than PAM, the plastic viscosity of modified PAM was lower at pH 10.

The gel strength of modified PAM was inclined at pH 9 -10. Hence, modified PAM had higher gel strength than bare PAM. Higher gel strength can promote better cutting in drilling [23], [27]. Further, modified PAM and PAM performed a flat gel strength at pH 10 due to a negligible difference in gel strength between 10 seconds and 10 minutes. Therefore, the drilling fluid performed with better rheological behaviour at pH 10.

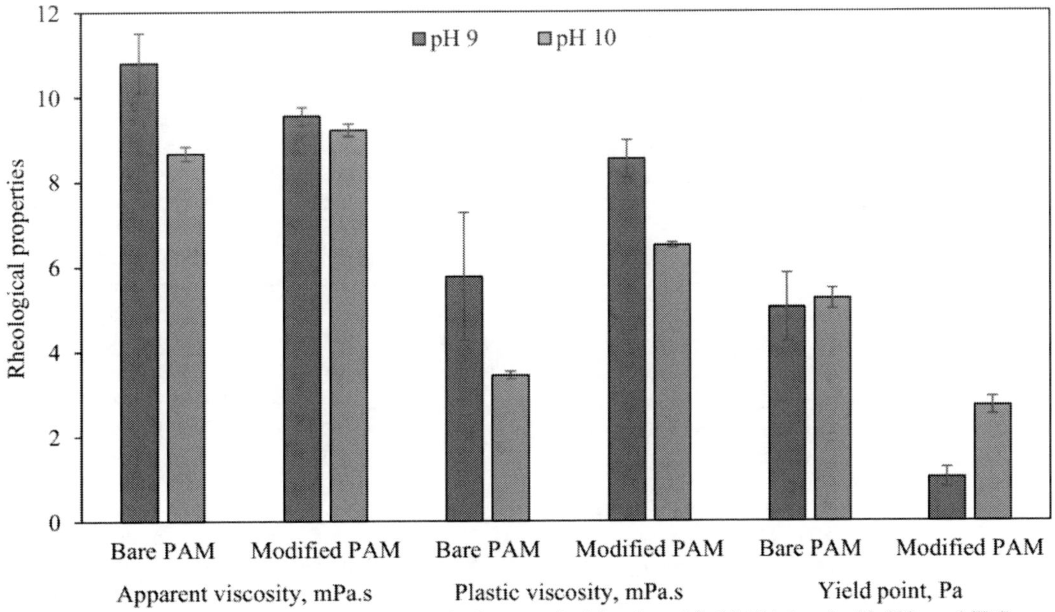

Fig. 5. Effect of pH on the rheological properties between PAM and modified PAM doped with SiO_2 and SDS.

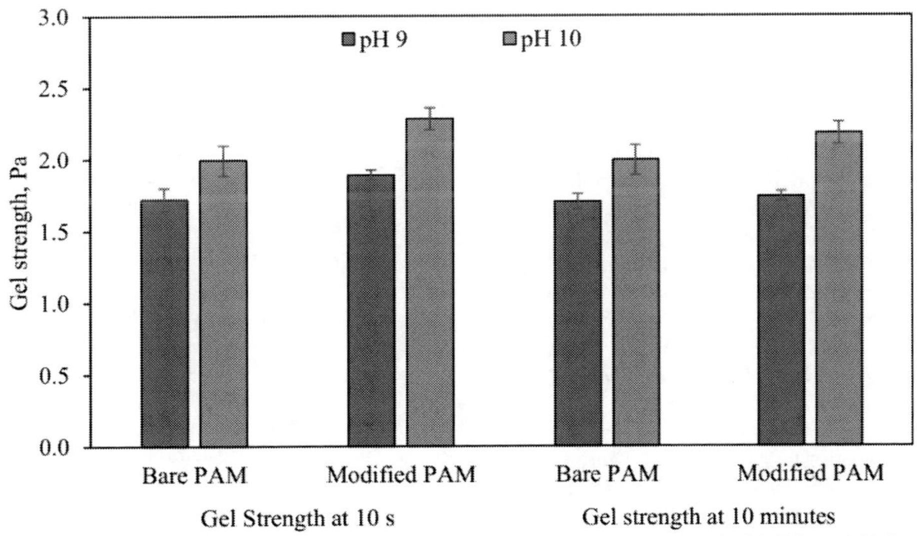

Fig. 6. Effect of pH on the gel strength between PAM and modified PAM doped with SiO_2 and SDS.

* Corresponding author: cwlai@um.edu.my.

4 Conclusion

In this paper, PAM and modified PAM were synthesised, characterised, and utilised in drilling fluid at a specific temperature (ambient to 80 °C) and pH (9 – 10). The modified PAM was investigated to enhance drilling fluids' performance, especially in bored pile drilling. The surface characterisation of modified PAM showed more hydrophilic than that of PAM. Besides, the rheological investigation showed that modified PAM performed better than PAM at 40 °C. Further, modified PAM had a better rheological performance at pH 10. In the future, modified PAM can be further tested in the bored pile construction to validate the frictional resistance of drilling fluid.

Acknowledgement

This research study was financially supported by Synergy Lite Sdn. Bhd. (No. PV027-2021).

References

[1] O. S. Onwuka, O. Igwe, S. I. Ifediegwu, and C. S. Uwom, "An assessment of the effectiveness of drilling waste treatment process in X-gas field, Niger Delta, Nigeria," *Geol. Ecol. Landscapes*, **vol. 2**, no. 4, (2018): 288–302

[2] A. R. Ismail, A. H. Alias, W. R. W. Sulaiman, M. Z. Jaafar, and I. Ismail, "Drilling fluid waste management in drilling for oil and gas wells," *Chem. Eng. Trans.*, **vol. 56**, (2017): 1351–1356

[3] J. K. Koh, C. W. Lai, M. R. Johan, S. S. Gan, and W. W. Chua, "Recent advances of modified polyacrylamide in drilling technology," *J. Pet. Sci. Eng.*, **vol. 215**, part A, (2022): 110566

[4] C. Lam, S. A. Jefferis, T. P. Suckling, and V. M. Troughton, "Effects of polymer and bentonite support fluids on the performance of bored piles," *Soils Found.*, **vol. 55**, no. 6, (2015): 1487–1500

[5] S. Jefferis, V. Troughton, and C. Lam, "Polymer systems for fluid supported excavations," in *Geotechnical issues in construction: short paper series and 2nd conference held on 2 November 2009, at CIRIA, London* (2011): 7–12.

[6] W. Europe, M. Doble, and A. Kumar, "Degradation of Polymers," in *Biotreatment of industrial effluents*, Elsevier Ltd, (2005): 101–110.

[7] B. Xie, J. Ma, Y. Wang, A. P. Tchameni, M. Luo, and J. Wen, "Enhanced hydrophobically modified polyacrylamide gel for lost circulation treatment in high temperature drilling," *J. Mol. Liq.*, **vol. 325**, (2021): 115155

[8] H. Gamal, S. Elkatatny, S. Basfar, and A. Al-Majed, "Effect of pH on rheological and filtration properties of water-based drilling fluid based on bentonite," *Sustainability*, **vol. 11**, no. 23, (2019): 6714

[9] H. Gudarzifar, S. Sabbaghi, A. Rezvani, and R. Saboori, "Experimental investigation of rheological & filtration properties and thermal conductivity of water-based drilling fluid enhanced," *Powder Technol.*, **vol. 368**, (2020): 323–341

[10] S. Medhi, S. Chowdhury, D. Kumar, and G. Aryab, "An investigation on the effects of silica and copper oxide nanoparticles on rheological and fluid loss property of drilling fluids," *J. Pet. Explor. Prod. Technol.*, **vol. 10**, (2019): 91–101

[11] E. Kusrini, F. Oktavianto, A. Usman, D. P. Mawarni, and M. I. Alhamid, "Synthesis, characterization, and performance of graphene oxide and phosphorylated graphene oxide as additive in water-based drilling fluids," *Appl. Surf. Sci.*, **vol. 506**, (2020): 145005

[12] T. Sharma and J. S. Sangwai, "Silica nanofluids in polyacrylamide with and without surfactant: Viscosity, surface tension, and interfacial tension with liquid paraffin," *J. Pet. Sci. Eng.*, **vol. 152**, (2017): 575–585

[13] S. K. Inturi, G. R. Dash, and J. Kodavaty, "Role of silica Nano particles in altering rheological properties of drilling fluid in Enhanced Oil Recovery," *Mater. Today Proc.*, **vol. 17**, (2019): 354–361

[14] L. M. Corredor, M. M. Husein, and B. B. Maini, "Impact of PAM-Grafted Nanoparticles on the Performance of Hydrolyzed Polyacrylamide Solutions for Heavy Oil Recovery at Different Salinities," *Ind. Eng. Chem. Res.*, **vol. 58**, no. 23, (2019): 9888–9899

[15] L. M. Corredor-rojas, A. H. Sarapardeh, M. M. Husein, P. M. Dong, and B. B. Maini, "Rheological behavior of surface modified silica nanoparticles dispersed in Partially Hydrolyzed Polyacrylamide and Xanthan Gum solutions : Experimental measurements , mechanistic understanding , and model development," *Energy Fuels*, **vol. 32**, no. 10, (2018): 10628–10638

[16] R. S. Kumar, K. R. Chaturvedi, S. Iglauer, J. Trivedi, and T. Sharma, "Impact of anionic surfactant on stability, viscoelastic moduli, and oil recovery of silica nanofluid in saline environment," *J. Pet. Sci. Eng.*, **vol. 195**, (2020): 107634

[17] T. Sharma, G. S. Kumar, B. H. Chon, and J. S. Sangwai, "Thermal stability of oil-in-water Pickering emulsion in the presence of

nanoparticle, surfactant, and polymer," *J. Ind. Eng. Chem.*, **vol. 22**, (2015): 324–334

[18] M. A. Haruna, S. Pervaiz, Z. Hu, E. Nourafkan, and D. Wen, "Improved rheology and high-temperature stability of hydrolyzed polyacrylamide using graphene oxide nanosheet," *J. Appl. Polym. Sci.*, **vol. 136**, no. 22, (2019): 47582

[19] M. S. Muhamad, M. R. Salim, and W.-J. Lau, " Surface modification of SiO 2 nanoparticles and its impact on the properties of PES-based hollow fiber membrane ," *RSC Adv.*, **vol. 5**, no. 72, (2015): 58644–58654

[20] G. N. ochuwe Abang, Y. S. Pin, and N. Ridzuan, "Application of silica (SiO_2) nanofluid and Gemini surfactants to improve the viscous behavior and surface tension of water-based drilling fluids," *Egypt. J. Pet.*, **vol. 30**, no. 4, (2021): 37–42

[21] B. Xiong *et al.*, "Polyacrylamide degradation and its implications in environmental systems," *npj Clean Water*, **vol. 1**, no. 17, (2018)

[22] Y. Zhang, Z. Miao, and J. Zou, "A new cation-modified Al-polyacrylamide flocculant for solid-liquid separation in waste drilling fluid," *J. Appl. Polym. Sci.*, **vol. 132**, no. 11, (2015): 41641

[23] E. U. Akpan, G. C. Enyi, G. Nasr, A. A. Yahaya, A. A. Ahmadu, and B. Saidu, "Water-based drilling fluids for high-temperature applications and water-sensitive and dispersible shale formations," *J. Pet. Sci. Eng.*, **vol. 175**, (2019): 1028–1038

[24] M. K. Fokuo *et al.*, "Tannin-Based Deflocculants in High Temperature High Pressure Wells: A Comprehensive Review," *Adv. Chem. Eng. Sci.*, **vol. 11**, no. 04, (2021): 263–289

[25] A.-G. Guezennec *et al.*, "Transfer and degradation of polyacrylamide based flocculants in hydrosystems: a review Transfer and degradation of polyacrylamide based flocculants in hydrosystems: a review. Environ-mental Science and Pollution Title: Transfer and degradation of polyacry," *Research*, **vol. 22**, no. 9, (2015) [Online].

[26] F. A. Makinde, A. D. Adejumo, C. T. Ako, and V. E. Efeovbokhan, "Modelling the effects of temperature and aging time on the rheological properties of drilling fluids," *Pet. Coal*, **vol. 53**, no. 3, (2011): 167–182.

[27] Z. Vryzas and V. C. Kelessidis, "Nano-based drilling fluids: A review," *Energies*, **vol. 10**, no. 4, (2017)

 https://doi.org/10.1051/e3sconf/202235502006

Improving the mechanical properties of jute fiber woven mat reinforced epoxy composites with addition of zinc oxide filler

T P Sathishkumar[1,*], *L* Rajeshkumar[2] , *G* Rajeshkumar[3], *M R* Sanjay[4], *Suchart* Siengchin[4] *and Navanee* Thakrishnan[1]

[1]Department Mechanical Engineering, Kongu Engineering College, Perundurai, Erode, Tamilnadu, India.
[2]Department of Mechanical Engineering, KPR Institute of Engineering and Technology, Coimbatore, Tamil Nadu, India
[3]Department Mechanical Engineering, PSG Institute of Technology and Applied Research, Coimbatore, Tamil Nadu, India
[4]King Mongkut's University of Technology North Bangkok (KMUTNB), Bangkok, Thailand.

Abstract. The present work is to prepare the jute fiber reinforced epoxy composites by changing the number of layers of jute fiber mat and adding the various content of zinc oxide filler. The composites were prepared by compression molding techniques and investigates the tensile, flexural and impact properties according to ASTM standard. The results reveal that increasing the number of jute fiber mat layer was increased the mechanical properties. The maximum strength was obtained in double-layer composites. By incorporation of zinc oxide filler, the filler content gradually increased the mechanical strengths which increased the bonding effect between fiber and matrix. The maximum mechanical strengths were obtained for composites containing the 25% of zinc oxide filler. More than the 25% of zinc oxide filler was decreased the mechanical properties due to lower the resin content and bonding between fiber and matrix.

Keyword. Jute fiber composites; Zinc oxide filler; mechanical properties; compression molding technique

1 Introduction

The natural fiber reinforced polymer composites are widely used in various applications, such as automobile parts, aerospace, construction and aviation [1]. These materials fit to the concept of circular economy to convert the biomass waste to value-added products. Changing of the weight content of the natural fiber in the composites manufacturing was improving the mechanical, dynamic and thermal properties. The natural fibers, such as jute, sisal, snake grass [2], Sansevieria ehrenbergii [3], pineapple leaf fiber, coir fiber, etc, are widely used to prepare the polymer composites for characterization.

The fillers have been used to mix with polymer to develop the filler and fiber-reinforced polymer composites for improving the properties. The polypropylene composites were made with continuous and short jute fiber by varying weight ratios from 20 to 40%. The Maleic Anhydride grafted Polypropylene (MAPP) coupling agent was mixed with resin before making composites. The tensile and flexural strength of continuous fiber and coupling agent mixed composite were found to be higher than the neat resin [1]. The polypropylene composite made up of chopped jute fiber showed higher mechanical properties at a fiber content of 30 wt% [3]. The chemically treated jute fiber composites showed higher mechanical properties than the untreated fiber composites. The NaOH and silane treated jute fiber showed higher bending strength [4]. The polypropylene composites reinforced with jute fabric showed higher tensile strength at 40 % fiber volume fraction [5]. The polyvinylchloride coated plane weave jute fiber in polyester composite showed higher tensile and flexural strength at a coating concentration of 200 g/m2 [6]. The epoxy composite was made up of woven jute mat with various coconut shell fillers. The composite with a 30 % weight content of filler showed better mechanical properties [8]. The Bi woven Jute fabrics were used to prepare the epoxy composite with *Portunus sanguinolentus* shell powder. The maximum mechanical properties of 55 MPa (Tensile strength), 5.45 MPa (Flexural strength) and 6.21 J (Impact strength) were obtained for chemically treated powder [9].

The unidirectional jute fiber was used to prepare the epoxy composite at 52 % of fiber content and compared the mechanical properties with bamboo fiber composites. The higher mechanical properties were found for jute fiber composite. The unidirectional jute fiber was used to prepare the epoxy composite at 52 % of fiber content and compared the mechanical properties with bamboo fiber composites. The improved mechanical properties were found for jute fiber composites, such as tensile strength of 216 MPa and flexural strength of 158 MPa [10]. The length of jute fiber, such as 10, 20, 30 and 40 mm was used to prepare the epoxy composites with and without NaOH treated fiber and at various weight fractions 0%, 5%, 10%, 15%, 20% and 25% [11]. The jute fiber was treated and coated with NaOH and polylactide for composites development. The mechanical properties were improved by increasing the coating thickness [12].

The injection molding technique was used to prepare the jute and flax fiber polypropylene composite. The jute fiber reinforced polypropylene composite showed higher tensile strength than the flax fiber reinforced polypropylene composite [13]. The rose and

* Corresponding author: atpsathish@kongu.ac.in

padauk wood dust was mixed with epoxy resin for preparing the jute epoxy composites. The 10 percent padauk wood filler mixed with jute epoxy composite showed higher mechanical properties [14]. The hot press technique was pursued to prepare the Polypropylene-MAPP/flax composites. The maximum tensile strength was found in longitudinal and transverse directions at 60% fiber content [15]. The fillers of zinc oxide and titanium dioxide were used to prepare the jute epoxy composites. The addition of filler showed in increasing the thermal stability and mechanical properties [16]. The present work is to investigate the mechanical properties of zinc oxide filler reinforced jute epoxy composite at various filler and fiber content.

2 Experimental

2.1 Fabrication of Composites

The jute fiber mat composite laminate was developed by hand lay up method followed by compression cold press technique. The jute mat was cut into 220 mm length × 220 mm width according to mold size. The epoxy and hardener were mixed in a ratio of 10:1 ratio for 5 minutes. The polyethylene thin sheet was placed on the mold for easy removal of the plate. The jute mat was dipped in the resin and placed on the female die mold. The steel roller was rolled over each layer for removing the air bubbles in the resin. The closed mold was kept in a hydraulic press at 15 bar pressure for 8 hours. The casted composite was removed from the mold and it was placed in a hot air oven for 1 hour at 45°C. The single, double and tribble layer jute mat was used to prepare the composites, and addition of zinc oxide filler of 5, 10, 15, 20, 25 and 30% weight content was mixed with double layer composites for preparing the filler with fiber-reinforced epoxy composites.

2.2 Mechanical Properties

The mechanical properties of tensile, flexural and interlaminar shear strength (ILLS) were studied according to ASTM standards. The tensile test was performed according to the ASTM D 638. The size of the specimen was 165 mm in length, 13 mm in width and 3 mm in thickness. The gauge length was 50 mm. The universal tenting machine (Model: DTRX-5 kN, Make: Deepak poly-Plast Pvt. Ltd, Load cell: 30kN) was used for the experiment with a cross head speed of 2 mm/min. The three-point flexural testing sample size was 120 mm (length), 12.7 mm (width) and 3 mm (thickness) [1-2]. The testing speed of the sample was 2 mm/min. The ILLS test was performed in UTM and the size of the samples was cut into ASTM D2344. The five identical specimens were tested for all mechanical tests and the average values were calculated for further discussion.

3 Results and discussion

3.1 Tensile Properties

Figure 1a shows the tensile stress versus strain curve of zinc oxide and jute fiber mat reinforced epoxy composites. Increasing the jute fiber mat increased the stress with strain. The epoxy composites containing two layers of jute fiber mat showed a higher tensile stress curve with strain. The three-layer composites showed less tensile stress than double-layer composites. The double-layer can absorb more stress during experiments due to the required level of matrix presented in the composites. However, increasing of the jute mat was more than double to reduced the quantity of matrix to lower the tensile stress. The double-layer composites with zinc oxide filler showed that increasing the filler content increased the tensile stress curve. The maximum tensile stress versus strain curve was obtained for double layer composite with 25% of zinc oxide filler content. The filler acted as a bonding agent between jute fiber and epoxy, which increase the tensile stress. However, with more filler content i.e 30%, the epoxy acted as more brittle and also less epoxy content [1, 3] that reduced the stress. Figure 1b showed the tensile properties of jute mat epoxy composites with and without zinc oxide. The percentage differences between the tensile stress of the composites are 3.5 (Single to double layers), 22.87 (Double to triple layers), 29.6 (25 to 5 % of zinc oxide), 23 (25 to 10 % of zinc oxide), 18.6 (25 to 15 % of zinc oxide), 3.8 (25 to 20 % of zinc oxide), and -14.8 (25 to 30 % of zinc oxide). Figure 1c shows the tensile modus of jute mat epoxy composites with and without zinc oxide. The percentage differences between the tensile modulus of the composites are 9.7 (Single to double layers), 16.08 (Double to triple layers), 27.7 (25 to 5 % of zinc oxide), 23 (25 to 10 % of zinc oxide), 11.18 (25 to 15 % of zinc oxide), 11.18 (25 to 20 % of zinc oxide), and -17 (25 to 30 % of zinc oxide).

3.2 Flexural properties

Figure 2a shows the flexural stress versus strain curve of zinc oxide and jute fiber mat reinforced epoxy composites. Increasing the jute fiber mat increased the flexural stress with strain. The double-layer jute fiber composite showed higher flexural stress than the single and triple layers. The more bending stress was transferred uniformly in the double-layer composite than in the other two. The double-layer composites with zinc oxide filler showed that increasing the filler content increased the flexural stress curve. The maximum flexural stress versus strain curve was obtained for double-layer composite with 25% of zinc oxide filler content. The filler acted as a bonding agent between jute fiber and epoxy, which increased the flexural stress. The more filler in epoxy composites led to brittle [1, 3, 7], which reduced the flexural stress. During the bending of the composite sample, the loading surface had compressive stress formation and bottom surface had tensile stress. The more filler acted as a load barrier member in the composites. Figure 2b showed the flexural stress of jute mat epoxy composites with and without zinc oxide. The percentage differences between the tensile stress of the composites are 27.97 (Single to double layers), 6.96 (Double to triple layers), 31.67 (25 to 5 % of zinc oxide), 23.62 (25 to 10 % of zinc oxide), 14.49 (25 to 15 % of zinc oxide), 5.17 (25 to 20 % of zinc oxide), and -21.68 (25 to 30 % of zinc oxide).

 https://doi.org/10.1051/e3sconf/202235502006

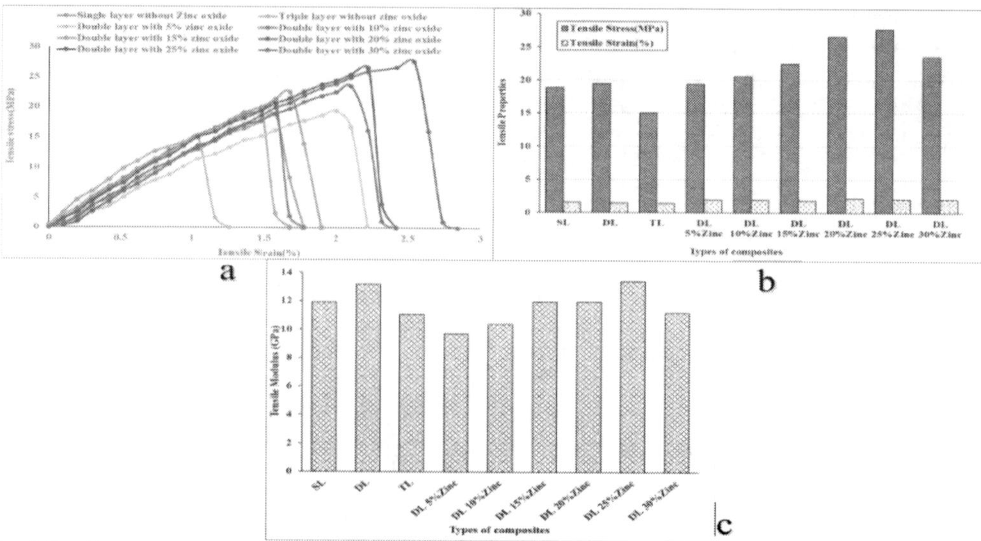

Figure 1. a) Tensile stress versus strain curve, b) Tensile stress and strain, and c) Tensile modulus of jute fiber mat reinforced epoxy composites with and without zinc oxide

Figure 2c shows the tensile modus of jute with and without zinc oxide. The percentage differences between the tensile modulus of the composites are 9.7 (Single to double layers), 16.08 (Double to triple layers), 27.7 (25 to 5 % of zinc oxide), 23 (25 to 10 % of zinc oxide), 11.18 (25 to 15 % of zinc oxide), 11.18 (25 to 20 % of zinc oxide), and -17 (25 to 30 % of zinc oxide).

3.3 Inter-laminar shear strength

Figure 3a shows the impact of zinc oxide and jute fiber mat reinforced epoxy composites. The impact strength of the composites was improved by increasing the number of jute fiber layers. The composites containing three layers of mat showed less impact strength than the double layers, and litter higher than the single layer jute mat composites. The double-layers composites had higher impact strength. The zinc oxide-filled double-layers epoxy composite showed an increase in the impact than the unfilled composite. Increasing the zinc oxide filler increased the impact strength due to the better bonding of fiber in epoxy with help of filler. The maximum impact strength was found in 25% of zinc oxide-filled composites. The percentage differences between the tensile stress of the composites are 50.3 (Single to double layers), 25 (Double to triple layers), 50.44 (25 to 5 % of zinc oxide), 40.27(25 to 10 % of zinc oxide), 40.29 (25 to 15 % of zinc oxide), 3.39 (25 to 20 % of zinc oxide), and -20.11 (25 to 30 % of zinc oxide).

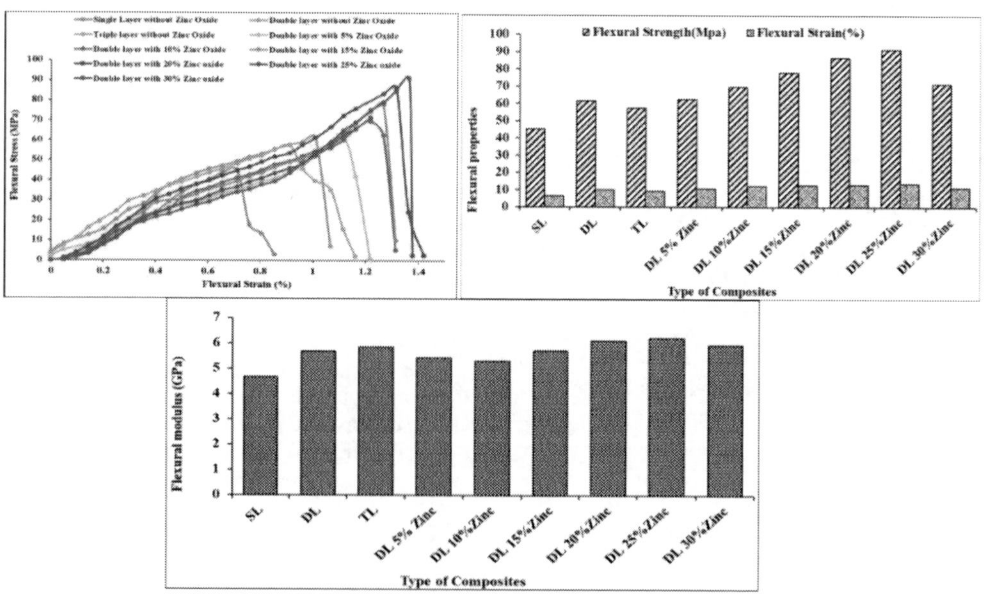

Figure 2. a) Flexural stress versus strain curve, b) Flexural stress and strain, and c) Flexural modulus of jute fiber mat reinforced epoxy composites with and without zinc oxide

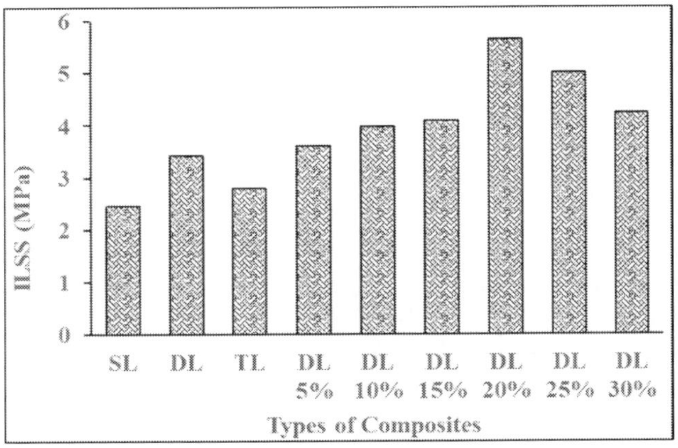

Figure 3. Impact strength of the jute fiber mat reinforced epoxy composites with and without zinc oxide

4 Conclusion

The jute fiber mat reinforced epoxy composite was prepared by compression molding techniques by changing the numbers of layers and, with or without zinc oxide filler. The tensile, flexural and impact strength were increased by increasing the jute fiber mat layers. Compare with single, double and triple layers, the double-layers composite showed higher mechanical properties due to the optimum level of matrix and fiber in the composite. Considering double-layer composites, the various percentagess of zinc oxide fillers were added to reinforced with epoxy for developed the composites. The more filler content increased the mechanical properties. The maximum strength was found for composites containing 25% of zinc oxide filler content. By increasing the filler content by more than 25%, there was a litter drop in mechanical properties due to the less resin content, which affects the bonding areas between fiber and matrix. Therefore, the double-layer jute fiber mat reinforced epoxy composites showed maximum mechanical properties at 25% of 25% of zinc oxide filler content.

References

[1] T. P. Sathishkumar, P. Navaneethakrishnan, S. Shankar, R. Rajasekar, N. Rajini, Characterization of natural fiber and composites – A review, Journal of Reinforced Plastic and Composites, 32(19), 1457-1476.

[2] T. P. Sathishkumar, P. Navaneethakrishnan, and S. Shankar. Tensile and flexural properties of snake grass natural fiber reinforced isophthallic polyester composites, Composite Science and Technology, 72 (10):1183-1190.

[3] T. P. Sathishkumar, P. Navaneethakrishnan, S. Shankar and R. Rajasekar, Characterization of new cellulose sansevieria ehrenbergii fibers for polymer composites, Composite Interfaces, 20(8):575-593.

[4] Sandhyarani Pailoor, H. N. Narasimha Murthy, Prakash Hadimani, T. N. Sreenivasa. Effect of chopped/continuous fiber, coupling agent and fiber ratio on the mechanical properties of injection-molded jute/polypropylene composites, Journal of natural fibers 2018, 16(1):126-136.

[5] Rezaur Rahman, Monimul Huque, Nazrul Islam, Mahbub Hasana. Improvement of physico-mechanical properties of jute fiber reinforced polypropylene composites by post-treatment, Composites: Part A, 39(2008):1739–1747.

[6] Anna Dilfi, Aiswarya Balan, Hong Bin, Guijun Xian, Sabu Thomas. Effect of surface modification of jute fiber on the mechanical properties and durability of jute fiber-reinforced epoxy composites, Polymer Composites, 39(S4)(2018): E2519-E2528.

[7] Sahadat Hossain, Muhammad Uddin, Razzak, Sarwaruddin Chowdhury, Ruhul Khan, Fabrication and characterization of jute fabrics reinforced polypropylene-based composites: effects of ionizing radiation and disaccharide(sucrose), Radiation effects & Defects in solids, 172(11-12)(2017): 1-11.

[8] Devarun Nath, Sayandeep Debnath, Mallika Datta, Sumona Chakrabarti, Shankha Ghosh, Biman Ghosh,. Poly(vinyl acetate)-coated jute fabric reinforced polyester composite with enhanced mechanical performance: Interfacial hydrogen bond and autohesion mechanism, Journal of Industrial Textiles, (2019):1–22.

[9] T P Sathishkumar, S Ramakrishnan, Mechanical properties of nanococonut shell filler mixed jute mat-reinforced epoxy composites for structure application, Fiber-Reinforced Nanocomposites: Fundamentals and Applications, (2020):459-476.

[10] P. Kumaran, S. Mohanamurugan, M. Madhu, R. Vijay, D. Lenin Singaravelu, A. Vinod, M. R. Sanjay, Suchart Siengchin, Investigation on thermo-mechanical characteristics of treated/untreated Portunus sanguinolentus shell powder-based jute fabrics reinforced epoxy composites, Journal of Industrial Textiles, 50(4)(2019): 1–33.

[11] Subhankar Biswasa, Sweety Shahinura, Mahbub Hasana, Qumrul Ahsana, Physical, mechanical and thermal properties of jute and bamboo fiber reinforced unidirectional epoxy composites, Procedia Engineering, 105(2015): 933 – 939.

[12] S Ramakrishnan, K Krishnamurthy, R Rajasekar, G. Rajeshkumar., An experimental study on the effect of nano-clay addition on mechanical and water absorption behaviour of jute fibre reinforced epoxy composites, Journal of Industrial Textiles, 49(5)(2019): 1–24.

[13] MK. Gupta, Investigations on jute fibre-reinforced polyester composites: Effect of alkali treatment and poly(lactic acid) coating, Journal of Industrial Textiles, 49(7)(2020): 1–20.

[14] S. Dinesh, P. Kumaran, S. Mohanamurugan, R. Vijay, D. Lenin Singaravelu, A. Vinod, M R. Sanjay, Suchart Siengchin, K. Subrahmanya Bhat, Influence of wood dust fillers on the mechanical, thermal, water absorption and biodegradation characteristics of jute fiber epoxy composites, Journal of Polymer Research, 27(9)(2020): 1-13.

[15] Morgane Tanguya, Alain Bourmauda, Johnny Beaugrandc, Thierry Gaudryb, Christophe Baleya, Polypropylene reinforcement with flax or jute fibre; Influence of microstructure and constituents properties on the performance of composite, Composites Part B, 139(2018): 64–74.

[16] P A Prasob, M Sasikumar. Static and dynamic behavior of jute/epoxy composites with ZnO and TiO2 fillers at different temperature conditions, Polymer Testing, 69(2018): 52-62.

Study of the Titanium Dioxide Nanoparticles Used as UV absorber on Natural Herbal Extract Sunscreen Products

Supamas Wirunchit[1], *Natchayaporn* Sakulpeeb[1], *Wantana* Koetniyom[2,3*]

[1]College of Materials Innovation and Technology (CMIT), King Mongkut's Institute of Technology Ladkrabang, Ladkrabang, Bangkok 10520, Thailand
[2]Department of Industrial Physics and Medical Instrumentation, Faculty of Applied Science, King Mongkut's University of Technology North Bangkok, Bangkok 10800, Thailand
[3]Lasers and Optics Research Center (LANDOS), King Mongkut's University of Technology North Bangkok, Bangkok 10800, Thailand

Abstract. This research focuses on the effects of UV absorber on natural herbal extracts including Wan Nga-chang (Ivory) and Wan Tan-diao containing Glycoside and Tannin with solvent prepared via the maceration method. The concentrations of UV absorber in this study, Titanium Dioxide (TiO_2), were varied from 5, 10, 15 % by weight and compared to the unadded reference material. The characteristics and properties of sunscreen products were analyzed by UV-Vis spectroscopy, X-ray diffraction technique (XRD) and Fourier Transform-Infrared spectrometer (FT-IR). The results of the experiment showed that the optimal concentration of UV absorber is herbal cream and TiO_2 at concentrations of 10%wt. From XRD and FT-IR techniques confirmed the elements and functional group of herbal extract and TiO_2, respectively.

Keyword. Titanium Dioxide Nanoparticles, UV-absorber, Natural Herbal Extract, Sunscreen Products

1 Introduction

At present, the greenhouse effect results in global warming, the penetration of sunlight, especially UV rays can enter Earth's atmosphere increase, therefore overexposure to UV rays will result in premature skin aging, wrinkles, uneven skin tone, dullness, freckles, skin cancer and destroy the genetic system [1-3]. Due to such problems from the melanin process under the skin through the enzyme Tyrosinase. To reduce those problems can be done by inhibiting the Tyrosinase enzyme [4-5] by natural extracts in the group of Glycoside and Tannin. It can inhibit the Tyrosinase enzyme. In addition, glycosides and tannins are also antioxidants and can be anti-inflammatory and antibacterial activity [6-8]. There are many methods of extracting natural substances from herbs, such as maceration, ultrasonic extraction, microwave extraction, and continuous extraction [6,8]. This research is focused on natural herbal extracts obtained from Wan Nga-chang (Ivory) and Wan Tan-diao in the group of glycosides and tannins, respectively. Using the natural extract is the method aiming to the sustainable production and fit to the bioeconomy and green economy. Glycoside is a molecule in which a sugar is bound to another functional group via a glycosidic bond. Glycosides play numerous important roles in living organisms [9]. Tannin is a class of astringent, polyphenolic biomolecules that bind to and precipitate proteins and various other organic compounds including amino acids and alkaloids [9]. Wan Nga-chang or Ivory, the scientific name: *Sansevieria stuckyi* is a plant in the family Agavaceae. It is an herbaceous plant. There are underground stems, no leaves, and green stems emerging from the soil. Shaped like ivory. The inflorescences or florets are white and fragrant [9] as shown in Figure 1. Wan Tan-diao, the scientific name: *Hypoxis aurea* Lour, is a plant in the family Hypoxidaceae. It is a tuberous herb with underground stems and round roots. The leaves are slender, elongated like blades of grass in a straight line, about 10-30 cm long, about 2-3 mm in width, and the flowers are single or paired with golden yellow flowers. For flower stalks about 2.5-10 cm long [9] as shown in Figure 2. This research is focusing the maceration method. Maceration is the process of fermentation of herbs with solvent until the tissue of the herb is soft and the solvent penetrates and dissolves the herb to produce an extract. It is also the most popular method by which the solvent suitable for the herb can be selected. The advantage of this method is that the extract is not heated and can reduce the chance of decomposition of the extract. Whereas, it will be wasted because many solvents are required [6]. Using herbal extracts to make cosmetic products, such as sunscreen is an interesting alternative for people allergic to chemical makeup.

This research focused on the assessment of the influence of Titanium dioxide nanoparticles (TiO_2) as a UV absorber material on sunscreen products from natural herbal extracts including Wan Nga-chang (Ivory) and Wan Tan-diao. These compounds are classified in the group of glycoside and tannin with solvent through the maceration method. After that, the influence of nano-titanium dioxide concentration on the light absorption of sunscreen products from natural herbal extracts was

* Corresponding author: wantana.k@sci.kmutnb.ac.th

studied. The characteristics and properties of sunscreen products with optical techniques by UV-Vis spectroscopy, elements in herbal extracts by X-ray diffraction technique (XRD) and functional group by FT-IR were analyzed.

Figure 1. The appearance of Wan Nga-chang (Ivory)

Figure 2. The appearance of Wan Tan-diao

2 Materials and methods

2.1 Extraction method of Ivory and Wan Tan Diao

Herb extract was comprised of Ivory and Wan Tan Diao by the solvent extraction or maceration technique. Maceration used the solvent that was cold pressed coconut oil in a solvent to herbal ratio of 2:1 as shown in Figure 3. The extracts from Wan Nga-chang (Ivory) and Wan Tan-diao were mixed with a ratio of 1:1 and solvent and the samples were stirred for 30 minutes.

2.2 Herbal cream preparation

The herbal cream preparations have three main components: water base, oil base and co-solvent. The mixed extracts from section 2.1 used oil base, whereas TiO_2 at the concentrations of 0, 5, 10 and 15 %w/w used RO water base. For co-solvent, viscolam AT 100P was used. The ratio of the three main components between water base, oil base and co-solvent are 75:23:2. These components were mixed with a homogenizer for 30 minutes, then rested for 10 minutes. This step was repeated for 3 times, and leaved it for 1 night for setting the cream as shown in Figure 4.

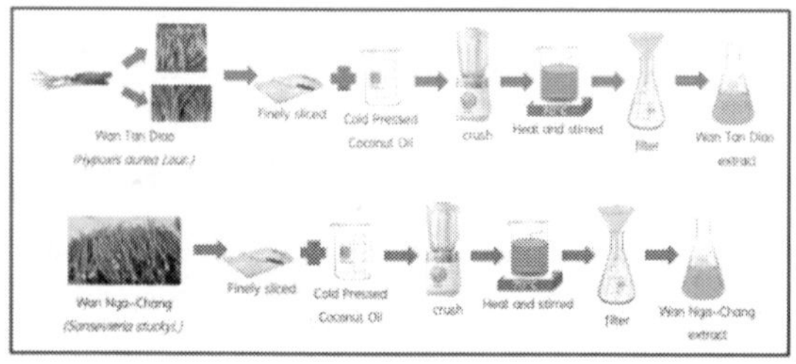

Figure 3. Herbal extraction method

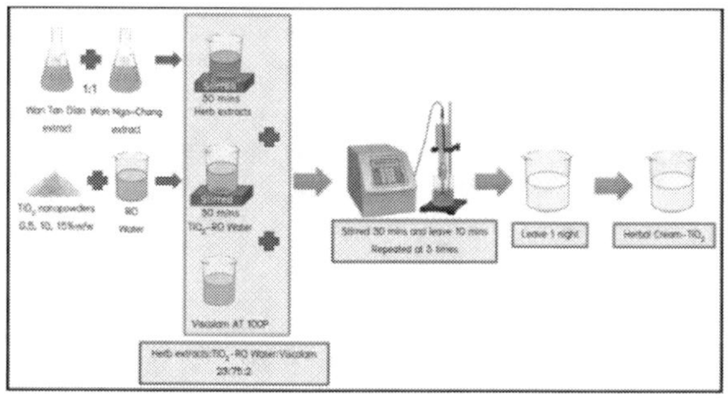

Figure 4. Herbal cream preparation

2.3 Films preparation

The glass substrates were cut in size of 2×2 cm^2, then cleaned with detergent, ultrasonicated in DI water, acetone, and isopropanol for 15 min in each step, and dried with nitrogen gas. The extracts that were mixed with TiO_2 with varying concentrations of 5, 10, 15%w/w were coated by doctor blade technique on the glass substrate compared with herbal cream, then substrate was dried for 30 min at 80 °C. The method was summarized as seen in Figure 5.

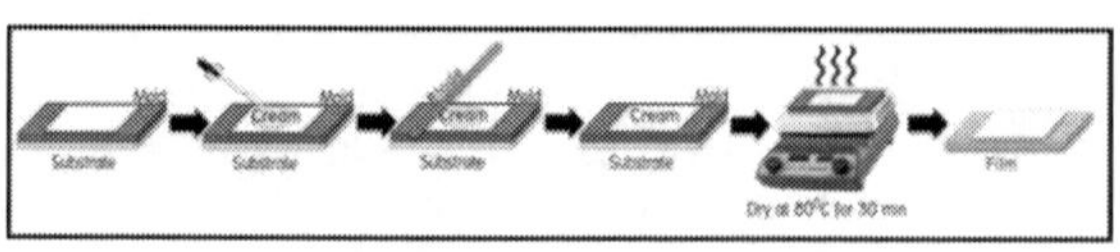

Figure 5. Films preparation

2.4 Characterization and measurements

Fourier transform infrared (FTIR) spectroscopy was used to identify the chemical structure of the films and possible interactions between their components. The FT-IR spectra of the films were measured from the wave number 500 – 4,000 cm^{-1} by Perkin Elmer UATR Two spectrophotometer. UV absorbance was characterized by UV-Vis spectroscopy PG Instrument with model T92+

Spectrophotometer in absorbance mode from the wavelength 200 - 800 nm. The crystalline properties of films were analyzed from X-ray diffraction (XRD) patterns obtained from scanning 2-theta from 10-80° at a fixed incident angle of 0.4° (Rigaku Smartlab using Cuk-alpha as the X-ray source) using X-ray diffractometer with Cu Kα radiation (λ=0.15406 nm).

3 Results and discussion

3.1 UV absorption

Figure 6 presents UV absorbance from 250 to 800 nm of herbal cream and TiO_2 at various concentrations compared with herbal cream. Especially, the UV absorption range from 300-400 nm, it was shown that the intensity of herbal cream and TiO_2 at concentrations of 5 and 10%wt increased, however, at TiO_2 15%wt, the intensity was decreased. The trend of this graph shows the optimal concentration of UV absorbers is herbal cream and TiO_2 at concentrations of 10%wt.

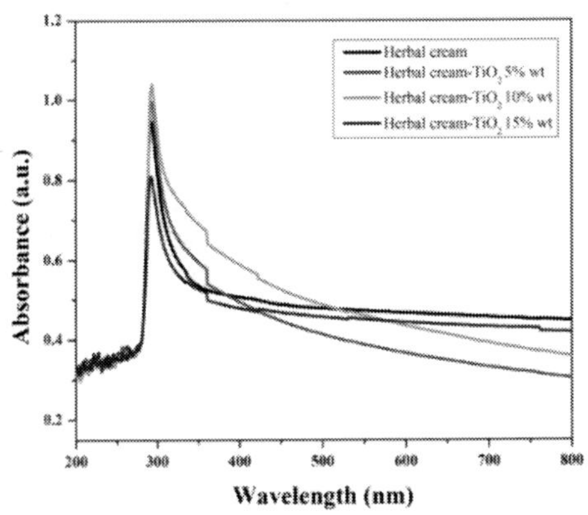

Figure 6. UV absorbance of herbal cream and TiO_2 at various concentrations

3.2 X-ray diffraction characterization

Figure 7 shows X-ray diffraction pattern of a commercial grade Titanium (IV) oxide nanoparticle (TiO_2 P25). It presents a mixed phase between anatase and rutile structure by comparing with the data for Titanium (IV) oxide (TiO_2) in anatase phase (A) from the JCPDS files no. 21-1272 and it has the highest intensity peaks at the (101) (004) (200) (105) (211) plane that corresponds to the position of the 2θ values are 25.31°, 37.80°, 48.04°, 53.89°, 55.06°, respectively. The Titanium (IV) oxide in rutile phase (R) data from the JCPDS file no. 21-1276 has the highest intensity peak at the (110) plane that corresponds to the position of the 2θ value is 27.44° [3]. Each phase is a tetragonal structure.

Figure 8 shows a typical XRD pattern of mixed herbal cream and TiO_2 at various concentrations. The herbal cream reference presents that herbal extracts did not show any pronounced structure due to the uniqueness of the organic compounds. For the mixtures where titanium dioxide was added at different concentrations, it was seen that the peak was formed due to the mixing of the rutile phase and the anatase phase of the titanium dioxide and the graph intensity was the highest at a concentration of titanium dioxide of 15% by weight.

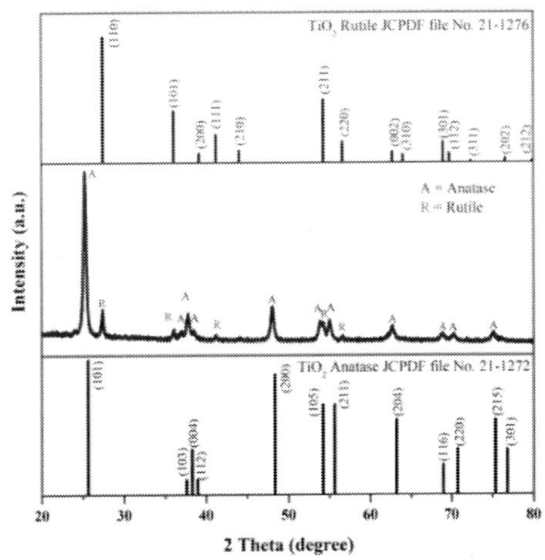

Figure 7. X-ray diffraction patterns of Titanium (IV) oxide nanoparticles commercial grade (TiO_2 P25)

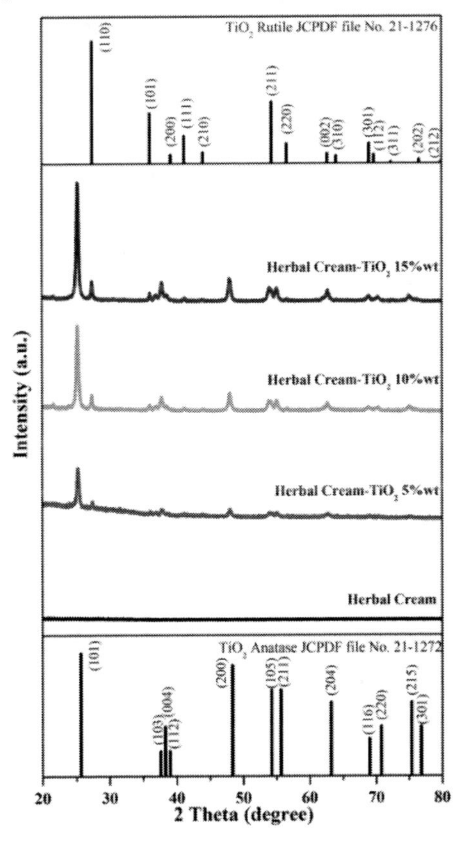

Figure 8. X-ray diffraction pattern of herbal cream and TiO_2 at various concentrations

3.3 Fourier Transform-Infrared spectrometer (FT-IR)

Figure 9 shows FT-IR curves of TiO_2 powder, herbal cream and herbal cream with TiO_2 absorbent material at the ratios of 5%, 10%, and 15% by weight, respectively. For TiO_2 powder, the oscillations between the titanium atom and the oxygen atom inside the molecule are observed at the wave number 523 cm^{-1} representing by the titanium and oxygen oscillations as a stretching oscillation (Ti-O Stretching) [3]. FT-IR spectra of herbal cream show oscillatory characteristics of tannin and glycoside extracts. Tannin shows a wide peak in the region 3550–3100 cm^{-1} that is characteristic of the OH stretching vibration of benzene nucleus and methylol group of tannin. A small peak around 2900 cm^{-1} is aromatic CH stretching vibration of both methyl and methylene groups. The carboxyl-carbonyl group appears at 1732 cm^{-1} in the spectrum of tannin. The bands at 1324 and 1037 cm^{-1} are symmetrical and asymmetrical C–O valence vibration. The peaks around 910-740 cm^{-1} in all spectra are deformation vibrations of the C-H bond in the benzene rings [10]. Glycoside observed the hydroxyl stretching between the wavenumbers 3319 - 3379 cm^{-1}, alkane stretching was observed between 2918 - 2947 cm^{-1}, carbonyl groups between 1719 - 1728 cm^{-1}, alkene stretching between 1641 - 1662 cm^{-1}, ether groups. between 1054 - 1070 cm^{-1} and alkene bending between 870 - 895 cm^{-1}, respectively [11]. FTIR spectrum of blended sample between herbal cream and TiO_2 at different concentrations showed that the mixed peaks of tannin and glycoside from herbal cream and TiO_2. Therefore, it can be concluded that mixed TiO_2 into herbal creams is possible without interaction with each other until the structure of TiO_2 changes.

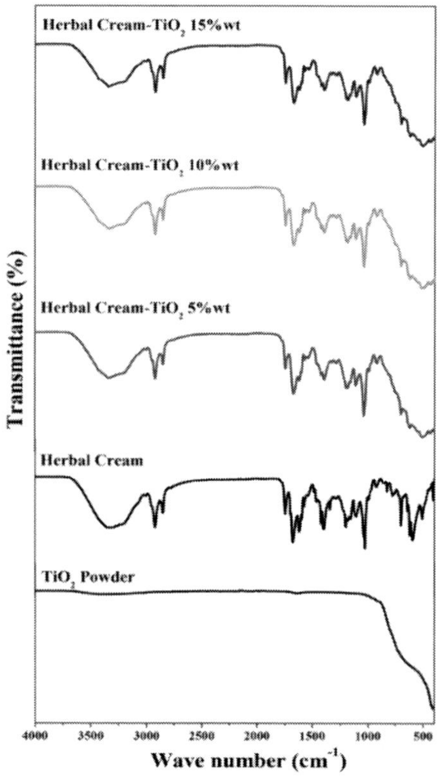

Figure 9. FTIR spectrum of herbal cream and TiO2 at various concentrations

4 Conclusion

Since most of the sunscreens in the market are chemically synthesized, it is not suitable for some people with sensitive skin. This research is therefore interested in developing a sunscreen from an easy-to-find herbal precursor. It is cheap, mass-produced in Thailand, and researched to add UV absorbers and find the right ratio between herbal extracts and UV absorbers. From this research, it was found that the optimal ratio was the UV absorbent ratio at 10% by weight, which can absorb UV rays in the range of 300-400 nm with the most effectively. This project will study more in the extracts of other medicinal plants for further comparison.

Acknowledgements

Thanks to Department of Industrial Physics and Medical Instrumentation (IMI) of King Mongkut's University of Technology North Bangkok (KMUTNB) and College of Materials Innovation and Technology (CMIT), King Mongkut's Institute of Technology Ladkrabang (KMITL), Thailand. This research was funded by Faculty of Applied Sciences at budget in 2022, King Mongkut's University of Technology North Bangkok, contract no. 653098.

References

[1] J. D'Orazio, S. Jarrett, A. Amaro-Ortiz, T. Scott, UV Radiation and the Skin, nternational journal of molecular sciences, 14(6) (2013):12222-12248.

[2] A. Amaro-Ortiz, B. Yan, J.A. D'Orazio, Ultraviolet radiation, aging and the skin: prevention of damage by topical cAMP manipulation, Molecules, 19(5) (2014):6202-6219.

[3] S. Wirunchit, C. Apivitcholchat, T. Chodjarusawad, W. Koetniyom, The Study of UV Protection Materials, AIP Conference Proceedings, 2010 (2018):020024-1–020024-10.

[4] Y.S.C. Bae-Harboe, H.Y. Park, Tyrosinase: a central regulatory protein for cutaneous pigmentation, Journal of Investigative Dermatology, 132(12) (2012): 2678-2680.

[5] R. Sarkar, S. Chugh, V.K. Garg, Newer and upcoming therapies for melisma, Indian journal of dermatology, venereology and leprology, 78(4) (2012): 417-428.

[6] A. Srisopa, *Chemical and biology properties of marigold extracts obtained by organic solvent-free extraction* (Pibulsongkram Rajabhat University, 2016).

[7] K. Boonpisuttinant, In vitro Anti-Melanogenesis on Murine Melanoma Cell Line (B6F10) and Tyrosinase Inhibittion Activeity of Hypoxis aurea Lour. Leave Extracts, Proceeding of ASTC (2013): 5-8.

[8] E.Y.A. Salih, R. Julkunen-Tiitto, O. Luukkanen, M.K.M. Fahmi, P. Fyhrquist, Hydrolyzable tannins (ellagitannins), flavonoids, pentacyclic triterpenes

https://doi.org/10.1051/e3sconf/202235502007

and their glycosides in antimycobacterial extracts of the ethnopharmacologically selected Sudanese medicinal plant Combretum hartmannianum Schweinf, Biomedicine & Pharmacotherapy, **144** (2021): 112264.
[9] https://th.wikipedia.org/
[10] O. Gonultas, M.B. Ucar, *Chemical composition of some commercial tannins produced in Turkey*, Proceedings of the 55th International Convention of Society of Wood Science and Technology, (2012):1-9.
[11] V.S.P.Chaturvedula, C. Mubarak, I. Prakash, IR Spectral Analysis of Diterpene Glycosides Isolated from Stevia rebaudiana, Food and Nutrition Sciences, 3(10) (2012): 1467-1471.

* Corresponding author: wantana.k@sci.kmutnb.ac.th

Formation of thermal oxide scale and its adhesion to hot-rolled low carbon steels with different final strip thicknesses

Sun Naipinij, *Sasapan* Sukieum, *Ravinupha* Namprai, and *Thanasak* Nilsonthi*

High Temperature Corrosion Research Centre and Department of Materials and Production Technology Engineering, Faculty of Engineering, King Mongkut's University of Technology North Bangkok, 1518, Pracharat 1 Road, Wongsawang, Bangsue, Bangkok, 10800, Thailand

Abstract. Currently, the steelmaking industry produces iron oxide waste scale resulting in pollution to the environment. It was necessary to have a good understanding of the composition, characteristics and properties of the oxide scale. This study focused on the formation of scale and its adhesion to the hot-rolled steel strip with different thicknesses. The oxide scale formed on an as-received hot-rolled steel strip was investigated by X-ray diffraction (XRD), scanning electron microscopy equipped with energy dispersive X-ray (SEM-EDS). Magnetite, hematite and iron were found from the XRD results of all samples, which had the thickness strip of 8, 10 and 12 mm. The scale was thinner for the thinner strip. The adhesion test was conducted by a tensile testing machine adapted with an observation set. The strain initiating the first spallation and mechanical adhesion energy was lowest for the sample with the highest thickness (12 mm). These results indicate that the waste scale produced by hot rolled steel industry can be controlled by the final strip thickness. There was a need to control the scale of waste in a reasonable way to protect the environment.

Keyword. Waste scale, Iron oxide, Rolling mill steel, Hot-rolled steel, Strip thickness, Adhesion

1 Introduction

The thermal oxide scale formed on the steel surface because of the process temperature in the hot rolling line, which is around 600-1250°C. It is de-scaled both, first before the slab enters the roughing mill and again before the steel bar enters the finishing mill, using a high-pressure water jet. Thermal oxide scale continues to rapid form during steel strip coiling and storage after the second de-scaling, until the temperature decreases below 400°C. The existence of this scale has a significant impact on the product's quality.

A slab is a raw material for making carbon steel strips in a hot rolling line, and it can be made in one of two ways: blast furnace (BF) or electric arc furnace (EAF). A blast-furnace route can be used to make a slab from iron ore, coal, and limestone. For the electric arc furnace route, steel scraps are the main raw material for making a slab, tramp elements in steel were present in the slab and might affect the thermal oxide scale structure. This may have an impact on the pickling behaviour in the following pickling line.

During the hot rolling process, steel is oxidised and a scale of iron oxide formed on the steel surface at process temperature comprises three layers a thin outermost hematite (Fe_2O_3), an intermediate magnetite (Fe_3O_4), and a thick inner wustite (FeO) [1-13]. For the Fe-O system, the scale layer can vary via concentration gradients of oxygen. An iron oxide formed always grows external scale rather than forming internal oxidation. Due to eutectoid decomposition during cooling, magnetite can be occasionally found in the wustite layer. The hematite layer may also be absent in some cases.

In the hot rolling process, a slab is rolled at high temperatures in order to obtain a strip with desired thickness [14]. At high temperatures, the steel reacts with oxygen in the ambient atmosphere and gives a thermal oxide scale [1-13]. The oxide formed is a waste of the process and it should be minimised. Many factors can affect the amount of the scale formation, such as alloying elements in the steel like C [15], Si [16], Cu [17], Sn [18] or Ti and Nb [19] or the parameters set in the process [20-22]. The objective of this work is first to investigate if the final strip thickness relates to the amount of oxide formation.

In addition, the amount of oxide formation can affect the adhesion of scale to steel. This property is important. If the hot-rolled steel is delivered to the customer for the direct application in this form, good scale adhesion is required. If the hot-rolled has to be successively cold-rolled, the bad scale adhesion is preferred. This is because the scale can be easily removed from the steel surface before further cold rolling. The adhesion of oxide scale on steel substrate can be assessed in many ways, such as the indentation test [23], the inverted blister test [24, 25], the bending test [26-29] or the tensile test [30-34]. Our group developed the tensile test [35-39] to evaluate the scale adhesion of low carbon steels [15, 19,

* Corresponding author: thanasak.n@eng.kmutnb.ac.th

22, 35-39] and stainless steels [40-42]. It will be used in this study. The structure and adhesion of the oxide scale after the hot rolling process are focused on in this study. The effect of final strip thickness on scale formation and adhesion is emphasised. During the industrial production of steel, tons of iron-rich scale are produced as waste materials [43]. The waste contains large amounts of iron oxides, heavy metals and other different contaminants. It affects the environment when disposed of in landfills [44]. The environment is polluted. The scale as a process waste should be kept to a minimum.

For preventing scale formation in steel industries, since the scale formation during the hot rolling process at high temperatures. Several hot rolling conditions have been proposed to prevent the scale problem. Control of slab heating temperature and alloying elements content are some of those proposed processes. However, these conditions are not sufficient in preventing scale defects in the commercial hot rolling process. The thick scale formed during heating for a long time in the furnace and remains even after water jet descaling. A short time holding in the furnace might control this scale. The existence of fayalite (Fe_2SiO_4) for high Si steel is an additional important factor to prevent scale formation. This is due to scale thickness decreasing with the increasing silicon content. However, the strong bonding of fayalite on the scale-steel interface is considered to descaling process. The scale generated by the steel plant is called waste, but now this term has been replaced with a by-product. Mill scale is one of the by-products produced during steel processing. Mill scale is used for magnetic storage, polishing, chemical manufacturing, pigment manufacturing, and biomedical application.

2 Materials and methods

2.1 Materials

The study used hot-rolled steel which is available from Sahaviriya Steel Industries Public Company Limited as a strip with a thickness of 8, 10 and 12 mm. The position of the hot-rolled coil at the head, middle and tail are used for examination as shown in Figure 1. The sample is cut from a strip obtained from a slab produced by the blast-furnace route. In the hot rolling process, the finishing and coiling temperatures of the sample, which had a strip thickness of 8 mm are 790°C and 580°C respectively, for the strip 10 mm is 780°C and 590°C, for strip 12 mm is 760°C and 560°C.

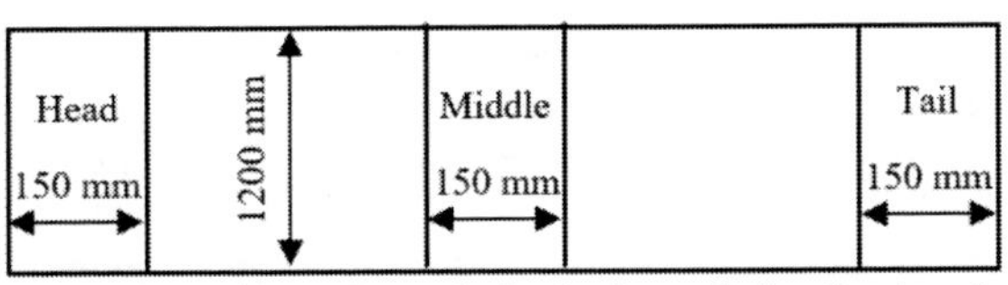

Figure 1. Position of sample in the hot-rolled coil taken for examination.

The hot-rolled steel is used for several products e.g. pipe and tube, automotive structural, machine structures, and gas cylinders. The final strip thickness is considered by product specification. Considered to phase transformation when exiting the hot rolling mill, the choice of finishing and coiling temperatures have a significant influence on achieving product property. Table 1 shows the chemical composition of the steel. Table 2 shows the information from mechanical tensile testing.

Table 1. Chemical composition of the as-received hot-rolled steel strip (wt.%).

Sample	C	Si	Cu	Mn	P	S
8 mm	0.138	0.006	0.010	0.441	0.017	0.009
10 mm	0.156	0.007	0.009	0.454	0.016	0.002
12 mm	0.158	0.009	0.007	0.440	0.020	0.006

Table 2. The tensile mechanical properties of a material.

Strip thickness	Mechanical properties		
	Yield strength (MPa)	Tensile strength (MPa)	Elongation (mm)
8 mm	444	502	18
10 mm	348	441	29
12 mm	385	463	27

2.2 Characterisation

The oxide scale morphology is observed by the scanning electron microscope (SEM, QUANTA 450). The energy-dispersive X-ray spectroscopy (EDS, OXFORD INSTRUMENTS, Model X-Max) was equipped with SEM for elemental analysis. The oxide phase is determined by the X-ray diffraction technique (XRD, SmartLab) using the Cu Kα line (k = 0.15406 nm) with a step size of 0.02 degree/step and a step time of 0.5 second/step.

2.3 Mechanical adhesion

The tensile testing machine (Instron, Model 5566) with a load of 10 kN is used. The strain rate of 0.04 s^{-1} at room temperature is operated. A high-magnification lens with a CCD camera has been used to observe the evolution of scale failure. The video processing is performed at a resolution of 640 × 480 pixels. Image framework programming is used to acquire the image. The CCD camera has a frame rate of 7.5 frames per second. The specimen is prepared according to the ASTM E8M standard. Figure 2 shows the tensile testing machine with the scale observation set and sample shape.

3 Results and discussion

Figure 3 presents cross sections of the 8-mm strip at the head, middle and tail positions. The XRD pattern for each scale was also shown in Figure 4. It was seen that the thickness of scale at the head position was 12.82 ± 2.57 µm. They were 16.42 ± 2.29 µm and 10.77 ± 1.15 µm at the middle and tail positions respectively. Fe_2O_3, Fe_3O_4 and Fe were found in the XRD patterns of the scale formed in these three positions. The literature [4] was reported the oxidation of iron and steel in oxygen or air at high temperatures. At temperatures above 700°C, iron oxidation follows the parabolic law, resulting in the formation of three-layered hematite, magnetite, and wustite. Carbon steel oxidation was generally slower than iron oxidation. The scale structure of carbon steel was similar to those formed on iron after very short-term oxidation.

100 mm

Figure 2. Tensile machine set up with observation set (a) and sample shape (b).

The oxidation behaviour of carbon steel at high temperatures has been extensively studied in the previous. Oxidation behaviour of steel and scale structure was more difficult to understand than iron oxidation due to the presence of various alloying and impurity elements. However, literature [45] has proposed that the steam partial pressure was unaffected by the rate of oxidation. The hematite layer was very thin, which affects the oxygen activity at the magnetite-hematite interface. From this research, hematite and magnetite layers were observed without the wustite layer. According to the Fe-O phase diagram as shown in Figure 5, wustite will fully decompose undergoes eutectoid transformation and turns into magnetite and iron at 570°C or below during the cooling after hot rolling. Literature [20] suggestion was a direct magnetite-formation mechanism. The presence of an iron substrate was not essential for the formation of the magnetite layer. The mechanism of forming the magnetite layer was via the following primary reaction without the involvement of the iron base.

$$Fe_{1-x}O \rightarrow [(x-y)/(1-4y)]Fe_3O_4 + [(1-4x)/(1-4y)]Fe_{1-y}O \quad (1)$$

where $Fe_{1-y}O$ is an iron-rich wustite, which subsequently decomposed into a eutectoid product via the following reaction.

$$4Fe_{1-y}O \rightarrow Fe_3O_4 + (1-4y)Fe \quad (2)$$

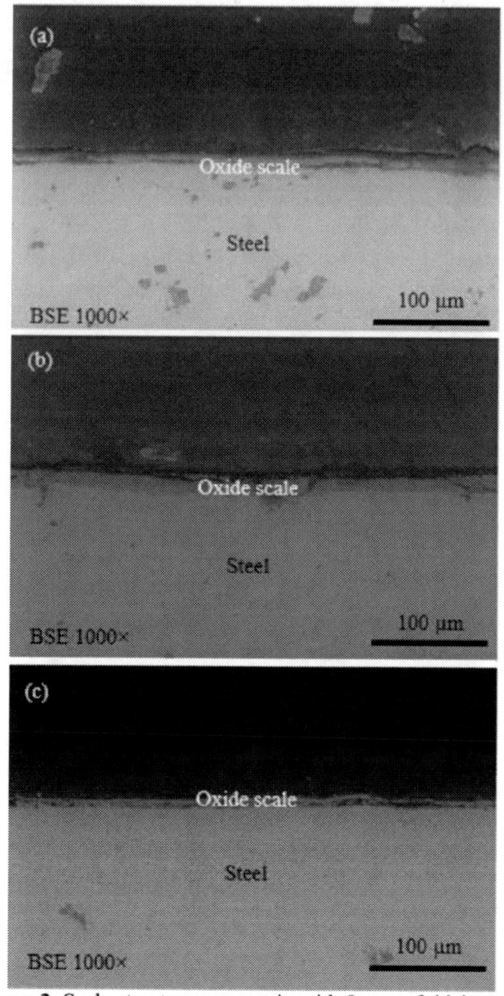

Figure 3. Scale structures on a strip with 8 mm of thickness at head (a), middle (b) and tail (c) positions.

During wustite decomposition, primary magnetite precipitation inside the wustite phase was readily formed and then followed by magnetite precipitation at the wustite-iron interface. Finally, eutectoid forming magnetite and iron. In our view, high temperature promoted the development of the wustite layer which was possible at temperatures higher than 570°C. At temperatures below 570°C, the layered magnetite and iron eutectoid were continuously formed. Conversely, wustite could be found at temperatures below 570°C when the cooling rate was high enough. In addition, wustite might undergo a eutectoid reaction to form a mixture of magnetite and iron when oxidation was preceded under a continuous cooling process. The oxide scale microstructure formed via phase transformation of wustite affects the surface quality of steel products. However, this study was focused on the effect of final strip thickness in

 https://doi.org/10.1051/e3sconf/202235502008

the simplest way for understanding the adhesion of iron oxide waste scale from the rolling mill steel industry. Therefore, it was necessary to have a good understanding of oxide scale composition. Table 3 shows the composition of the oxide scale obtained from hot-rolled steel. The EDS mapping spectra for the sample shows that the main elemental components were Fe, O and C. This indicates that the iron oxide as the main phase was presented, and the carbon-rich originate from the hot-rolled steel. It can be noted that the presence of Si was found on a scale of a 10 mm strip thickness at the middle position. It was due to Si-rich precipitates in the oxide scale of their steel.

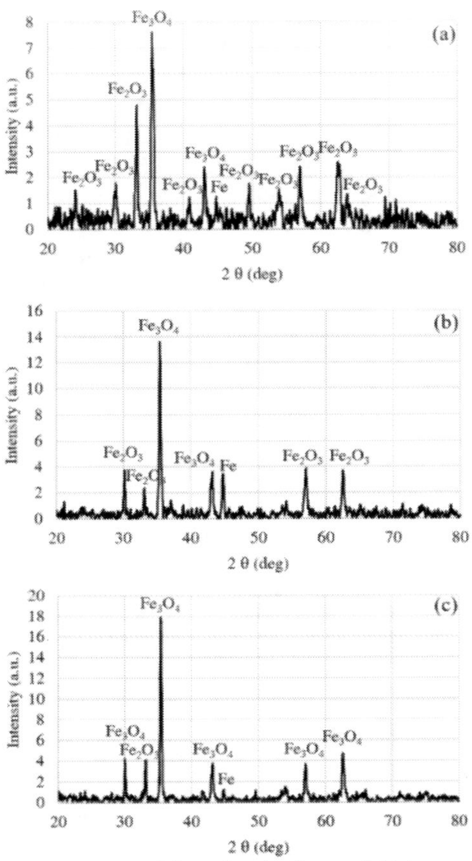

Fig. 4. XRD patterns of the strip with 8 mm of thickness at the head (a), middle (b) and tail (c) positions.

Table 3. Elemental quantification in percentage weight (wt.%) obtained by EDS analysis for the scale as-received hot-rolled steel strip.

Thick	Position	Fe	O	C	Si
8 mm.	Head	40.26	23.74	36.00	
	Middle	37.59	30.22	32.20	
	Tail	34.75	31.43	33.81	
10 mm.	Head	21.64	33.64	42.45	
	Middle	38.76	38.08	22.02	1.13
	Tail	37.49	35.37	25.65	
12 mm.	Head	43.26	23.31	33.44	
	Middle	40.33	34.00	25.67	
	Tail	39.91	37.26	22.82	

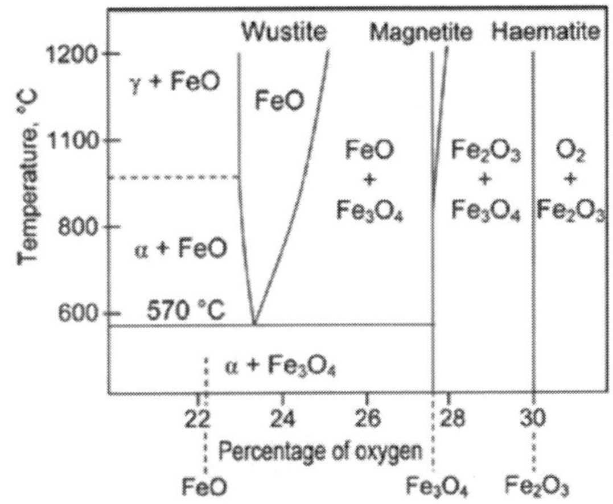

Figure 5. Fe-O phase diagram [46].

Figure 6 illustrates cross sections of the 10-mm strip at the head, middle and tail positions. The XRD pattern for each scale was also shown in Figure 7. It was seen that the thickness of scale at the head position was 10.41 ± 3.89 μm. They were 37.44 ± 7.61 μm and 13.34 ± 3.34 μm at the middle and tail positions respectively. Fe_2O_3, Fe_3O_4 and Fe were found in the XRD patterns of the scale formed in these three positions.

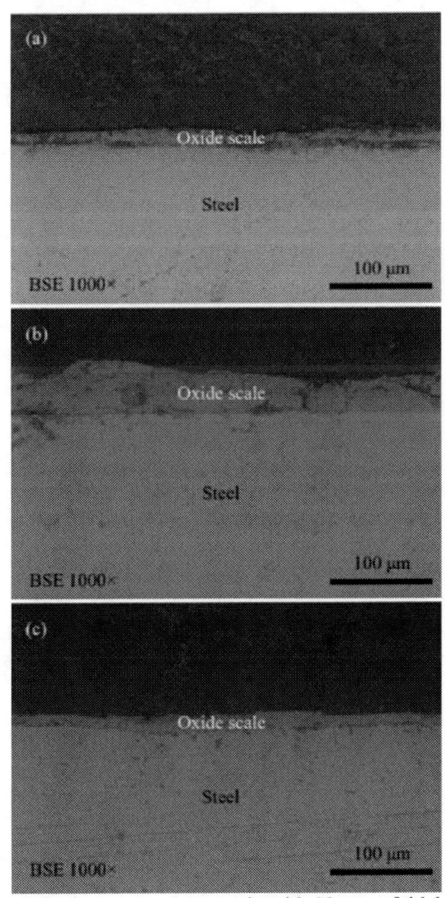

Figure 6. Scale structures on a strip with 10 mm of thickness at the head (a), middle (b) and tail (c) positions.

 https://doi.org/10.1051/e3sconf/202235502008

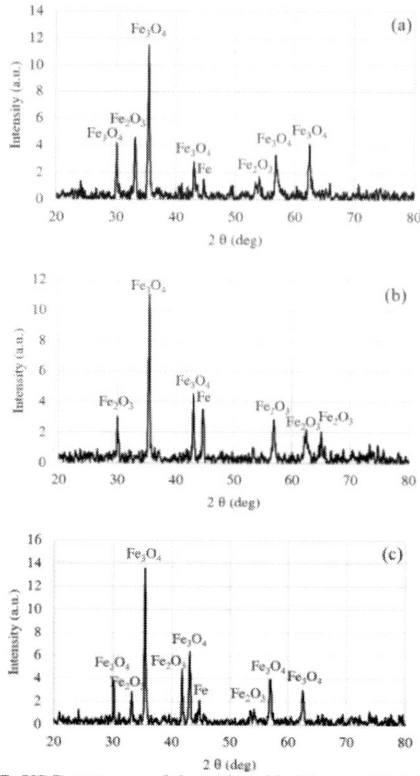

Figure 7. XRD patterns of the strip with 10 mm of thickness at head (a), middle (b) and tail (c) positions.

Figure 8 shows cross sections of the 12-mm strip at the head, middle and tail positions. The XRD pattern for each scale was also shown in Figure 9. It was seen that the thickness of scale at the head position was 29.44 ± 2.63 µm. They were 45.13 ± 8.99 µm and 74.36 ± 6.54 µm at the middle and tail positions respectively. Fe_2O_3, Fe_3O_4 and Fe were found in the XRD patterns of the scale formed in these three positions.

From XRD results, the oxide scale formed at rolling temperatures comprises hematite and magnetite with iron as shown in all samples. Qualitative phase analysis provides oxide scale information gleaned from an X-ray diffraction pattern. Each phase of a crystalline oxide will produce a unique diffraction pattern. The diffracted peak positions and intensities from a particular oxide phase serve as a fingerprint that can be compared to a database of reference patterns to identify a phase. A pattern plot of diffracted X-ray intensity with Bragg angle. The peak positions of the hematite phase shown at 24.17°, 33.19°, 35.66°, 40.90°, 49.51°, 54.13°, 62.49°, and 64.05° were indexed to (012), (104), (110), (113), (024), (116), (214), and (300) respectively. While the peak positions of the magnetite phase shown at 30.07°, 35.42°, 43.05°, 56.93°, and 62.51° were indexed to (220), (311), (400), (511), and (440) respectively. The peak positions of the iron phase shown at 44.68°, 65.03°, and 82.34° were indexed to (110), (200), and (211) respectively. It can be noted that the peak positions and relative intensities of the sharp peaks in XRD patterns indicate the hematite, magnetite and iron were well crystallised.

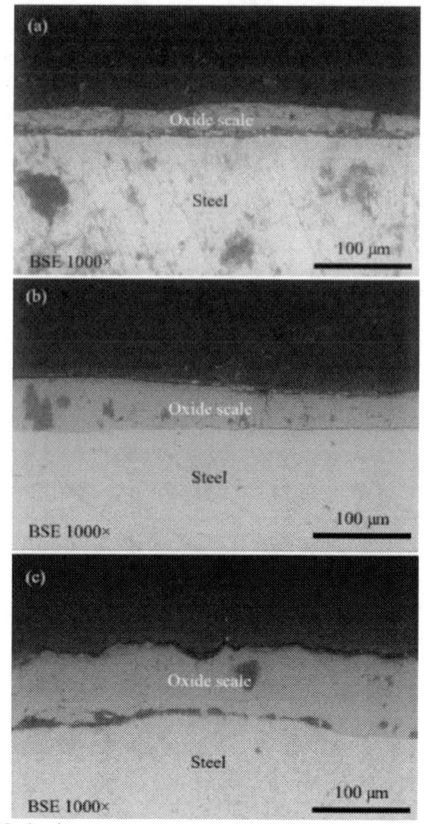

Figure 8. Scale structures on a strip with 12 mm of thickness at the head (a), middle (b) and tail (c) positions.

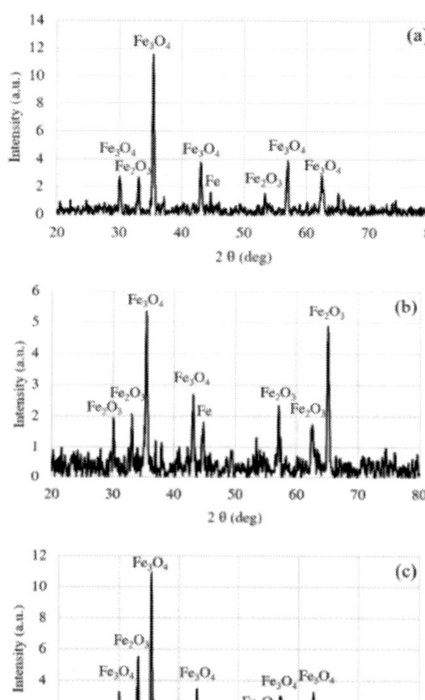

Figure 9. XRD patterns of the strip with 12 mm of thickness at head (a) position middle (b) and tail (c) positions.

https://doi.org/10.1051/e3sconf/202235502008

A comparison of the oxide scale thickness as a function of final strip thickness for specimens at the position of the head, middle and tail was shown in Figure 10. The thickness was measured using cross-sectional metallographic images. It was found that the scale thickness of the studied 12-mm thick strip was significantly increased at all positions. It was well known that the thickness of oxide scale was linear with the weight gain of the scale due to the absorption of oxygen. With the increase in exposure time, the internal and external thickness of the oxide scale gradually thickens. If the strip thickness is high, provide more time for cooling after the hot rolling process. Scale continued to grow as a result. It was also seen that the thickness at the tail position on a strip of 12 mm was higher than that head and middle positions. This might be due to the accumulation of heat at the strip tail during the hot rolling process, corresponding to using a longer cooling time. The results indicate a drop in scale thickness when the final strip thickness was decreased. This can be attributed to the higher cooling rate of the thinner strip. While the scale thickness at the middle position on strips of 8mm and 10mm seems to be higher than the head and tail positions. Literature [47], this study was to improve the longitudinal performance uniformity of hot-rolled coils. The results show that the average cooling rate of the head and tail parts were higher than that of the middle part during coil cooling. This might cause lower scale thickness on strips of 8-mm and 10-mm at head and tail positions. However, accumulation of heat at the strip tail was going to have an effect on the strip of 12 mm. It can be observed that the global scale thickness of the strip of 12 mm seems to be higher.

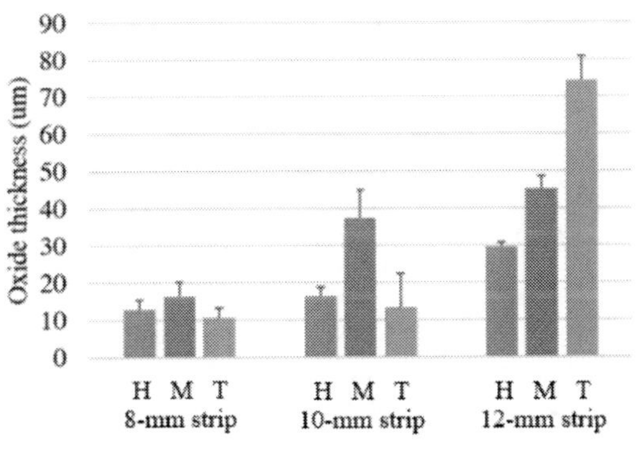

Figure 10. Oxide scale thickness of strip with 8, 10 and 12 mm of thickness at the head (H), middle (M) and tail (T) positions.

Comparison literature [48], this research shows the thicknesses of oxides formed on AISI 1045 were 2.92 µm and 6.22 µm for oxidation of 168 and 720 hours respectively. Oxidation of steel was performed in the heated air inside the furnace at 673 K (400°C). After the oxidation process was completed, the sample was kept in a desiccator for cooling down to room temperature. Note that before oxidation the steel was cut into 20 mm in diameter with a thickness of 2 mm. The results show that the scale thickness was found in the range of 3-6 µm, this lower than that scale thickness was observed on as-received hot-rolled steel with a final strip thickness of 8-12 mm. This was due to the difference between the thickness of the sample, oxidation temperature and oxidation time.

The oxide adhesion to a steel substrate will be discussed formally in terms of oxide-steel separation. The adhesion of scale on steel was previously considered [49]. Many failures occurred in the oxide near the interface, involving oxide growth stresses, oxide and steel plasticity, and thermal cycle effects caused by differential thermal contraction. Cracks normal to the interface, cracks parallel to the interface or spalling were all examples of local failures. Attention was focused here on the scale spallation and adhesion. Figure 11 shows strain initiating the first spallation of the strip with 8, 10 and 12 mm of thickness at the head, middle and tail positions. It was seen that the strain resulting induced the first spallation of a strip with a thickness of 12 mm was lower at all positions. This indicates that during the hot rolling process oxide scale was easier to flake off. In terms of mechanical adhesion energy, it was determined using information from the literature [26, 50-54]. The values reveal that the strip with a thickness of 12 mm has lower mechanical adhesion energy at all positions as shown in Figure 12.

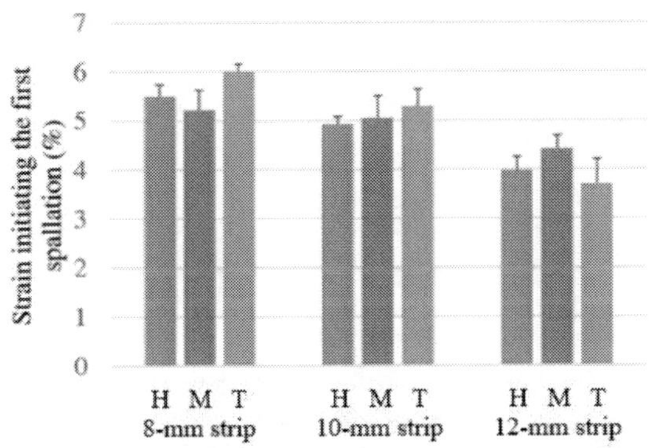

Figure 11. Strain initiating the first spallation of the strip with 8, 10 and 12 mm of thickness at the head (H), middle (M) and tail (T) positions.

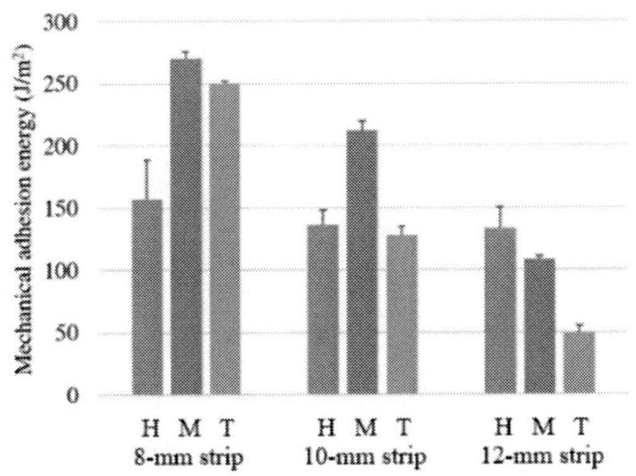

Figure 12. Mechanical adhesion energy of strip with 8, 10 and 12 mm of thickness at the head (H), middle (M) and tail (T) positions.

The scale adhesion was increased with decreasing final strip thickness. This was confirmed by the result in Figure 11. It was seen that the strain initiating the first spallation tends to be lower for the 12-mm sample. It was reported in the literature that the higher scale thickness tends to be lower strain initiating the first spallation [25, 52-53]. This was because if metals diffuse from the internal steel-scale interface to the external scale-gas interface, voids can be formed at the scale-steel interface [55]. The thinner scale had a lower diffusing rate of metal to the external interface, and therefore voids should be less. Then the adhesion should be better. The thinner strip should be pressed more during the rolling. This process might change the properties of the scale and the steel-scale interface strength in a way that increased the scale adhesion. During the hot rolling process, oxide scale spallation was always a possibility. Among the waste from the steelmaking industry, iron oxide produces millions of tons resulting in pollution to the environment. To protect the environment, it was necessary to treat steel waste carefully. Literature [56] was reported the characterisation of iron oxide waste scale obtained by the rolling mill steel industry. The oxide sample revealed α-Fe_2O_3 (hematite) and Fe_3O_4 (magnetite) as principal and secondary phases. It was important to have a good understanding of the relationship between scale structure and strip thickness conditions because the oxide structure had a significant impact on the descaling performance of hot rolled steel strip [57]. It was well known that the pickling process was easier on wustite than that on magnetite [57-59].

4 Conclusion

Formation of thermal oxide scale and its adhesion to hot-rolled low carbon steel with different final strip thicknesses was studied. The purpose of this study was on scale failure in relation to waste scale. The following conclusions could be drawn:

4.1 The strain initiation of the first spallation and adhesion energy of oxide formed on 12 mm strip thickness was found to be lower, causing oxide scale to spall easily.

4.2 More oxide thickness and lower adhesion of scale on hot-rolled steel with 12 mm strip thickness resulted in increased iron oxide waste scale obtained by the rolling mill steel industry.

4.3 The present work studied the only effect of final strip thickness on oxide scale formation and its adhesion to hot-rolled steel. The effect of final strip thickness on adhesion of scale on steel substrate should further be investigated by the pickling test or other methods e.g. the indentation test for comparison.

Acknowledgements

This research was funded by National Science, Research and Innovation Fund (NSRF), and King Mongkut's University of Technology North Bangkok with Contract no. KMUTNB-FF-65-09. Sahaviriya Steel Industries Public Company Limited for the provision of the hot-rolled steel strip.

References

1. P. Kofstad, *High temperature corrosion* (Elsevier Applied Science, London, UK, 1988)
2. R.Y. Chen, W.Y.D. Yuen, Oxide-scale structures formed on commercial hot-rolled steel strip and their formation mechanisms, Oxidation of Metals, 56, 1/2, (2001): 89-118
3. R.Y. Chen, W.Y.D. Yuen, Oxidation of low-carbon, low-silicon mild steel at 450-900°C under conditions relevant to hot-strip processing, Oxidation of Metals, 57, 1/2, (2002): 53-79
4. R.Y. Chen, W.Y.D. Yuen, Review of the high-temperature oxidation of iron and carbon steels in air or oxygen, Oxidation of Metals, 59, 5/6, (2003): 433-468
5. R.Y. Chen, W.Y.D. Yuen, Examination of oxide scales of hot rolled steel products, Iron and Steel Institute of Japan International, 45, 1, (2005): 52-59
6. P. Sarrazin, A. Galerie, J. Fouletier, *Les mécanismes de la corrosion sèche: une approche cinétique* (EDP Science, France, 2000)
7. P. Sarrazin, A. Galerie, J. Fouletier, *Mechanisms of high temperature corrosion: a kinetic approach* (Materials Science Foundations, Trans Tech Publications, Stafa-Zürich, Switzerland, 2008)
8. M.J.L. Gines, G.J. Benitez, T. Perez, E. Merli, M.A. Firpo, W. Egli, Study of the picklability of 1.8 mm hot-rolled steel strip in hydrochloric acid, Latin American Applied Research, (2002): 281-288
9. Z.Y. Jiang, A.K. Tieu, W.H. Sun, J.N. Tang, D.B. Wei, Characterisation of thin oxide scale and its surface roughness in hot metal rolling, Materials Science and Engineering A, 435-436, (2006): 434-438
10. L. Suárez, R. Petrov, L. Kestens, M. Lamberigts, Y. Houbaert, Texture evolution of tertiary oxide scale during steel plate finishing hot rolling simulation tests, Materials Science Forum, 550, (2007): 557-562
11. M. Zhang, G. Shao, Characterization and properties of oxide scales on hot-rolled strips, Materials Science and Engineering A, 452-453, (2007): 189-193
12. Y.-L. Yang, C.-H. Yang, S.-N. Lin, C.-H. Chen, W.-T. Tsai, Effects of Si and its content on the scale formation on hot-rolled steel strips, Materials Chemistry and Physics, 112, (2008): 566-571
13. A. Segawa, Reproduction and deformation characteristics of oxide scale in hot rolling using vacuum rolling mill, Materials Science Forum, 696, (2011): 150-155
14. V.B Ginzburg, *Flat-rolling fundamentals* (New York: Marcel Dekker, Inc. 2000)
15. S. Chandra-ambhorn, T. Phadungwong, K. Sirivedin, Effects of carbon and coiling temperature on the adhesion of thermal oxide scales to hot-rolled carbon steels, Corrosion Science, 115, (2017): 30-40
16. S. Chandra-ambhorn, T. Nilsonthi, Y. Wouters, A. Galerie, Oxidation of simulated recycled steels with 0.23 and 1.03 wt.% Si in Ar–20% H_2O at 900°C, Corrosion Science, 87, (2014): 101-110

17. R.Y. Chen, W.Y.D. Yuen, Isothermal and step isothermal oxidation of copper-containing steels in air at 980-1220°C, Oxidation of Metals, 63, 3/4, (2005): 145-168
18. S. Chandra-ambhorn, A. Jutilarptavorn, T. Rojhirunsakool, High temperature oxidation of irons without and with 0.06 wt.% Sn in dry and humidified oxygen, Corrosion Science, 148, (2019): 355-365
19. S. Chandra-ambhorn, K. Ngamkham, High temperature oxidation of micro-alloyed steel and its scale adhesion, Oxidation of Metals, 88, (2017): 291-300
20. R.Y. Chen, W.Y.D. Yuen, A study of the scale structure of hot-rolled steel strip by simulated coiling and cooling, Oxidation of Metals, 53, 5/6, (2000): 539-560
21. R.Y. Chen, W.Y.D. Yuen, Copper enrichment behaviours of copper-containing steels in simulated thin-slab casting processes, Iron and Steel Institute of Japan International, 45, 6, (2005): 807-816
22. S. Chandra-ambhorn, K. Ngamkham, N. Jiratthanakul, Effects of process parameters on mechanical adhesion of thermal oxide scales on hot-rolled low carbon steels, Oxidation of metals, 80, 1, (2013): 61-72
23. J.L. Beuth, N. Dhanaraj, J. Hammer, S. Laney, F.S. Pettit, G.H. Meier, *ASM Materials Solutions Conference and Show* (Columbus, OH, October 18-24, 2004)
24. J. Mougin, M. Dupeux, A. Galerie, L. Antoni, Inverted blister test to measure adhesion energy of thermal oxide scales on metals or alloys, Materials Science and Technology, 18, 10, (2002): 1217-1220
25. J. Mougin, M. Dupeux, L. Antoni, A. Galerie, Adhesion of thermal oxide scales grown on ferritic stainless steels measured using the inverted blister test, Materials Science and Engineering A, 359, (2003): 44-51
26. A. Galerie, F. Toscan, E. N'Dah, K. Przybylski, Y. Wouters, M. Dupeux, Measuring adhesion of Cr_2O_3 and Al_2O_3 scales on Fe-based alloys, Materials Science Forum, 461-464, (2004): 631-638
27. M.M. Nagl, W.T. Evans, D.J. Hall, S.R.J. Saunders, The failure of iron oxide scales at growth temperature under tensile stress, Journal de Physique IV, 3, (1993): 933-941
28. M.M. Nagl, S.R.J. Saunders, W.T. Evans, D.J. Hall, The tensile failure of nickel oxide scales at ambient and at growth temperature, Corrosion Science, 35, 5-8, (1993): 965-977
29. M.M. Nagl, W.T. Evans, D.J. Hall, S.R.J. Saunders, An in situ investigation of the tensile failure of oxide scales, Oxidation of Metals, 42, (1994): 431-449
30. M. Krzyzanowski, J.H. Beynon, The tensile failure of mild steel oxides under hot rolling conditions, Steel Research, 70, (1999): 22-27
31. M. Krzyzanowski, J.H. Beynon, Finite element model of steel oxide failure during tensile testing under hot rolling conditions, Materials Science and Technology, 15, 10, (1999): 1191-1198
32. M. Krzyzanowski, J.H. Beynon, C.M. Sellars, Analysis of secondary oxide-scale failure at entry into the roll gap, Metallurgical and Materials Transactions B, 31, 6, (2000): 1483-1490
33. D. Geneve, D. Rouxel, P. Pigeat, M. Confente, Descaling ability of low-alloy steel wires depending on composition and rolling process, Corrosion Science, 52, (2010): 1155-1166
34. Y.T. Chiu, C.K. Lin, J.C. Wu, High-temperature tensile and creep properties of a ferritic stainless steel for interconnect in solid oxide fuel cell, Journal of Power Sources, 196, (2011): 2005-2012
35. K. Ngamkham, S. Niltawach, S. Chandra-ambhorn, Development of tensile test to investigate mechanical adhesion of thermal oxide scales on hot-rolled steel strips produced using different finishing temperatures, Key Engineering Materials, 462-463, (2011): 407-412
36. S. Chandra-ambhorn, N. Klubvihok, Quantification of adherence of thermal oxide scale on low carbon steel using tensile test, Oxidation of Metals, 85, 1, (2016): 103-125
37. T. Nilsonthi, W. Issaard, S. Chandra-Ambhorn, Development of the scale adhesion assessment using a tensile testing machine equipped with a CCD camera, Oxidation of Metals, 88, (2017): 41-55
38. W. Issaard, T. Nilsonthi, Application of the tensile test with a CCD camera to assess the adhesion of scale to Si-containing hot-rolled steels, Key Engineering Materials, 728, (2017): 26-30
39. N. Na-kalasin, S. Yenchum, T. Nilsonthi, Adhesion behaviour of scales on hot-rolled steel strips produced from continuous casting slabs, Materials Today: Proceedings, 5, (2018): 9359-9367
40. S. Chandra-ambhorn, P. Saranyachot, T. Thublaor, High temperature oxidation behaviour of Fe-15.7 wt.% Cr-8.5 wt.% Mn in oxygen without and with water vapour at 700°C, Corrosion Science, 148 (2019): 39-47
41. S. Chandra-ambhorn, T. Thublaor, P. Wiman, High temperature oxidation of AISI 430 stainless steel in Ar-H_2O at 800°C, Corrosion Science, 167, (2020)
42. P. Wiman, T. Thublaor, T. Rojhirunsakool, M.H.S. Bidabadi, Z.G. Yang, T. Siripongsakul, W. Chandra-ambhorn, S. Chandra-ambhorn, Corrosion behaviour of AISI 430 stainless steel in O2-40%H_2O at 800°C, Corrosion Science, 203, (2022)
43. M.Z. Ruhiyuddin, D. Murizam, K.R. Ahmad, Synthesis and characterization of iron produced from iron mill scale, Key Engineering Materials, 594–595, (2014): 118-122
44. S.R. Prim, M.V. Folgueras, M.A. de Lima, D. Hotza, Synthesis and characterization of hematite pigment obtained from a steel waste industry, Journal of Hazardous Materials, 192, (2011): 1307-1313
45. N.J. Cory, T.M. Herrington, The location of hydrogen in the kinetics of oxidation of ferrous alloys

by superheated steam, Oxidation of Metals, 29, 1-2, (1988): 135-152
46. A.S. Khanna, *Introduction to high temperature oxidation and corrosion* (Ohio, ASM International, 2004)
47. H. Li, T. Li, C. Li, Z. Wang, G. Wang, Improvement of longitudinal performance uniformity of hot-rolled coils for cold-rolled DP980 steel, Metals, 10, (2020): 1-14
48. H. Krungkarnchana, C. Kongvarhodom, Low temperature corrosion: oxidation of carbon steel and stainless steel in air, Applied Science and Engineering Progress, 12, 1, (2019): 44-51
49. G.C. Wood, J. Stringer, The adhesion of growing oxide scales to the substrate, Journal de Physique IV, 3, (1993): 65-74
50. F. Toscan, L. Antoni, Y. Wouters, M. Dupeux, A. Galerie, Oxidation kinetics and scale spallation of iron-chromium alloys with different titanium contents, Materials Science Forum, 461-464, (2004): 705-712
51. S. Chandra-ambhorm, F. Roussel-Dherbey, F. Toscan, Y. Wouters, A. Galerie, M. Dupeux, Determination of mechanical adhesion energy of thermal oxide scales on AISI 430Ti alloy using tensile test, Materials Science and Technology, 23, 4, (2007): 497-501
52. G. Bamba, Y. Wouters, A. Galerie, F. Charlot, A. Dellali, Thermal oxidation kinetics and oxide scale adhesion of Fe-15Cr alloys as a function of their silicon content, Acta Materialia, 54, (2006): 3917-3922
53. S. Chandra-ambhorn, Y. Wouters, L. Antoni, F. Toscan, A. Galerie, Adhesion of oxide scales grown on ferritic stainless steels in solid oxide fuel cells temperature and atmosphere conditions, Journal of Power Sources, 171, (2007): 688-695
54. T. Nilsonthi, Determination of mechanical adhesion energy of thermal oxide scales on steel produced from medium and thin slabs using tensile test, Key Engineering Materials, 658, (2015): 106-110
55. H.E. Evans, Stress effects in high temperature oxidation of metals, International Materials Reviews, 40, 1, (1995): 1-40
56. J. Kargin, L.L.S. Valladares, L.E. Borja-castro, J. Xize, D.G. Mukhambetov, Y. V. Konyukhov, N.O. Moreno, A.G.B. Dominguez, C.H.W. Barnes, Characterization of iron oxide waste scales obtained by rolling mill steel industry, Hyperfine Interactions, 243, (2022): 1-10
57. J. Tominaga, K. Wakimoto, T. Mori, M. Murakami, T. Yoshimura, Manufacture of wire rods with good descaling property, Transactions of the Iron and Steel Institute of Japan, 22, 8, (1982): 646-656
58. S. Garber, Fundamental aspects of scale on mild steel strip, Journal of the Iron and Steel Institute, 192, (1959): 153-160
59. K. Sachs, G.T.F. Jay, A magnetic seam at the scale/metal interface on mild steel, Journal of the Iron and Steel Institute, 195, (1960):180-189

Effective solvents for proteins recovery from microalgae

Pau Loke Show[1,*], *Kit Wayne* Chew[2,*], *Kuan Shiong* Khoo[3,*], and *Atthasit* Tawai[4]

[1]Department of Chemical and Environmental Engineering, Faculty of Science and Engineering, University of Nottingham Malaysia, Jalan Broga, Semenyih 43500, Selangor Darul Ehsan, Malaysia.
[2]School of Energy and Chemical Engineering, Xiamen University Malaysia, Jalan Sunsuria, Bandar Sunsuria, 43900 Sepang, Selangor, Malaysia.
[3]Faculty of Applied Sciences, UCSI University, UCSI Heights, 56000 Cheras, Kuala Lumpur, Malaysia.
[4]Biorefinery and Process Automation Engineering Center, Chemical and Process Engineering, The Sirindhorn International Thai-German Graduate School of Engineering, King Mongkut's University of Technology North Bangkok, Bangkok, 10800, Thailand

Abstract. From an industrial perspective, the exploitation of microalgae as protein source is of great economical and commercial interest due to several attractive characteristics. However, the protein extraction efficiency is limited by the multiple layers of rigid and thick cell walls that are enriched with recalcitrant structure of cellulose. Therefore, an efficient method of cell disruption is necessary to disintegrate the cell wall and promote the release of protein contents. The conventional methods for downstream processing, e.g. disruption, isolation, extraction, concentration and purification, are energy-intensive and costly because they typically compose of several operational units. To reduce the overall process cost and establish an economical feasible process for the large-scale production of microalgae derived products, a more cost-effective and eco-friendly technique in downstream processing is in critical demand. One of the main challenges for protein extraction from microalgae cells is the recalcitrant structure of microalgae cell wall. This work aims to provide a guideline on the selection of the solvent to facilitate the proteins release during the cell disruption process. The influences of various solvent types (methanol, ethanol, 1-propanol and water) were evaluated and compared based on the protein yields. It was found that water solvent released the highest protein concentration from microalgae compared to the other tested solvents.

Keyword. Liquid biphasic system technologies, Microalgae, Protein, Solvents.

1 Introduction

The mass production of protein-derived products from microalgae cultures offers significant advantages as it is considered to be renewable biomass that fits to the concept of sustainable production [1]. Microalgal proteins obtained from aquaculture industry are one of the most promising and sustainable substitute sources to conventional proteins, for example fishmeal and soybean [2]. This is mainly due to the microalgal-enriched protein content, nutritional profile and quality of microalgae's proteins [3]. Microalgae cells have the ability to synthesize all types of essential amino acids, which are mostly equivalent or even better with that of other higher organisms, such as plants. Furthermore, their amino acid compositions are not significantly affected by changes or stresses in environmental conditions [4]. Additionally, microalgae possess high efficiency photosynthetic activity, high growth rate, short harvesting cycle, high disease resistance and high biomass density [5-7].

However, one of the main difficulties in the mass production of protein products in commercial and industrial process is the low efficiency in protein extraction and recovery because of the presence of multiple layers of thick and tough cell wall [8, 9]. The process development of simple and rapid cell disruption and cell lysis in large scale industry is necessary to facilitate the release of proteins [10]. One of the vital factor for promoting the protein release is to choose a suitable extraction solvent for the cell disruption process [11]. This study aims to evaluate the influence of extraction solvent and provides a guidance on the selection of the effective solvent to facilitate the proteins release during the cell disruption process of microalgae.

The relevancy of this research work also involves the green environmental initiative by creating high rates of photosynthesis in mitigation of carbon dioxide (CO2) adopted from the cultivation of *Chlamydomonas sp.* Tai-03 microalgae. In regards to reserve energy and sustainable solutions, the concept of using liquid biphasic system combines the pretreatment and extraction step into one-system, thereby creating an efficient and effective bioprocessing of protein extraction from microalgae cultures in aquaculture industries.

* Corresponding author: showpauloke@gmail.com; kitwayne.chew@xmu.edu.my; kuanshiong.khoo@hotmail.com

2 Materials and methods

2.1 Chemicals

Food grade alcohols of ethanol, 1-propanol, 2-propanol (R&M Chemicals, Selangor, Malaysia) were purchased and used as the extraction solvents. Salts for the bottom phase during extraction that are utilized in this study were dipotassium hydrogen phosphate (K_2HPO_4), magnesium sulphate ($MgSO_4$), ammonium sulphate ($(NH_4)_2SO_4$) and potassium dihydrogen phosphate (KH_2PO_4) and all chemicals were purchased from R&M Chemicals (Selangor, Malaysia).

2.2. Setting up of liquid biphasic flotation system

Liquid biphasic flotation (LBF) unit of 50 mL volume capacity was used as the separation system, and it was provided by Donewell Resources (Puchong, Selangor, Malaysia). A 50 mL glass tube was connected from the bottom to a gas compressor. The bottom of the glass tube was drilled and fitted with a rubber tube to connect to the gas compressor. A sintered glass disk (Grade 4 porosity) was fitted at the bottom of the glass tube to allow the air bubbles that are generated during passing of the compressed air flow. The air flow rate that was supplied to the LBF unit is monitored and controlled by a flowmeter (model: RMA-26-SSV, Dwyer, Michigan, IN, USA) with a range of 50 to 200 cc/min.

2.3 Cultivation of *Chlamydomonas sp.* Tai-03 microalgae

The optimal culture conditions for the growth of *Chlamydomonas sp.* Tai-03 had been determined by our previous study [12]. Therefore, in this study, the culture conditions use BG-11 medium (containing 25% initial nitrate concentration (0.375 g/L $NaNO_3$). The microalgal culturing is conducted in the indoor 1 L photobioreactor. The light intensity is set up at 200 μmol/m².s. The initial concentration of cell inoculum is 0.12 g/L. The culture reactor is controlled in the mode of continuous stirring at 300 rpm, and is supplied by continuous sparging of 5% CO_2 gas at a flow rate of 0.1 vvm [12].

2.4. Protein Assay

The protein concentration is measured by using Bradford Reagent. The 2 mL of various dilutions of protein samples obtained from LBF unit at either top or bottom solutions are prepared with a serial dilution method (10x dilution) and mixed with 0.2 mL of Bradford Reagent, and the mixture was incubated for 10 min before the absorbance reading using UV-Vis spectrophotometer at a reading wavelength of 595 nm. The absorbance of the protein concentration was based on the calibration curve between Bovine serum albumin (BSA) concentrations and A595 values. BSA is commonly used as standard for measurement of protein concentration because of its commercial availability, high purity and low cost. The protein concentration of the microalgae samples can be calculated using the Equation (1) generated from the standard curve (Figure 1).

$$y = 0.0014x \quad (1)$$

Where y is the absorbance value at 595 nm and x is the protein concentration (μg/mL) equivalent to BSA

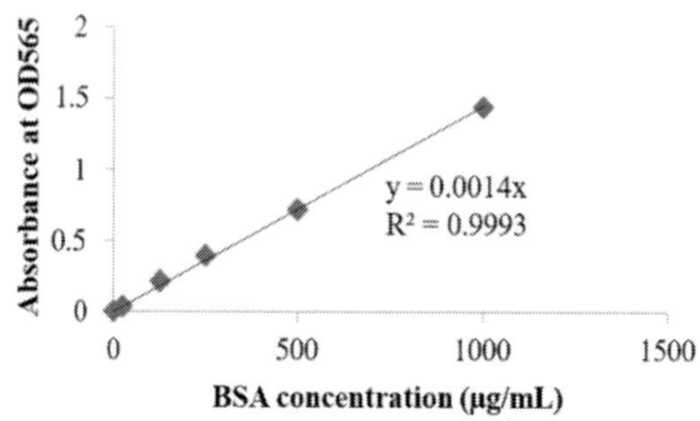

Figure 1. Standard curve for measurement of protein concentration using BSA as the standard based on the Bradford Reagent.

3. Results and discussion

In this study, four types of different solvents, including methanol, ethanol, propanol and water were used in extraction experiments (Figure 2). The protein concentrations (%) per dried weight of microalgae cells were measured by using Bradford protein assay. The experimental results showed that water showed the highest efficiency in protein extraction of *C. Tai-03* cultures because it provided the highest proteins concentration with about 3% of protein concentration per dry weight, and followed by 1-propanol (2.75%), ethanol (1.80%) and methanol (1.50%), respectively.

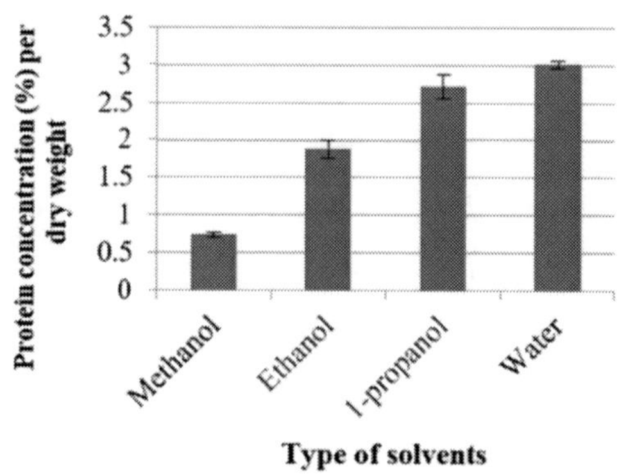

Figure 2. Total proteins contents of *Chlamydomonas Tai-03* obtained from extraction by LBF unit by using different types of solvents. Each bar is an average of three replicates.

In these recent years, liquid biphasic system (LBS) has become a proven tool for separation and purification process that fit to the concepts of circular economy and bioeconomy in microalgae industries. The application of controlling system based on Internet of Things (IoT) in LBSs in clarification, partitioning and partial purification of biomolecules and bioproducts, in this case, protein products, had showed the rapid development. This LBS technology helps to successfully achieve the high recovery yield and the high purity of targeted products in a single separation step. Additionally, the LBS shows the other interesting characteristics, such as high selectivity and easily to process scale up. Thus, LBS is proposed and proved to be an attractive technology that meets the requirements of the high throughput in commercial and industrial processes. Furthermore, it has good benefit in terms of economic and environmental protection.

Rapid progress and improvements in biorefining technology, new regulatory directives, product quality constraints, and the production efficiency have necessitated the development of more advanced and powerful downstream bioprocesses for various industries, such as biotechnology, biopharmaceuticals, as well as bioenergy. This situation motivates to intensely improvements in conventional biorefining processes, and drives the development of entirely new approaches, for example, numerous available extractive technology, including extractive fermentation, extractive bioconversion liquid biphasic system, liquid biphasic flotation system, and newly developed liquid biphasic flotation assisted with ultrasound and microwave, electricity. Combinations of the alcohol/salt LBF system and the ultrasonication have the ability of killing two birds with one stone, because it is not only able to execute the cell rupturing, it also capable to recover targeted bioproducts simultaneously and continuously.

Some other process parameters in LBS, such as crude feedstock concentration, flotation time, type of salt, concentration of salt, type of alcohol, concentration of alcohol, initial volumes of salt and alcohol were experimentally investigated. As demonstrated in this study, the small molecular weight aliphatic alcohols, methanol, ethanol, 1-propanol and 2-propanol can be combined with various types of salts, such as dipotassium hydrogen phosphate (K_2HPO_4), magnesium sulphate ($MgSO_4$), ammonium sulphate ($(NH_4)_2SO_4$) and potassium dihydrogen phosphate (KH_2PO_4). Under the optimal condition of LBF operation at 250 g/L of $(NH_4)_2SO_4$, 60% (v/v) of 2-propanol, initial VR of 1.0, 20 g/L of crude algal biomass load, 4 mm^3/min of air flowrate and 10 min of flotation time.

The recycling of phase components in LBF was also tested and evaluated to minimize the use of alcohol solvent and salt. It was demonstrated that top phase (containing alcohol) of recycling sample can achieve improved performance of extraction for at least three consecutive recycling runs. Under this optimized process conditions, the proportion of protein recovered in the top phase was 88.86% for the third recycling run in microalgae recovery studies.

4. Conclusions

In this work, comparing four types of solvents in LBF system, it was found that water could be an excellent solvent for microalgal cell disruption with the indication of higher proteins yield than the other solvents. Further investigation should be done on different strains of microalgae to evaluate the general effectiveness of water in improving the release of proteins yield from microalgae with different composition of cell wall.

References

[1] L. Moreno-Garcia, K. Adjallé, S. Barnabé, G. Raghavan, Microalgae biomass production for a biorefinery system: recent advances and the way towards sustainability, Renewable and Sustainable Energy Reviews, 76 (2017): 493-506.

[2] F.A. Ansari, A. Guldhe, S.K. Gupta, I. Rawat, F. Bux, Improving the feasibility of aquaculture feed by using microalgae, Environmental Science and Pollution Research, 28 (2021): 43234-43257.

[3] C. Acquah, S.M. Tibbetts, S. Pan, C. Udenigwe, Nutritional quality and bioactive properties of proteins and peptides from microalgae, Handbook of Microalgae-Based Processes and Products, Elsevier2020, pp. 493-531.

[4] P. Spolaore, C. Joannis-Cassan, E. Duran, A. Isambert, Commercial applications of microalgae, Journal of bioscience and bioengineering, 101 (2006): 87-96.

[5] D.Y.Y. Tang, K.S. Khoo, K.W. Chew, Y. Tao, S.-H. Ho, P.L. Show, Potential utilization of bioproducts from microalgae for the quality enhancement of natural products, Bioresource technology, 304 (2020): 122997.

[6] K.S. Khoo, K.W. Chew, G.Y. Yew, W.H. Leong, Y.H. Chai, P.L. Show, W.-H. Chen, Recent advances in downstream processing of microalgae lipid recovery for biofuel production, Bioresource technology, 304 (2020): 122996.

[7] N.S.M. Aron, K.S. Khoo, K.W. Chew, A. Veeramuthu, J.-S. Chang, P.L. Show, Microalgae cultivation in wastewater and potential processing strategies using solvent and membrane separation technologies, Journal of Water Process Engineering, 39 (2021): 101701.

[8] M. Alhattab, A. Kermanshahi-Pour, M.S.-L. Brooks, Microalgae disruption techniques for product recovery: influence of cell wall composition, Journal of Applied Phycology, 31 (2019): 61-88.

[9] V. Timira, K. Meki, Z. Li, H. Lin, M. Xu, S.N. Pramod, A comprehensive review on the application of novel disruption techniques for proteins release from microalgae, Critical Reviews in Food Science and Nutrition, 62 (2022): 4309-4325.

[10] K.S. Khoo, K.W. Chew, G.Y. Yew, S. Manickam, C.W. Ooi, P.L. Show, Integrated ultrasound-assisted liquid biphasic flotation for efficient extraction of astaxanthin from Haematococcus pluvialis, Ultrasonics sonochemistry, 67 (2020): 105052.

[11] P.-L. Show, M. Sriariyanun, Prospect of liquid biphasic

https://doi.org/10.1051/e3sconf/202235502009

system in microalgae research, Applied Science and Engineering Progress, 14 (2020): 1-2.
[12] C.H. Tan, C.-Y. Chen, P.L. Show, T.C. Ling, H.L. Lam, D.-J. Lee, J.-S. Chang, Strategies for enhancing lipid production from indigenous microalgae isolates, Journal of the Taiwan Institute of Chemical Engineers, 63 (2016): 189-194.

Pasting Property of Jasmine Rice Flour, Quality and Sensory Evaluation of Gluten-Free Rice Penne as Affected by Protein and Hydrocolloid Addition

Pakkawat Detchewa[1], *Viboon* Pongkanpai[2], *Chutamas* Maneewong[3], *Chanthima* Phungamngoen[4] and *Anuchita* Moongngarm[5]*

[1] Department of Agro-Industry Technology and Management, Faculty of Agro-Industry, King Mongkut's University of Technology North Bangkok, Prachinburi Campus, Thailand
[2] Department of Food Science and Technology, Faculty of Innovative Agriculture and Fishery Establishment Project, Prince of Songkla University, Surat Thani Campus, Thailand
[3] Department of Biotechnology, Faculty of Science, Maejo University, Chiang Mai, Thailand
[4] Department of Food Technology, Faculty of Technology, Khon Kaen University, Khon Kaen, Thailand
[5] Department of Food Science and Nutrition, Faculty of Technology, Mahasarakham University, Maha Sarakham, Thailand

Abstract. Jasmine rice flour (JMRF) is one of the most popularly consumed rice products in Thailand but applications to prepare gluten-free pasta are limited as JMRF amylose content is low. This research was carried out to develop JMRF suitable for gluten-free rice pasta (penne) (GFRP) by adding soy protein isolate (SPI, 5%), egg white protein powder (EW, 5%) and hydrocolloids (guar gum, GG and xanthan gum, XG, 1%). JMRF blended with proteins and hydrocolloids was investigated for pasting properties. JMRF blended with EW and SPI with GG showed the highest peak viscosity, whereas setback values observed in JMRF blended with EW and GG or XG were not significantly different. GFRP was prepared and determined for protein content, color and sensory evaluation. The protein content of GFRP with SPI blended with GG and XG ranged from 10.50 to 10.95% is higher than GFRP with EW and penne without SPI or EW (control). GFRP with SPI had higher yellowness than GFRP with EW. Adding SPI, EW and gums reduced cooking loss, while sensory evaluation showed a higher liking score for GFRP with SPI and GG than GFRP with EW. Results suggested that adding SPI and GG to JMRF improved pasting properties, protein content, color and acceptance. GFRP showed promise as a new alternative sustainable source to replace wheat in pasta products. However, characteristics, such as product chemical composition, texture analysis and nutritional benefits require further evaluation.

Keyword. Gluten-free pasta, penne, rice flour, hydrocolloid, protein

1. Introduction

Penne is a type of pasta that is widely consumed around the world, especially in European countries and North America. Penne is made from semolina or wheat flour (*Triticum durum* L.) mixed with water to form a dough. During kneading, a gluten network is developed that is necessary for pasta production [1]. The dough is extruded into cylinder-shaped pieces, with ends cut at an angle. However, pasta contains gluten, which causes allergies as a celiac disease [2]. Gluten-free pasta lacks the gluten network and shows poor sensory and cooking quality [3]. Rice is considered a gluten-free cereal grain and rice flour are used to produce gluten-free products, such as rice cookies [4], gluten-free rice spaghetti [5] and rice noodles [6]. Gluten-free rice spaghetti and rice noodles are produced from high amylose rice flour (> 22 %), which provides a satisfactory texture with low cooking loss and sensory acceptance [6, 7]. Jasmine rice (Khao Dawk Mali 105, *Oryza sativa*) is one of the most famous and popularly consumed rice varieties in Thailand. It has a unique fragrant smell and soft texture when cooked with low amylose content (< 20%). Jasmine rice is commonly consumed as whole grain or polished grain cooked rice. Broken Jasmine rice is a byproduct of the rice milling process and is generally sold as animal feed at a low price. Production of rice flour from broken rice can add value, reduce the production of waste and promote sustainable rice production. Jasmine rice flour has low amylose content and its application is limited to rice noodles and gluten-free pasta products. The amylose content in rice is related to the quality of rice noodles, and low amylose content produces rice noodles with low cooking quality. Previous studies have reported the development of gluten-free pasta using proteins and hydrocolloids as

* Corresponding author: anuchitac@yahoo.co.th

several gluten-free products using varieties of flour (e.g., rice, chickpea and maize) [4, 9]. Protein is a desirable food ingredient that promotes textural properties and nutrition enhancement. Gluten-free rice spaghetti incorporation with egg white protein improved textural properties [10]. The main protein in egg white is ovalbumin (54%), which has a free sulfhydryl (SH) group that enables gelation [11]. Detchewa and Naivikul [10] found that adding 5% egg white protein powder improved the texture, cooking quality and sensory evaluation of gluten-free rice spaghetti.

Soy protein is widely utilized in food products as a gelling agent. Detchewa *et al.* [5] improved the texture of gluten-free rice spaghetti by adding soy protein isolate (5%). Scanning electron microscopy showed that adding soy protein isolate to gluten-free rice spaghetti increased porosity at the surface and decreased cooking time. Rachman *et al.* [11] reported that the addition of egg white protein or soy protein improved the textural characteristics of banana pasta. Hydrocolloids are used in food products as thickeners, stabilizers, gelling agents and emulsifiers. Adding hydrocolloids improved the texture of gluten-free noodles. Silva *et al.* [12] found that adding 1% xanthan gum (XG) or 1% guar gum (GG) improved the strength of sweet potato starch noodles with 4% broccoli powder.

However, previous studies using protein and hydrocolloids to improve the quality of low amylose rice flour and preparation of GFRP using Jasmine rice are limited. This study investigated the effect of adding proteins, such as egg white (EW), soy protein isolate (SPI) and hydrocolloids including GG and XG on the pasting properties of Jasmine rice flour and the sensory evaluation of gluten-free rice penne.

2 Materials and Methods

2.1 Materials

Jasmine rice flour or Khao Dawk Mali 105 rice variety (*Oryza sativa* cv.) was purchased from Limited Partnership Charoenworrakit. Semolina or wheat flour (Divella S.p.A, Italy) was purchased from a local market. Soy protein isolate (SPI), egg white protein powder (EW), guar gum (GG) and xanthan gum (XG) were purchased from Krungthepchemi Co., Ltd., Bangkok, Thailand.

2.2 Proximate composition and amylose content

Moisture, protein (Nx6.25), fat and ash contents of the samples were analyzed according to the AOAC (2000) [13]. Protein contents of gluten-free rice penne samples were determined using the Kjeldahl method [13], while amylose contents of JMRF and semolina were determined according to the previous study by Juliano [14] using the iodine-based colorimetric method. Absorbance was measured at 620 nm wavelength (Biochem Libra S32, UK). Amylose content was estimated based on the standard curve of prepared potato amylose.

2.3 Pasting properties

Pasting properties of JMRF with proteins and hydrocolloids, and semolina were determined using a Rapid Visco Analyzer (model 4S, Newport Scientific, Australia) as pasting temperature, peak viscosity, hot paste viscosity, breakdown, final viscosity and setback [15].

2.4 Gluten-free rice penne production

Rice flour was blended with proteins and hydrocolloids following the mixture levels listed in Table 1.

The raw materials of each formula were well mixed before extrusion using a single-screw extruder (model CT, Chareon Tut, Thailand) to prepare pasta as penne type. Barrel temperatures were set at 50°C, 90°C and 90°C for zones 1-3 and the die temperature was 80°C. The screw speed was set at 220 rpm. The extruded GFRPP was dried at 50°C using a tray dryer until moisture content was below 12% [16].

Table 1 Nomenclature of GFRP

Designated name	% Mixture				
	JMRF	EW	SPI	GG	XG
JMRF	100	-	-	-	-
JMRFEW5	95	5	-	-	-
JMRFSPI5	95	-	5	-	-
JMRFEW5GG1	94	5	-	1	-
JMRFEW5XG1	94	5	-	-	1
JMRFSPI5GG1	94	-	5	1	-
JMRFSPI5XG1	94	-	5	-	1

JMRF= Jasmine rice flour, JMRFEW5= Jasmine rice flour with egg white protein 5%, JMRFSPI5= Jasmine rice flour with soy protein isolate 5%, JMRFEW5GG1= Jasmine rice flour with egg white protein 5% and GG 1%, JMRFEW5XG1= Jasmine rice flour with egg white protein 5% and XG 1%, JMRFSPI5GG1= Jasmine rice flour soy protein isolate 5% and GG 1%, JMRFSPI5XG1= Jasmine rice flour soy protein isolate 5% and XG 1%

2.5 Color measurement

Cooked GFRP samples were measured for color using a Hunter Lab Color (Colex, USA). The color L* value indicates lightness, while a* and b* are redness and yellowness, respectively.

2.6 Cooking quality measurement

Cooking qualities of the GFRP samples and wheat penne were determined using the optimum cooking time, water absorption index and cooking loss according to the method of AACC [17] and Detchewa *et al.* [16].

2.7 Sensory evaluation

Gluten-free rice penne samples were evaluated after cooking by 30 untrained panelists for appearance, color, texture, flavor, taste and overall liking. The sensory evaluation was performed using a 9-point hedonic scale, where 9= like extremely, 5= neither like or dislike and 1= dislike extremely.

2.8 Statistical analysis

All experiments were conducted in triplicate. Results were analyzed by one-way analysis of variance (ANOVA) using SPSS version 22 (SPSS Inc, Chicago, USA). Duncan's multiple range test (DMRT) was used for multiple comparisons with statistical significance set at $p<0.05$.

3 Results and Discussion

3.1 Chemical compositions and amylose content

Chemical compositions of JMRF, semolina, EW and SPI are shown in Table 2. Moisture, protein, fat and ash contents of JMRF were 10.50%, 6.37%, 0.67% and 0.56%, respectively, while semolina contained 10.56% moisture, 12.92% protein, 0.97% fat and 0.80% ash content. Protein content in wheat flour was higher than JMRF. Sissons *et al.* [18] reported that protein content of semolina higher than 12-13% gave good pasta quality. SPI presented higher protein content (93.13%) than EW (82.20%). SPI has a minimum protein content of 90% dry weight basis, while egg white protein powder concentrate contains a minimum protein content of 65%. Amylose content was 14.10% in JMRF and 25.41% in semolina (Table 2). Bhattachary *et al.* [6] reported that noodles with high amylose content (> 22%) showed a positive correlation with hardness chewiness, gumminess and tensile strength.

Table 2 Chemical compositions of major raw materials

Raw material	Chemical composition (%)				
	Moisture	Protein	Fat	Ash	Amylose
JMRF	10.50 ±0.15^a	6.37 ±0.15^d	0.67± 0.12^b	0.56 ±0.10^d	14.10 ±0.26^b
Semolina	10.56 ±0.15^a	12.92 ±0.50^c	0.97± 0.08^a	0.80 ±0.10^c	25.41 ±0.11^a
EW	5.80 ±1.00^b	82.20 ±0.80^b	0.02 ±0.00^c	3.00± 0.20^b	-
SPI	4.70 ± 0.06^c	93.13 ±1.03^a	0.75± 0.09^b	3.80 ±0.14^a	-

Mean values with different small letter superscripts within the same column are significantly different at $p<0.05$.

3.2 Pasting properties

Pasting properties of JMRF and JMRF with proteins and hydrocolloids, and semolina are shown in Figure 1. Pasting temperature or gelatinized temperature of semolina, JMRF and JMRF with proteins and hydrocolloids are shown in Figure 1 (A). Wheat flour gelatinization temperature was higher than JMRF. The gelatinized temperature of JMRF added with egg white protein 5% (JMRFEW5) or soy protein isolate 5% (JMRFSPI5) was not significantly different from JMRF. Adding GG and XG as well as EW or SPI affected the gelatinization temperature. The gelatinized temperatures of JMRFEW5GG1, JMRFEW5XG1, JMRFSPI5GG1 and JMRFSPI5XG1 were lower than JMRF. This result concurred with Zhang *et al.* [19] who reported that the addition of hydrocolloids to high amylose starches showed initial onset of viscosity or gelatinization at low temperature.

Peak viscosity showed high viscosity of maximum starch swelling before disintegration. Peak viscosities of JMRFEW5 and JMRFSPI5 were lower than JMRF. Lower starch content and high protein content in this sample resulted in peak viscosity decreasing, as indicated in JMRF with egg white protein or SPI with XG 1%. Adding 1% XG gave a lower peak viscosity than JMRF, JMRFEW5 and JMRFSPI5. This result was similar to Weber *et al* [20]. The addition of XG to normal corn starch decreased peak viscosity at 95°C. XG is highly hygroscopic and reduces water availability for starch swelling, resulting in decreased peak viscosity. Adding GG 1% to JMRF with EW or SPI showed higher peak viscosity. This result was similar to Chauban *et al.* [21] who reported that amaranth blended with 1% GG showed higher peak viscosity than 0.5% GG. GG interacted with the starch granules and the entanglement led to an increase in peak viscosity.

Breakdown viscosity reflects starch granule stability during the heating process with water. The low breakdown value indicated higher integrity of the starch granules or higher shear resistance of starch paste [22]. JMRF showed the highest breakdown while adding proteins and hydrocolloids decreased breakdown. Breakdown values of JMRFEW5GG1, JMRFEW5XG1, JMRFSPI5GG1 and JMRFSPI5XG1 were not significantly different from semolina. These results suggested that proteins and hydrocolloids improved the heat and shear resistance of starch.

Setback viscosity indicates the degree of starch retrogradation or recrystallization after the heating and cooling process [22]. Setback viscosities of JMRF and JMRFEW5 were higher than JMRF. Adding protein and gums (JMRFEW5GG1 and JMRFEW5XG1) showed the highest setback, while addition of SPI and GG or XG (JMRFSPI5GG1 and JMRFSPI5XG1) gave higher setback than JMRF only (control).

Proteins contain many hydrophilic groups, such as -NH_2, -OH-, -COOH- and -NH-, which form crosslinks with starch [23]. These crosslinks generated higher setback viscosity compared to JMRF only. High setback or high amylose retrogradation imparted good rice noodle quality through reinforcement of the amylose network [24, 25].

https://doi.org/10.1051/e3sconf/202235502010

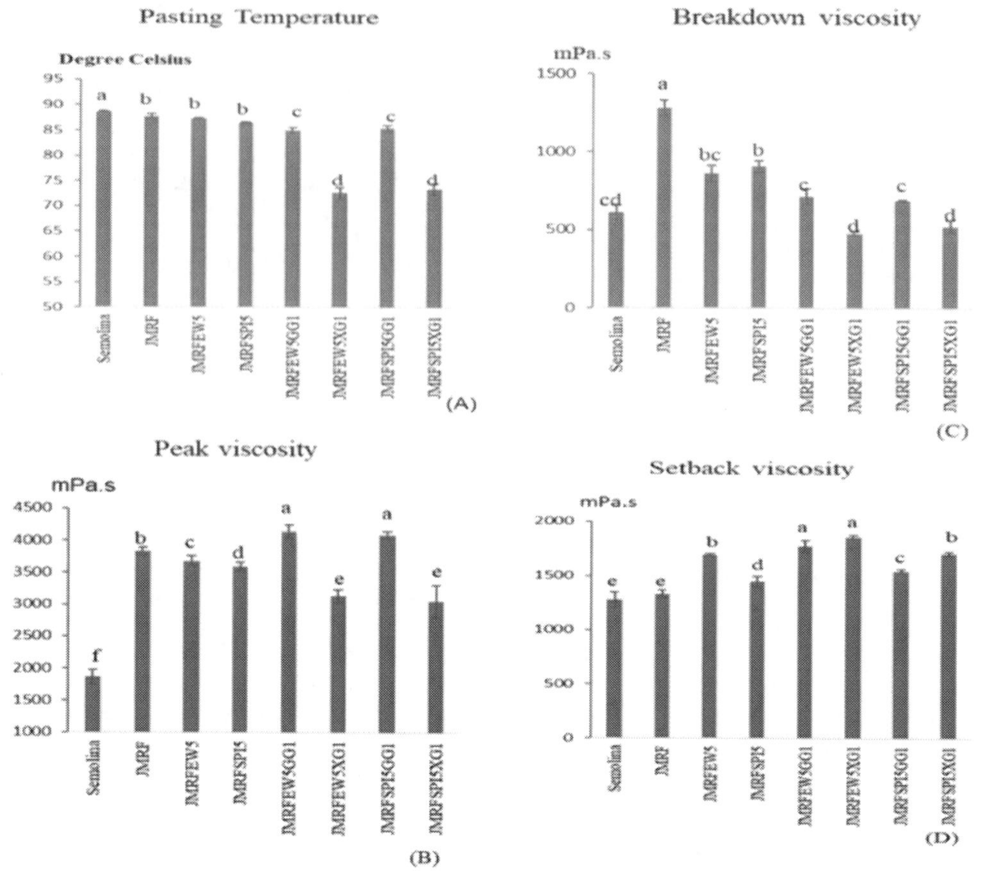

Figure 1. Pasting properties of JMRF, JMRF blended with protein and hydrocolloid, and semolina. Pasting temperature (A), peak viscosity (B), breakdown (C) and setback viscosity (D). JMRF= Jasmine rice flour, JMRFEW5= Jasmine rice flour with egg white protein 5%, JMRFSPI5= Jasmine rice flour with soy protein isolate 5%, JMRFEW5GG1= Jasmine rice flour with egg white protein 5% and GG 1%, JMRFEW5XG1= Jasmine rice flour with egg white protein 5% and XG 1%, JMRFSPI5GG1= Jasmine rice flour soy protein isolate 5% and GG 1%, JMRFSPI5XG1= Jasmine rice flour soy protein isolate 5% and XG 1%

3.3 Gluten-free rice penne

After the mixed rice flour was extruded to obtain GFRP, several quality aspects were evaluated. A photo of GFRP was compared with commercial wheat penne (Figure 2.)

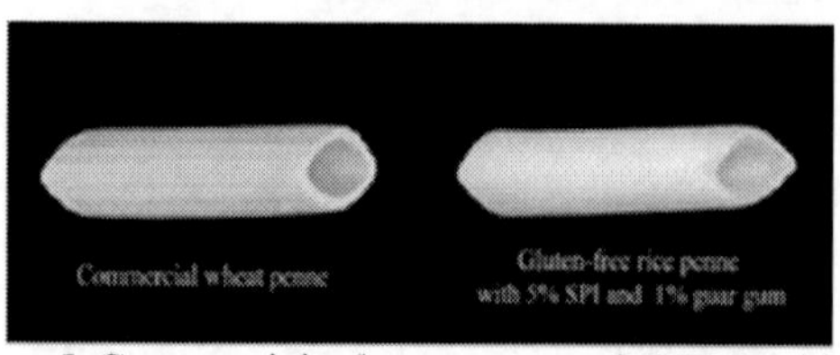

Figure 2 Commercial wheat penne and GFRP with 5% soy protein isolate and 1% GG

3.4 Color of gluten-free rice penne

The L*, a* and b* values of GFRP and wheat penne are shown in Table 3. The L* value of gluten-free rice penne varied from 67.21 to 70.33, while wheat penne was 62.56. The a* and b* values of GFRP ranged from 0.13 to 1.40 and from 11.87 to 17.41, respectively. GFRP added with egg white protein and gums (JMRFEW5, JMRFEW5GG1 and JMRFEW5XG) showed lower lightness (*L), redness(a*) and yellowness (b*) than GFRP made from JMRF only, while GFRP with soy protein isolate showed higher yellowness (b* value) with increasing SPI level than GFRP added with egg white protein and GFRP without egg white protein. Increasing SPI decreased the a* value. This result was similar to Detchewa *et al.* [5], who reported that increasing soy protein isolate content in gluten-free rice spaghetti resulted in increased yellowness (*b), while Akesowan [26] reported that adding soy protein isolate to ice cream increased the yellowness.

3.5 Protein content in penne

Protein contents of wheat penne and GFRP with proteins and hydrocolloids are shown in Figure 3. The protein content of wheat penne was 12.8%, significantly higher than the GFRP samples. Protein content higher than 12% gives good pasta quality, whereas flour with protein content below 10% is undesirable [27]. Protein content in GFRP with SPI (10.50-10.95%) was higher than GFRP with egg white protein (9.75-9.90 %). These results were similar to Rachman *et al.* [11] who reported that soy protein fortification in banana pasta gave higher protein content than banana pasta with egg white protein.

Table 3 Color of wheat penne and GFRP with proteins and hydrocolloids

Gluten-free rice penne	L*	a*	b*
Wheat penne	62.56±0.01^{h}	3.32±0.02^{a}	29.90±0.07^{a}
JMRF	70.33±0.26^{a}	1.40±0.03^{b}	11.87±0.02^{h}
JMRFEW5	67.87±0.12^{e}	0.13±0.02^{f}	14.02±0.02^{e}
JMRFSPI5	66.17±0.18^{g}	0.73±0.03^{c}	17.41±0.03^{b}
JMRFEW5GG1	69.90±0.17^{b}	0.17±0.06^{f}	12.57±0.03^{g}
JMRFEW5XG1	68.85±0.07^{c}	0.13±0.04^{f}	12.92±0.05^{f}
JMRFSPI5GG1	67.21±0.21^{f}	0.21±0.05^{e}	15.20±0.10^{d}
JMRFSPI5XG1	68.28±0.24^{d}	0.44±0.01^{d}	16.55±0.06^{c}

Mean values with different small letter superscripts within the same column are significantly different at $p<0.05$.

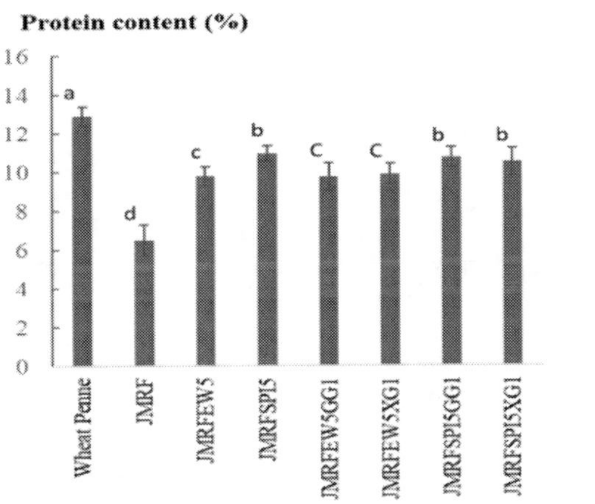

Figure 3 Protein content of wheat penne and GFRP with proteins and hydrocolloids

3.6 Cooking quality

The optimum cooking time, cooking loss and water absorption of GFRP and wheat penne samples are shown in Table 4. The optimum cooking time of GFRP was significantly higher (14.23-22.04 min) than wheat penne (10.02 min). Cooking loss in GFRP ranged from 22.05 to 34.39 %, higher than wheat penne (5.34%). When wheat penne was boiled in water, the gluten network entrapped the starch resulting in lower cooking loss [28]. Cooking losses of GFRP with 5% SPI and 1% GG (22.05%) were the lowest among the GFRP samples, while GFRP without protein and gum added showed the highest cooking loss (34.39%). Rachman *et al.* [11] reported similar results using SPI and egg white protein to decrease the cooking loss of banana pasta, while Chauhan *et al.* [21] found that 1% GG in gluten-free amaranth pasta gave the minimum cooking loss (8.5%) due to matrix formation between starch and gum. Water absorption expresses penne hydration after cooking and relates to the eating quality. Water absorption of wheat penne was 94.49%, whereas absorption of GFRP varied from 54.54 to 84.34%. Adding hydrocolloids increased water absorption due to their hydrophilic groups. GFRP with 5% SPI and 1% GG had higher water absorption than GFRP with 5% SPI and 1% XG. This result agreed with Chauhan *et al.* [20] who found that gluten amaranth pasta with 1% GG had higher water absorption (180.60%) than the control (144.78%). These results were also in agreement with Sutheeves *et al.* [29] who reported that the addition of GG in gluten-free instant noodles (297.26%) gave higher water absorption than XG (228.79%).

Table 4 Cooking quality of GFRP samples and wheat penne

Sample	Cooking time (min)	Cooking loss (%)	Water adsorption (%)
Wheat penne	10.2 ± 0.14^{g}	5.34 ± 0.48^{g}	94.49 ±0.32^{a}
JMRF	14.23 ± 0.16^{f}	34.39 ±0.55^{a}	54.54 ±0.05^{g}
JMRFEW5	15.08 ±0.03^{e}	32.40 ±0.57^{b}	56.44 ±0.12^{f}
JMRFSPI5	17.09 ±0.12^{d}	30.17 ±0.24^{c}	56.60 ±0.31^{f}
JMRFEW5GG1	19.27 ±0.38^{c}	22.25±0.07^{f}	60.66 ±0.38^{d}
JMRFEW5XG1	20.25 ±0.35^{b}	27.23 ±0.32^{d}	58.71 ±0.06^{e}
JMRFSPI5GG1	19.22 ±0.31^{c}	22.05 ± 0.35^{f}	88.23 ±0.59^{b}
JMRFSPI5XG1	22.04 ±0.06^{a}	25.20 ±0.28^{e}	74.34 ±0.48^{c}

Mean values with different small letter superscripts within the same column are significantly different at $p<0.05$.

3.7 Sensory evaluation

Sensory evaluation of GFRP with EW or SPI with hydrocolloids and wheat penne was evaluated by 30 untrained panelists who were familiar with penne consumption. The sensory evaluation reflects the quality of penne by consumer satisfaction through liking scores. GFRP was characterized by the mean scores of the sensory parameters including appearance, color, texture, flavor, taste and overall liking, as indicated in Table 4. GFRP with 5% SPI and 1% GG had the highest score for sensory attributes including appearance (8.0), color (7.6), texture (7.9), flavor (7.5), taste (7.3) and overall liking (7.8). The color of GFRP with SPI showed a higher score than GFRP with/without egg white protein because the color of soy protein was more yellow, as observed by the b* value in Table 3. The control GFRP received a high flavor score because it was made from Jasmine rice flour only, resulting in a more intense rice aroma than samples added with protein and gums. Daygon *et al.* [30] reported that Jasmine rice flour (Khao Dawk Mali 105) contained acetyl-1-pyrroline (2-ACPY), which has a fragrant or aromatic flavor and sweet aroma attribute.

Table 5 Sensory evaluation of wheat penne and gluten-free rice penne with proteins and hydrocolloids

Sample	Appearance	Color	Texture	Flavor	Taste	Over all liking
Wheat penne	8.4±0.6^{a}	8.7±0.4^{a}	8.2±0.5^{a}	8.0±0.7^{a}	7.6±0.8^{a}	8.5±0.5^{a}
JMRF	5.6±0.4^{e}	5.7±0.7^{e}	5.1±0.6^{f}	7.6±0.5^{b}	7.1±0.8^{b}	6.3±0.3^{e}
JMRFEW5	6.2±0.6^{d}	5.8±0.6^{e}	6.1±0.5^{e}	7.4±0.4^{b}	7.0±1.0^{b}	6.4±0.2^{e}
JMRFSPI5	7.5±0.5^{c}	8.1±0.4^{b}	6.5±0.6^{d}	7.8±0.7^{b}	7.0±0.9^{b}	7.2±0.4^{c}
JMRFEW5GG1	7.8±0.2^{bc}	7.3±0.6^{cd}	7.2±0.7^{c}	6.5±0.6^{c}	7.2±0.4^{ab}	7.5±0.1^{c}
JMRFEW5XG1	7.5±0.5^{c}	7.2±0.7^{d}	7.1±0.7^{c}	7.0±0.2^{d}	7.0±0.2^{b}	7.0±0.1^{d}
JMRFSPI5GG1	8.0±0.5^{b}	7.6±0.4^{c}	7.9±0.6^{b}	7.5±0.3^{b}	7.3±0.4^{ab}	7.8±0.2^{b}
JMRFSPI5XG1	7.7±0.4^{bc}	7.4±0.6^{cd}	7.8±0.6^{b}	7.1±0.1^{d}	7.2±0.4^{b}	7.1±0.1^{d}

Mean values with different small letter superscripts within the same column are significantly different at $p<0.05$.

Figure 2 shows the appearance of wheat penne and GFRP with 5% soy protein isolate and 1% GG. Both penne samples revealed a good appearance. Thus, overall liking showed that GFRP with 5% soy protein isolate and 1% GG gave the highest scores, while the commercial wheat penne gave the highest score for all sensory attributes.

4 Conclusions

This research focused on GFRP production based on Jasmine rice flour. The addition of SPI or EW with GG or XG into JMRF affected the pasting property of rice flour. Results suggested that the quality of GFRP including protein content of penne, color, cooking quality and sensory acceptance was improved using SPI or EW with GG or XG. The incorporation of 5% SPI and 1% GG were the most suitable percentages to obtain good quality and acceptable GFRP. However, further studies are required to investigate product chemical compositions, texture and nutritional quality.

Acknowledgments

This research was funded by the National Science Research and Innovation Fund (NSRF) and King Mongkut's University of Technology North Bangkok, under Contract no. KMUTNB-FF-65-74.

References

[1] R.R. Matsuo, J.E. Dexter, B.L. Dronzek, Scanning electron study of spaghetti processing, *Cereal Chemistry,* 55 (1978): 744-753.

[2] M. Atteno, L. Costa, A. Cozzolino, R. Tortora, R. Caso, A. Del Puente, C. Ciacci, The enthesopathy of celiac patients: Effects of gluten-free diet, *Clinical Rheumatology*, 33 (2014): 537-541.

[3] L. Padalino, A. Conte, M. A. Del Nobile, Overview on the general approaches to improve gluten-free pasta and bread, *Foods*, 5(87) (2016): 1-18.

[4] S. Sasanam, B. Thunthanaruk, V. Rungsardthong, J. Laohavijitjan, S. I. Mussatto, D. Uttapap, Physicochemical and pasting properties of rice flour, banana flour, and job's tears flour: flour blends and application in gluten-free cookies, *Applied Science and Engineering Progress*, 16(2) (2023): 5992.

[5] P. Detchewa, M. Thongngam, J-L Jane, O. Naivikul, Preparation of gluten-free rice spaghetti with soy protein isolate using twin-screw extrusion, *Journal of Food Science and Technology*, 53(9) (2016): 3485-3494.

[6] M. Bhattacharya, S. Y. Zee, H. Corke, Physicochemical properties related to quality of rice noodle, *Cereal Chemistry,* 76(1999): 861-867.

[7] D.E. Kohlwey, J. H. Kendall, R. B. Mohindra, Using the physical properties of rice as a guide to formulation, *Cereal Food World*, 40 (1995): 728-732.

[8] L. Padlino, M. Mastromatteo, L. Lecce, S. Spinelli, A. Conte, M. A. Del Nobile, Optimization and characterization of gluten-free spaghetti enriched with chickpea flour, *International Journal of Food Science, Nutr.* 49 (2014): 1544-1556.

[9] P. Detchewa, O. Naivikul, Effect of egg white protein powder on the textural properties, cooking qualities and sensory evaluation of gluten-free rice spaghetti, *VRU Research and Development Journal Science and Technology,* 11 (2) (2016): 37-46.

[10] Y. Mine, Recent advances in the understanding of egg white protein functional, *Trends in Food Science and Technology*, 6 (1995): 225-232.

[11] A. Rachman, M. A. Brennan, J. Morton, C. S. Brennan, Effect of egg white protein and soy protein fortification on physicochemical characteristics of banana pasta, *Journal of Food Processing and Preservation*, 43.e14801 (2019): 1-7.

[12] E. Silva, M. Birkenhake, E. Scholten, L.M.C. Sagis, E. Van Der Linden, Controlling rheology and structure of sweet potato starch noodles with high broccoli powder content by hydrocolloids, *Food Hydrocolloids*. 30 (2012): 42-52.

[13] AOAC., *Official Method of Analysis. (International 17th ed, Williams, S) Association of Official Analytical Chemists*, Arlington, Virginia. (2000).

[14] B.O. Juliano, C. M. Perez, A. B. Blakeney, D. T. Castillo, N. Knonseree, B. Lainelet, E.T. Lapid, V. V. S. Murty, C. M. Paule, B.D. Webb, International cooperative testing on the amylose content of milled rice, *Starch – Stärke*, 33 (1981): 157-162.

[15] P. Detchewa, P. Prasajak, W. Sriwichai, A. Moongngarm, The effects of unripe banana flour on resistant starch content and quality characteristics of gluten-free rice cookies, *Journal of Sustainability Science and Management*, 16(2) (2021): 67-78.

[16] P. Detchewa, P. Prasajak, C. Phugnamnagoen, W. Sriwichai, O. Naivikul, A. Moongngarm, Substitution of rice flour with rice protein improved quality of gluten-free rice spaghetti processed using single screw extrusion, *Lebensmittel-Wissenschaft & Technologie,* 153 (2022):112512.

[17] American Association of Cereal Chemists (AACC), Approved methods of The AACC, 10th edn. *American Association of Cereal Chemists* (AACC). St. Paul (2000).

[18] M. Sissons, S. Cutillo, I. Marcotuli, A. Gadaleta, Impact of durum wheat protein content on spaghetti in vitro starch digestion and technological properties, *The Journal of Cereal Science*, 98 (2021): 103156.

[19] Y. Zhang, M. Li, X. You, F. Fang, B. Li, Impacts of guar and XGs on pasting and gelling properties of high amylose corn starches, International *Journal of Biological Macromolecules*, 146 (2020): 1060-1068.

[20] F. H. Weber, M. T. P. S. Clerici, F. P. Collares-Queiroz, Y. K. Chang, Interaction of guar and XGs with starch in the gels obtained from normal, waxy and high-amylose corn starches, *Starch – Stärke*, 61 (2009): 28-34

[21] A. Chauhan, D. C. Saxena, S. Singh, Effect of hydrocolloids on microstructure, texture and quality characteristics of gluten-free pasta, *ournal of Food Measurement and Characterization*, 11 (2007): 1188-1195.

[22] L. Zhang, X. Wang, S. Li, J. Sun, X. Liu, Effect of inulin on the pasting, textural, and rheological properties of sweet potato starch, *CYTA- J. Food*, 17 (2019): 733-743.

[23] P.K. Goel, R.S. Singhal, P.R. Kulkarni, Studies on interactions of corn starch with casein and casein hydrolysates, *Food Chemistry*, 64 (1999): 393-389.

[24] B. Otegbayo, D. Oguniyan, O. Akinwumi, Physicochemical and functional characterization of yam starch for potential industrial application, *Starch – Stärke*, 66 (2014) :235-250.

[25] S. Kraithong, S. Sawdkuen, Quality attributes and cooking properties of commercial Thai rice noodles, *PeerJ*, 11113: 1-19.

[26] A. Akesowan, Influence of soy protein isolate on physical and sensory properties of ice cream, *Thai Journal of Agriculutral Science*, 42(1) (2009): 1-6.

[27] M. J. Sissons, N.E. Egan, M. C. Gianibelli, New insights into the role of gluten on durum pasta quality using reconstitution method, *Cereal Chemistry*, 82(5) (2005): 601-608.

[28] C. Zweifel, S. Handschin, F. Escher, B. Conde-Petit, Influence of high-temperature drying on structural and textural properties of durum wheat pasta, *Cereal Chemistry*, 80 (2003): 159-167.

[29] S. Sutheeves, P. Chai-Uea, D. Thirathumthavorn, Impact of hydrocolloids on the physico-chemical and sensory properties of gluten-free instant noodles from rice flour and mung bean starch, *Italian Journal of Food Science*, 32 (2020): 438-449.

[30] V. D. Daygon, S. Prakash, M. Calingacion, A. Riedel, B. Ovenden, P. Snell, J. Mitchell, M. Fizgerald, Understanding the Jasmine phenotype of rice through metabolite profiling and sensory evaluation, *Metabolomics*, 12(62) (2016): 1-15.

* Corresponding author: anuchitac@yahoo.co.th

The application of an analytic network process to select vegetarian food for chronic disease patients

Wilaitip Punyota [1,*], *Athakorn* Kengpol [2], *and Preecha* Rungsaksangmanee [3]

[1]Industrial Engineering, Faculty of Engineering, King Mongkut's University of Technology North Bangkok, Bangkok, Thailand

[2]Industrial Engineering, Faculty of Engineering, King Mongkut's University of Technology North Bangkok, Bangkok, Thailand
[3]Infirmary of Kasetsart University, Bangkok, Thailand

Abstract. Food is important for patients with chronic diseases. Food choice is a multiple-criteria decision problem because there are so many options and factors to consider. The number of chronic disease patients continues to increase. People with chronic diseases eat a lot of healthy food groups, such as vegetarian foods, because it can help prevent, treat, and lower the risk of chronic diseases. People with chronic diseases should eat vegetarian food. From previous research studies, no research has been done to prioritize the appropriate vegetarian food menus for people with chronic diseases. Therefore, the objective of this research is to prioritize the appropriate vegetarian food menu for patients with three chronic diseases as appropriate for each disease by using the analytic network process. The results show that the model for prioritizing food menus in patients with chronic disease is appropriate and nutritionally correct. The benefits of this research can help support food decisions for consumers with chronic diseases. This model encourages people who suffer from chronic diseases to eat nutritious foods that are appropriate for their condition. Vegetarian food that is appropriate for an individual is referred to as being a part of a sustainable food system.

Keyword. Vegetarian food, Analytic network process, Sustainable food system

1 Introduction

The change in the modern world has resulted in many people living with risky behaviors. Consumers eat foods that are very sweet, salty, and high in fat, lacking in fruits and vegetables, being less physically active, and living in the midst of pollution. Also, the selection of favorite food leads to the production of food waste, which contributed a major portion of greenhouse gas worldwide. Chronic diseases, such as diabetes, high cholesterol, high blood pressure, osteoporosis, cancer, and others are caused by these factors. These conditions can lead to more serious conditions, such as coronary artery disease, chronic kidney disease, and dementia. It also lowers the quality of life, makes people sick, and is a leading cause of death.

The most common noncommunicable diseases (NCDs) in Thai people are diabetes, obesity, high blood pressure, high cholesterol, chronic respiratory disease, and cancer. The population with noncommunicable diseases has been rising, and the average age of these patients is getting younger. The first obvious risk factor is the increase in body weight among the overweight, which increases the risk of developing NCD [1]. Additionally, the spread of coronavirus 2019 (COVID-19) has heightened the impact of NCDs. According to the World Health Organization (WHO), people with underlying diseases are more likely to develop a severe illness or die from COVID-19 than healthy people [2]. These long-term illnesses are caused by poor eating habits, not getting enough sleep, and overworking. These factors have a long-term negative effect on health. A healthy diet is the most important factor in preventing NCDs.

In Thailand, healthy food is becoming more popular, and consumption is increasing. Health-conscious consumers are looking for food that balances the body and lowers the risk of getting many different diseases. Entrepreneurs are competing to develop better products for the body to meet the needs of these people. Numerous healthy food groups, such as clean food, vegetarian food, or food made with organic ingredients, as well as vitamins and dietary supplements, are extremely popular [3]. As a result, eating healthy foods is a good way to care for the body. Vegetarian food is one of the healthiest and most popular foods. It is mainly composed of fruits and vegetables, with no meat. According to the World Health Organization, eating more than 400 g of fruits and vegetables per day can reduce the risk of contracting non-communicable diseases [4]. Therefore, vegetarian foods are suitable for reducing the risk of chronic diseases.

In the past, numerous factors influenced the desire to make healthy food choices. Consumers believe that

* Corresponding author: athakorn@kmutnb.ac.th

eating healthier foods can improve their overall health. These factors are essential for the selection of healthy foods that consumers can accept. There have been numerous types of restaurants that have expanded rapidly to meet consumer demand. As the number of restaurants increases, consumers have more food options to choose from. Consumers have to make dietary decisions on a regular basis but often lack the knowledge to use them. Most consumers choose food based on their own experiences, friends' advice, advertising, or their favorite food tastes. Therefore, if consumers lack experience or knowledge, they can make decisions with only a general understanding. Consumers are unable to select the most appropriate food for their individual needs, resulting in these limitations on their food choices. Food production is a resource-intensive effort with a significant environmental impact. Access to nutritious and healthy food remains a problem for many. Chronic diseases are related to consuming unhealthy foods and participating in unhealthy daily activities. This is a major public health issue that has a negative impact on the economy and society.

The majority of previous vegetarian food research studies have focused on consumer food consumption behaviors, factors that influence vegetarian food choices, and vegetarian diet satisfaction. Consumers are unaware of the nutritional value of their food [5]. If a patient with chronic disease is informed and consumes appropriate healthy foods, they can reduce their disease occurrences and medical costs. However, no research supports or helps solve the problem of proper menu provision for chronic disease patients. Therefore, the objective of this research is to determine the appropriate vegetarian food menu for patients with three chronic diseases by using the analytic network process (ANP). Appropriate food means food that is safe for the body and helps prevent, alleviate, or treat chronic diseases. The analytic network process framework represents network relationships to represent complex interrelationships between decision levels or clusters [6]. It is able to manage the relationship between the decision criteria and their impact on choice. The analytic network process uses a comparative ratio scale as a super matrix to reduce analytic hierarchy process (AHP) errors [7]. The analytic network process method is used in this study to prioritize appropriate food menus for people with chronic diseases. The benefits of this research encourage consumers with chronic diseases to consume appropriate vegetarian foods for their condition. Consumer-appropriate vegetarian food is considered to be part of a sustainable food system (SFS) [8]. Sustainable food systems need to be made into long-term food systems so that they are more productive and can provide food that is healthy and nutritious. Vegetarian food has health, environmental, global food security, and animal welfare benefits. As a result, vegetarian food is a logical means of promoting sustainable food development [9].

2 Literature review

2.1 Vegetarian food

Vegetarian food involves abstaining from eating meat, fish, and poultry. It contains various levels of fruits, vegetables, grains, pulses, nuts, and seeds, but no meat. Plant-based vegetarian food is becoming more popular, and consumers may find it to be a cost-effective and low-risk treatment for a variety of chronic diseases. According to a number of studies, vegetarian foods reduce the risk of diabetes, hypertension, and heart disease [9]. Vegetarianism is becoming more popular all over the world, in part because it is healthier. Vegetarian foods can prevent and treat heart disease, hypertension, diabetes, cancer, osteoporosis, renal disease, and dementia. The lower prevalence of chronic diseases among vegetarians may be partially attributable to dietary choices other than the absence of meat. According to Glenn *et al.,* vegetarianism is associated with a 22% reduction in cardiovascular mortality and a 28% reduction in cardiovascular disease [10]. For example, bitter lemon is a part of the Cucurbitaceae, which has long been used in traditional medicine to treat diseases such as cancer and diabetes. A lectin found in bitter melon lowers blood glucose levels by acting on peripheral tissues and suppressing appetite, similar to how insulin works in the brain [11]. Ginger decreases blood pressure by dilating blood vessels and blocking calcium channels [12]. Based on previous research studies, Mehta (2017) proposes that proper vegetarian diet planning can be beneficial to health. The most significant effect is a 25% reduction in the risk of death and morbidity from coronary heart disease [13]. Agnoli et al. (2017) study important nutrients in vegetarian diets (proteins, vitamin B12, iron, zinc, calcium, vitamin D, and n-3 fatty acids). Regular consumption of plant-based foods can provide health benefits and help prevent chronic diseases [14]. In summary, vegetarian food is healthy because it reduces risks, alleviates diseases, and prevents chronic diseases. Additionally, it reduces the cost of medical care and increases agricultural production [15].

2.2 The Analytic Network Process (ANP)

One of the multi-criteria decision-making solutions is the AHP and ANP methods. This method has the advantage of being applicable to both quantitative and qualitative decision-making criteria. AHP is used for advanced decision analysis. It can eliminate weight-related issues. AHP compares each pair of elements using a hierarchical structure. Occasionally, the structure involves non-hierarchical decision-making elements. This is a limitation of the AHP. Thus, the ANP can manipulate the relationship between decision criteria and non-hierarchical selections. ANP uses a comparative ratio scale and processes it as a super matrix, which reduces AHP errors [16]. The ANP creates a network of relationships between criteria and sub-criteria that indicate dependencies between the criteria and sub-criteria. Inner dependencies are between criterion components (nodes). External dependencies are between-sub-criteria dependencies [17]. From previous research studies, Xia and Cheng (2019) use an analytic network process (ANP) to organize multiple-criteria decision-making (MCDM) to determine sustainable rural built-up landscapes in rural planning. The results show that weights and relative importance have been appropriately assigned to each criterion [18]. Hu et al. (2019) use a multi-criteria ANP to make maintenance networks for clusters of agricultural machinery. The ANP can solve the problem of evaluating competing alternatives and implementing decision-making criteria. This decision-making process involves a limited number of options, each of which is given a priority or preference [19]. In this research, the ANP method is used to determine appropriate foods for consumers with chronic diseases.

3 Material and Method

This section describes five steps for prioritizing a vegetarian diet for patients with chronic disease. The beginning step is to study appropriate foods for chronic diseases. Step 2 defines the structure of the problem. Step 3 is paired comparison. Step 4 is to determine the importance. Finally, step 5 finds a consistency ratio.

3.1 Study information on foods suitable for chronic diseases

This study is a pilot study to prioritize vegetarian food in patients with chronic disease. This study focused on high blood pressure, diabetes, and dyslipidemia, which are three of the most common long-term diseases. Then select a total of 5 dishes, namely, menu1: spicy mixed vegetable soup, menu2: stir-fried tofu with ginger, menu3: bitter melon with mixed Chinese herbs soup, menu4: fried eggplant and basil with textured vegetable protein (TVP), menu5: sour soup with mixed vegetables. These foods are readily available for purchase and consumption on the market. Fig. 1 shows examples of appropriate foods for chronic diseases.

Figure 1. Examples of appropriate foods for chronic diseases.

3.2 Define the structure of the problem

The objective of defining the structure of the problem is to describe the relationship between the problem's goals and the different criteria. The section explains the structure of the food evaluation problem in chronic disease patients. The structure consists of two parts: 1) Goal: The objective of the decision is to select the appropriate food for patients with chronic diseases. 2) Criteria: A criterion is a factor that influences a decision. It is defined as a food menu with five options.

3.3 Paired Comparison

A paired comparison is used to determine the significance of each relationship using data from the Bureau of Nutrition, the Department of Health, and the Ministry of Public Health [20]. The raw ingredients of each dish are considered when determining the relationship between diet and chronic diseases. The right nutritional value for each chronic disease determines which foods are most important for that disease. The considered factor and the reference factor are compared using an equal-numbers comparison. As shown in Table 1, the Saaty rating scale is used for pair comparison, which has a value between 1 and 9 [21]. The pair comparison evaluation is represented in Table 1 by the pair comparison matrix. The objective comparison scores *i* and *j* is represented by a_{ij} as shown in Fig. 2.

Table 1. Priority for comparing decision criteria.

Importance Level	Numerical Value
Equally	1
Equally to Moderately	2
Moderately	3
Moderately to Strongly	4
Strongly	5
Strongly to Very Strongly	6
Very Strongly	7
Very Strongly to Extremely	8
Extremely	9

Criteria i

$A =$

	1	2	⋯	*i*
1	a_{11}	a_{21}	⋯	a_{i1}
2	a_{12}	a_{22}	⋯	a_{i2}
⋮	⋮	⋮	⋮	⋮
j	a_{1j}	a_{2j}	⋯	a_{ij}

Criteria j

Figure 2. Matrix of pair comparisons.

3.4 Prioritization

From the paired comparison in Step 3.3, the following step is to calculate the significance of the criteria from the A_{norm} matrix using equation 1. The priority weights for each goal are averaged in each row of the A_{norm} matrix.

$$A_{norm} = \begin{bmatrix} \frac{a_{11}}{\sum_{j=1}^{n} a_{1j}} & \cdots & \frac{a_{i1}}{\sum_{j=1}^{n} a_{ij}} \\ \vdots & \ddots & \vdots \\ \frac{a_{1j}}{\sum_{j=1}^{n} a_{1j}} & \cdots & \frac{a_{ij}}{\sum_{j=1}^{n} a_{ij}} \end{bmatrix} \quad (1)$$

3.5 Consider the Consistency Ratio (CR)

The consistency ratio from pair comparisons is used to determine whether or not pair comparison assessments are consistent with one another. The eigen vector value is used to determine the priority of the consistency check. It can be used to calculate the consistency ratio (CR), which is the ratio of the consistency index (CI) to the random consistency index (RI) as shown in equation 2. The consistency index value is determined by the variance of λ_{max} and the number of decision criteria. Table 2 shows the random index used to calculate the random consistency index [15].

$$CR = \frac{CI}{RI} = \frac{\frac{\lambda_{max} - n}{n-1}}{RI} \quad (2)$$

where n is the number of decision criteria.

λ_{max} is the sum of the product between the sum of the elements in each digit of the matrix and the significant weight.

Table 2. Random Index.

Order	1	2	3	4	5	6	7	8	9	10
R.I.	0	0	0.52	0.89	1.11	1.25	1.35	1.40	1.45	1.49

If the consistency ratio (CR) is less than 0.1, then the comparative evaluations of the pairs are consistent. It can bring weight values and comparative evaluation results in pairs for decision making in the next step. If the consistency ratio (CR) is greater than 0.1, it means that the pairs are not consistent with each other. The weights and results of the paired comparison cannot be used to consider making any more decisions.

4 Results and discussion

This ANP method is used to determine the priority of foods for patients with chronic diseases. The results and discussion are as follows:

4.1 The result of determining the structure of the problem

This step shows a diagram of the relationship between goals and decision criteria.

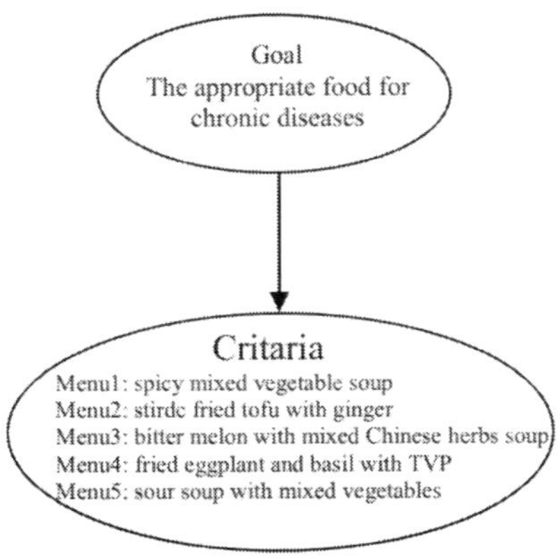

Figure 3. Relationship between goals and criteria

4.2 The results of the paired comparison assessment

A paired comparison of the importance of each food menu relationship is assessed by a group of nutrition experts. The decision criterion and each criterion are compared proportionally in a paired comparison. Table 3 presents an example of a paired comparison of a pair of foods for diabetes.

Table 3. Paired comparison with decision goals for diabetes.

Criteria	Menu 1	Menu 2	Menu 3	Menu 4	Menu 5
Menu 1	1	2	1/5	1/4	1/3
Menu 2	1/2	1	1/6	1/5	1/4
Menu 3	5	6	1	2	3
Menu 4	4	5	1/2	1	2
Menu 5	3	4	1/3	1/2	1

Table 3 shows an example of a comparison of the criteria used to make a decision for diabetic patients. The evaluation results can be explained by comparing the impact or importance of criteria on decision-making goals, as shown below: A score of 2 between criterion 1 and criterion 2 indicates that criterion 1 is twice as important as criterion 2 to the decision goal. The raw ingredients of each dish are considered when determining the relationship between food and chronic diseases. The right nutritional value for each chronic disease determines which foods are most important for that disease. On the other hand, if the significance comparison score is proportional, the score for criterion 2 versus criterion 5 is 1/4. This means that criterion 2 is four times less important to the goal of making a decision than criterion 5. From the paired comparison, the example evaluation results can be shown in the form of a metric A as follows:

$$A = \begin{pmatrix} 1 & 2 & 1/5 & 1/4 & 1/3 \\ 1/2 & 1 & 1/6 & 1/5 & 1/4 \\ 5 & 6 & 1 & 2 & 3 \\ 4 & 5 & 1/2 & 1 & 2 \\ 3 & 4 & 1/3 & 1/2 & 1 \end{pmatrix}$$

The A_{norm} matrix is obtained by performing the operation with equation 1 as follows:

$$A_{norm} = \begin{pmatrix} 0.07 & 0.11 & 0.09 & 0.06 & 0.05 \\ 0.04 & 0.06 & 0.08 & 0.05 & 0.04 \\ 0.37 & 0.33 & 0.45 & 0.51 & 0.46 \\ 0.30 & 0.28 & 0.23 & 0.25 & 0.30 \\ 0.22 & 0.22 & 0.15 & 0.13 & 0.15 \end{pmatrix}$$

The significance of each criterion is determined by averaging each row of the matrix A as shown in Table 4. An example of evaluating the significance of the criteria for diabetes is (0.07+0.11+0.09+0.06+0.05)/5 = 0.07.

Table 4. The weight of comparing pairs for diabetes.

Criteria	Menu	Weight
1	Spicy mixed vegetable soup	0.07
2	Stir-fried tofu with ginger	0.05
3	Bitter melon with mixed Chinese herbs soup	0.42
4	Fried eggplant and basil with TVP	0.27
5	Sour soup with mixed vegetables	0.17

4.3 Weight of the importance of decision-making criteria

The nutritional value of each menu item for a chronic disease is calculated by using nutrition data from the Bureau of Nutrition. The ANP method is used to analyze the results of food priorities in chronic disease patients using a Python programming language with a quick and easy library. Figure 4 shows ANP analysis by Python.

```
priority.pri_eigen(Diabetes, return_eigenval=True)

Xmax = priority.pri_eigen(Diabetes, return_eigenval=True)

priority.pri_eigen(Diabetes)

Priority = priority.pri_eigen(Diabetes)

priority.incon_std(Diabetes)

CR = priority.incon_std(Diabetes)
Paired Comparison Diabate =
 [[1.         2.         0.2        0.25       0.33333333]
 [0.5        1.         0.16666667 0.2        0.25      ]
 [5.         6.         1.         2.         3.        ]
 [4.         5.         0.5        1.         2.        ]
 [3.         4.         0.33333333 0.5        1.        ]]
Priority =
 [0.0762662  0.05056501 0.4271063  0.2728264  0.17323609]
SumXmax = 5.097532883557998
Consistency Ratio(CR) = 0.021770732937053037
```

Figure 4. ANP analysis by Python.

The results of food prioritization for high blood pressure are shown in Table 5, diabetes is shown in Table 6, and dyslipidemia is shown in Table 7, respectively.

Table 5. The priority of the criterion for high blood pressure.

No.	Criteria	Weight
1	Spicy mixed vegetable soup	0.31
2	Stir fried tofu with ginger	0.31
3	Bitter melon with mix Chinese herbs soup	0.18
4	Fried eggplant and basil with TVP	0.11
5	Sour soup with mixed vegetables	0.06

Table 6. The priority of the criterion for diabetes.

No.	Criteria	Weight
1	Bitter melon with mix Chinese herbs soup	0.42
2	Fried eggplant and basil with TVP	0.27
3	Sour soup with mixed vegetables	0.17
4	Spicy mixed vegetable soup	0.07
5	Stir fried tofu with ginger	0.05

Table 7. The priority of the criterion for dyslipidemia.

No.	Criteria	Weight
1	Sour soup with mixed vegetables	0.39
2	Spicy mixed vegetable soup	0.27
3	Bitter melon with mix Chinese herbs soup	0.19
4	Fried eggplant and basil with TVP	0.07
5	Stir fried tofu with ginger	0.05

4.4 The result of consider the consistency ratio (CR)

This procedure performs a consistency ratio analysis (Table 8). For example, the weight of the diabetes criteria is shown below.

$$A_{Diabetes} = \begin{pmatrix} 1 & 2 & 1/5 & 1/4 & 1/3 \\ 1/2 & 1 & 1/6 & 1/5 & 1/4 \\ 5 & 6 & 1 & 2 & 3 \\ 4 & 5 & 1/2 & 1 & 2 \\ 3 & 4 & 1/3 & 1/2 & 1 \end{pmatrix} \times \begin{pmatrix} 0.07 \\ 0.05 \\ 0.42 \\ 0.27 \\ 0.17 \end{pmatrix} = \begin{pmatrix} 0.39 \\ 0.25 \\ 2.19 \\ 1.40 \\ 0.89 \end{pmatrix}$$

$$\lambda_{max} = \frac{\frac{0.39}{0.07} + \frac{0.25}{0.05} + \frac{2.19}{0.42} + \frac{1.40}{0.27} + \frac{0.89}{0.17}}{5} = 5.09$$

Based upon the size of the matrix $n = 5$, the random index of Table 3 is 1.12, which when placed into equation **2** obtains the following consistency ratio:

$$CR = \frac{CI}{RI} = \frac{\frac{\lambda_{max} - n}{n-1}}{RI} = \frac{\frac{5.09-5}{5-1}}{1.12} = 0.02$$

The importance of appropriate food for patients with three chronic diseases: high blood pressure, diabetes, and dyslipidemia and five menus, namely, menu1: spicy mixed vegetable soup, menu2: stirred-fried tofu with ginger, menu3: bitter melon with mixed Chinese herbs soup, menu4: fried eggplant and basil with TVP, menu5: sour soup with mixed vegetables.

Table 8. Consistency ratio (CR).

Chronic diseases	Consistency Ratio (CR)
High blood pressure	0.008
Diabetes	0.021
Dyslipidemia	0.018

The results of the priority are discussed as follows: food evaluation results for high blood pressure patients show that spicy mixed vegetable soup and stir-fried tofu with ginger have the highest weight, 0.31. Therefore, both menus are most appropriate for patients. According to Table 5, the next most appropriate foods for high blood pressure patients are bitter melon with mixed Chinese herbs soup, fried eggplant and basil with TVP, and sour soup with mixed vegetables, weighing 0.18, 0.11, and 0.06, respectively. As shown in Table 6, the most appropriate food for diabetes patients is bitter melon with mixed Chinese herbs soup, which weights 0.42. The next most appropriate foods for diabetes are fried eggplant and basil with TVP, sour soup with mixed vegetables, spicy mixed vegetable soup, and stir-fried tofu with ginger, with respective weight values of 0.27, 0.17, 0.07, and 0.05. Finally, food evaluation results for dyslipidemia patients show that sour soup with mixed vegetables has the highest weight of 0.39 and is the most appropriate food for patients, as shown in Table 7. The next most appropriate foods for dyslipidemia are spicy mixed vegetable soup, bitter melon with mixed Chinese herbs soup, fried eggplant and basil with TVP, and stir-fried tofu with ginger, with respective weight values of 0.27, 0.19, 0.07, and 0.05. The results of the consistency ratio determination show that the CR for high blood pressure is 0.0081, the CR for diabetes is 0.0271, and the CR for dyslipidemia is 0.0189. Table 8 indicates that all of the CR values are less than 0.1, which means that the criteria for making decisions are consistent.

5 Conclusion

The objective of this research is to prioritize the appropriate vegetarian food menu for patients with three chronic diseases as appropriate for each disease by using the analytic network process. This is a decision made under quantitative and qualitative decision-making criteria. The goal of the decision is to identify appropriate foods for all three chronic conditions: high blood pressure, diabetes, and dyslipidemia. The decision criterion is all five menus: menu1: spicy mixed vegetable soup, menu2: stirred-fried tofu with ginger, menu3: bitter melon with mixed Chinese herbs soup, menu4: fried eggplant and basil with TVP, menu5: sour soup with mixed vegetables. The right nutritional value for each chronic disease determines which foods are most important for that disease. The results indicate that the analytic network process can prioritize nutritionally appropriate food for chronic disease patients. The two most appropriate food for high blood pressure patients is spicy mixed vegetable soup and stirred-fried tofu with ginger. The most appropriate food for diabetes patients is bitter melon with mixed Chinese herbs soup. The most appropriate food for dyslipidemia is a sour soup with mixed vegetables. The reliability of the evaluation is indicated by a C.R. value of less than 0.1. The benefits of this research contribute to the promotion of dietary guidelines for chronic disease patients. The recommendations of this analysis allow consumers to select foods that are nutritionally appropriate for their health. Vegetarian food that is appropriate for an individual is considered to be part of a sustainable food system (SFS). Sustainable food systems must be reformed in order to become more productive and capable of providing suitable, healthy, and nutritious food to all people on a sustainable basis. Vegetarian food has numerous advantages in terms of human health, especially for patients with chronic diseases. As a result, a vegetarian diet is a logical way to promote sustainable food development and consumption. More fruits and vegetables are grown to support a vegetarian diet for people with chronic diseases when there are more vegetarians. It is an increase in agricultural productivity that is sustainable. For future research, the number of chronic diseases and foods should be increased to better match the daily situation. The results of this research could be improved by studying weight scales that are both more reliable and more widely accepted. These weight scales should be based upon expert assessments of each disease. Furthermore, modern technologies such as artificial intelligence (AI) may be used to determine and compare the appropriate foods for chronic disease patients to the foods in this study.

Acknowledgements

This research is funded by the Faculty of Engineering, King Mongkut's University of Technology North Bangkok, Thailand.

References

[1] Division of Non-Communicable Diseases, Department of disease control, Ministry of Health in Thailand, NCD Situation Report 2019 Diabetes, Hypertension and Related Risk Factors, (2019)

[2] K.L. Hon, K. K. Y. Leung, J. W. Tang, A. K. C. Leung, Y. Li, COVID-19 in Hong Kong – public health, food safety, and animal vectors perspectives, Journal of Virological Methods, 290 (2020)

[3] V. Mehta, Vegetarian Diet: A Boon or Bane for Health, Journal of Medical Research and Innovation, 2 (2017)

[4] B. Chaniphun, Vegetables and Fruits Consumption at Least 400 g Per Day for Health: Evidence-based Approach, Journal of Nutrition Association of Thailand, 55(2020):53–65

[5] J. Saintila, Y. E. Calizaya-Milla, D. J. Aliaga, Knowledge of Vegetarian and Nonvegetarian Peruvian Dietitians about Vegetarianism at Different Stages of Life, Journal of Nutrition and Metabolic Insights, 14 (2021):1-9

[6] A. Kengpol, P. Neungrit, A decision support methodology with risk assessment on prediction of terrorism insurgency distribution range radius and elapsing time: An empirical case study in Thailand, Computers & Industrial Engineering, 75 (2014):55-67

[7] Y.-H. Chang, W.-M. Wey, and H.-Y. Tseng, Using ANP priorities with goal programming for revitalizations strategies in historic transport: A case study of the Alishan Forest Railway, Expert Systems with Application, 36 (2009):8682-8690

[8] H. D. Vries, M. Donner, M. Axelos, Sustainable food systems science based on physics' principles, 123 (2022):382-392

[9] A. Jabri, A. Kumar, E. Verghese, A. Alameh, A. Kumard, M. S. Khan, S. U. Khan, E. D. Michos, S. R. Kapadia, G. W. Reed, A. Kalra, Meta-analysis of effect of vegetarian diet on ischemic heart disease and all-cause mortality, American Journal of Preventive Cardiology, 7(2021)

[10] A. J. Glenn, E. Viguiliouk, M. Seider, B. A. Boucher, T. A. Khan, Relation of Vegetarian Dietary Patterns with Major Cardiovascular Outcomes: A Systematic Review and Meta-Analysis of Prospective Cohort Studies, Nutrition, 6(2019):1-12

[11] O. O. Elekofehinti, E. O. Ariyo, M. O. Akinjiyan, O. S. Olayeriju, A. O. Lawal, I. G. Adanlawo, J. B. T. Rocha, Potential use of bitter melon (Momordica charantia) derived compounds as antidiabetics: In silico and in vivo studies, Journal of Pathophysiology, 25,4(2018):327-333

[12] H. Hasani, A. Arab, A. Hadi, M. Pourmasoumi, A. Ghavami, M. Miraghajani, Does ginger supplementation lower blood pressure? A systematic review and meta-analysis of clinical trials, Journal of Phytotherapy, 33,1(2019)

[13] V. Mehta, Vegetarian Diet: A Boon or Bane for Health?, Jounal of Medical Research and Innovation, 2(2017)

[14] C. Agnoli, L. Baroni, I. Bertini, S. Ciappellano, A. Fabbri, M. Papa, S. Sieri, Position paper on vegetarian diets from the working group of the Italian Society of Human Nutrition. Journal of Nutrition, Metabolism and Cardiovascular Diseases, 27(2017):1037-1052

[15] A. U. Khank, Analytical hierarchy process (AHP) and analytic network process (ANP) methods and their applications: a twenty year review from 2000-2019, 12 (2020):369-459

[16] P. Chemweno, L. Pintelon, A. V. Horenbeek, P. Muchiri, Development of a risk assessment selection methodology for asset maintenance decision making: an analytic network process (ANP) approach, International Journal of Production Economics, 170(2015):663–676

[17] A. Kengpol, W. Punyota, Deep learning food assessment for the elderly with chronic diseases during the COVID-19 pandemic. 26th International Conference on Production Research (ICPR) 2021, (2021)

[18] L. Xia, W. Cheng, Sustainable development strategy of rural built-up landscapes in Northeast China based on ANP approach, Journal of Energy Procedia, 157(2019):844-850

[19] Y. Hu, S. Xiao, J. Wen, J. Li, An ANP-multi-criteria-based methodology to construct maintenance networks for agricultural machinery cluster in a balanced scorecard context, Journal of Computers and Electronics in Agriculture, 158 (2019): 1-10

[20] Bureau of Nutrition, Department of Health, Ministry of Public Health in Thailand. Homepage. http://nutrition.anamai.moph.go.th/image/file/food/images/files/foodCOVID-19.pdf

[21] T. L. Saaty, *Decision Making with Dependence and Feedback: The Analytic Network Process*. 2nd ed. Pittsburgh :RWS Publications, (2001)

Decision support system for consumer behavior of Elderly Chinese tourists on healthy beverages

Thanyathip Pichitkarnkar [1,*] and *Athakorn* Kengpol [2]

[1]Department of Industrial Business and Human Resource Development, Faculty of Business and Industrial Development, King Mongkut's University of Technology North Bangkok, Thailand
[2]Advanced Industrial Engineering Management Systems Research Center, Department of Industrial Engineering,Faculty of Engineering, King Mongkut's University of Technology North Bangkok, Thailand
* Corresponding author. E-mail: thanyathip.p@bid.kmutnb.ac.th

Abstract Based on tourist expenditure, food and beverages are a crucial market and healthy beverages are a dominant product developed to serve customer needs. At present, elderly tourist groups emerge with purchasing power in the tourism industry and related businesses. However, market and academic research on healthy beverages still need further knowledge and only a few decision support models exist. The objectives of this research are (1) to investigate major influential factors of consumer behavior (CB) of elderly Chinese tourists on healthy beverages; and (2) to structure a decision support system for CB. The research methodology employs qualitative and quantitative approaches. In-depth interviews are conducted by including 120 experts with 410 survey samples collected from Chinese tourists. Content analysis and structural equation modelling (SEM) are employed to analyze the data, including the sensitivity analysis for testing model robustness. The results show that marketing stimuli affect CB and product innovation at a significant level of 0.05. On the other hand, product innovation slightly affects CB at a significant level. The sensitivity analysis reveals that decreasing price and trial usage are influential CB. Also, the increasing and decreasing of attitude, relative advantage, and compatibility affect CB. This research aims to design an applicable decision support system of CB for Chinese tourists, which can identify influential factors of CB on healthy beverages when an entrepreneur launches a new product and be a guideline for SMEs to assess customer satisfaction by responding directly to customer needs. Moreover, the DSS model can also enable sustainable development and engage the right business strategies that resonate with customer needs.

Keywords: Healthy Beverages, Decision Support System, Structural Equation Modeling (SEM), Consumer Behavior, Hospitality Industry

1 Introduction

The COVID-19 crisis has affected every industrial sector in the world, especially the hospitality industry. The hospitality industry refers to a subsection within the service industry which comprises hotel businesses, restaurants, and tourism, all of which are a cornerstone of the Thai economy as they bring employment opportunities and lead to robust academic growth [1]. During the COVID-19 situation, the management of stocks of foods and beverages are difficult, due to unexpected stop/slow-down of customer's activities leading to food waste problem. Therefore, it is necessary to optimize the management of these supplies to the customer's demands. The sectors that generate the majority of revenue for the service industry are tourism and food & beverages (F&B). In order to optimize the F&B industry, entrepreneurs need to adapt themselves to overcome the shift in consumer behavior, e.g., new lifestyles, global trends, product innovation, health concerns, etc.

The food and beverage sector are at the lead based on tourists' expenditure in a destination country and obtains a large market share [2] whereby healthy beverages are one of the dominant products developed to serve customer needs. Healthy beverages are a kind of drink that contains vitamins, low sugar, and low caffeine; for example, vegetable juice, juice, herbal drinks, herbal tea, green tea, low-fat milk, and yogurt, among others. Generally, being healthy is the most desirable state for everybody and is a reason for people to be concerned about their longevity and quality of life, especially for aging populations. Therefore, to maintain good health, people consume healthy food and beverage. The healthy beverage market has its target groups classified by age, demographic, and gender. Preferences are upon customer perception towards taste, smell, and colors, such as orange juice, grape juice, apple juice, pomegranate juice, lime juice, coconut juice, green tea, etc.

Thailand is home to a large number of healthy beverage manufacturers, both market leaders and SMEs; for instance, Doi Kham, Gingen, Handyherb, QminC, Tipco, and Malee, which have a high production capacity, resulting in a low average cost per unit and competitive advantage compared to SME manufacturers.

As the number of aging populations is getting higher globally, this contributes to changes in various businesses, which is tourism where the group that owns high purchasing power is the elderly [3]. Chinese tourists

*Corresponding author: thanyathip.p@bid.kmutnb.ac.th

are a dominant group that travels to countless destination countries throughout the globe, including Thailand. In 2018, Thailand welcomed approximately 10.5 million Chinese visitors while, in 2019, the number achieved a sharp growth of 27.5%, before the outbreak of the COVID-19 pandemic [4]. In this regard, the increment contributed to the highest tourism revenue (roughly USD 19,357 million). Currently, the Chinese populations are approximately 240 million and based on a survey conducted by the National Working Commission on Aging, the elderly Chinese tourists accounted for 20%. Elderly Chinese tourists increasingly travel at a steady rate compared to other age groups and are indeed a group that has a very high expenditure on tourism. Moreover, this group prefer travelling with a travel agency, either in-bound or out-bound. As Thailand is a popular destination for elderly Chinese tourists, it is an opportunity for Thai entrepreneurs to study the behavior of this tourist group.

Furthermore, changes in tourist behavior are quite complicated and need to be aware of by entrepreneurs. Literature review reveals that most tourists alter their interests to local food and beverage. Hence, it is a good sign for small and medium enterprises (SMEs) to enlarge their sales volume. Nonetheless, healthy beverage manufacturers still lack basic knowledge and research backup to develop their products that can fulfil customers' preferences. They need know-how to understand customer needs, create innovation to enhance the market share, and increase their competitiveness. To achieve the targets and prepare effective marketing strategies, a Decision Support System (DSS) is requested to empower SMEs to compete in the red ocean market. The DSS is an effective tool to help them decide on product development while also minimizing the costs [1]. Consequently, a study of DSS for CB of elderly Chinese tourists on healthy beverages is a guideline for SMEs to be informed of customer satisfaction and able to touch customer needs. Considering industrial engineering, the DSS is an efficient tool that facilitates the decision-making process of operators. In addition, technological approaches and scientific ones were applied to this research in order to shape a DSS applicable to the field of social sciences. Therefore, the development of the DSS model for improvement of marketing strategies and product development can enable the operators to optimize their businesses and attain competitive advantage, which eventually leads to sustainable development.

The elderly tourist groups emerge with potential purchasing power in the tourism industry and related businesses. Thus, it is still in need of both market and academic research as DSS models for commercial purposes with respect to beverages are still rare. Moreover, there are still gaps in the empirical data obtained from academic research that need to be fulfilled. As a result, the research objectives are (1) to explore major influential factors for consumer behavior of elderly Chinese tourists on healthy beverages; and (2) to structure a decision support system of consumer behavior toward healthy beverages.

2 Materials and Methods

This research studies factors that affect consumer behavior to create a DSS that helps an entrepreneur to finalize a decision on product development and improve marketing strategies. The research method involves literature review, research design, collection of qualitative and quantitative data, and analysis of the findings through Structural Equation Modeling (SEM). Afterwards, the model robustness is tested in a sensitivity analysis.

2.1 Consumer behavior (CB)

Kotler and Armstrong [5] summarized about four key types of factors influencing consumer behavior, which comprised marketing stimuli, other stimuli, consumer psychology, and consumer characteristics. In general, a consumer would decide to purchase a product based on his/her brand consciousness and the trust he/she has in such a product [6]. Brand and quality expectation is what consumers always look for. The previous study [6] expressed that brand conceptualization affected CB and [7] confirmed that customers trust a brand if they are confident in product quality. Consequently, brand, trust, and product expectations are considered CB in this study.

2.2 Marketing stimuli and product innovation

There are four components of marketing stimuli [5]: price, product, promotion, and place, which affect consumer behavior. On the other hand, other stimuli considered in the CB model are economy, politics, technology, and culture. [8] expressed that marketing mix (price, product, promotion, and place) is an influential factor in brand consciousness and consumer satisfaction. Similarly, [9] confirmed that the marketing mix affected the quality expectation of a product and trust in the brand that leads to purchasing decisions. [10] and [11] found that attitude and reference group affected consumer purchasing intention. The model from [12] indicated that attitudes, reference groups, and culture are major factors for CB. Then, in this research, marketing stimuli also cover other stimuli.

Drucker [13] explained that innovation is an important tool for entrepreneurs to build their potential in business competition using existing resources or further developing based on new knowledge. Thus, a crucial success factor for marketing strategies is innovation, as it brings more customers and markets via the improvement of sustainable competitive advantage [14]. Innovations consist of three main attributes: 1) novelty, 2) economic benefits, and 3) knowledge and creativity. However, academic research classified innovation into three areas, which are 1) innovation target (product innovation and process innovation), 2) innovation change (radical innovation and incremental innovation), and 3) innovation impact (technological innovation and administrative innovation). Nowadays, the demand shift for product innovation is noticeably higher which impacts the marketing stimuli. [5] given that innovative products in the commercial arena required manipulation through customer adoption, which comprised five stages, i.e., awareness, interest, evaluation, trial, and adoption. The factors that influenced the adoption rate are 1) relative advantage – innovations that enjoy comparable products or substitutes; 2) compatibility – matching of customer values and experiences whereby a customer posits to purchase a specific innovative product; 3) complexity or degree of difficulty or easiness in using a specific innovative product; 4) divisibility or easiness with respect to the trial of a specific innovation; 5) communicability or how a specific innovation can be understandable or describable to others; 6) cost; 7) business risks generated by the innovation against the outcomes; 8) scientific credibility; 9) social approval. The study of [15] revealed that product innovation influenced purchasing decisions, similar to [16], which expressed that service innovation and product innovation affected marketing performance. Hence, companies that can combine both product innovation and service innovation can achieve greater revenue growth and more profitability.

In summary, marketing stimuli consist of price, product, promotion, place, attitude, reference group, and consumer culture while product innovation is composed of relative advantage, compatibility, trial usage, and communication.

2.3 Structural equation modeling

Structural Equation Modeling (SEM) is considered a robust multivariate approach that one employs to evaluate empirical data and causal assumptions of qualitative data. The SEM is applicable to several fields such as the study of tourism and tourist behavior [17], the study of the interrelationship between attitude, experiences, and overall customer satisfaction of elderly Australian tourists [18], the study of factors concerning attitudes of consumers that influence price shift and adoption [19], among others.

2.4 Decision support system (DSS)

A decision Support System can help an entrepreneur to engage in a decision-making process. A model is prepared to ensure businesses can respond to customer preferences. A DSS serves the management, operations, and planning levels to handle well-performed decision-making. Moreover, it is a useful appliance to increase consumer satisfaction and consumer confidence [20]. The DSS is applied in market research to study consumer purchasing behavior [21].

2.5 Research methodology

The research methodology is designed into three phases presented in Figure 1. Phase I: Literature Review, research and theories of CB concepts, marketing, product innovation, DSS, and healthy beverages are studied to identify appropriate factors. Phase II: Scenario Development consists of four steps. It starts with Phase II starts with engaging with the in-depth interview in step 1, collecting qualitative data, analyzing data using content analysis, and constructing a questionnaire in step 2. Next, the validity and reliability of the questionnaire are confirmed by experts in both academic and industrial fields and tested by Cronbach alpha coefficient (α) from 60 samples of the try out group in which the coefficient of the whole questionnaire is 0.946. Step 3, collecting quantitative data from 410 samples of Chinese tourists who visited Thailand between January – May 2022 in several tourist attractions around the country. Normally, SEM analysis requires only 280 samplings (14 latent variables × 20 multiply). Fourteen variables are price, product, promotion, place, attitude, reference group, consumer culture, relative advantage, compatibility, trial usage, communication, brand, trust, and product expectation. The data is analyzed by using SEM to create

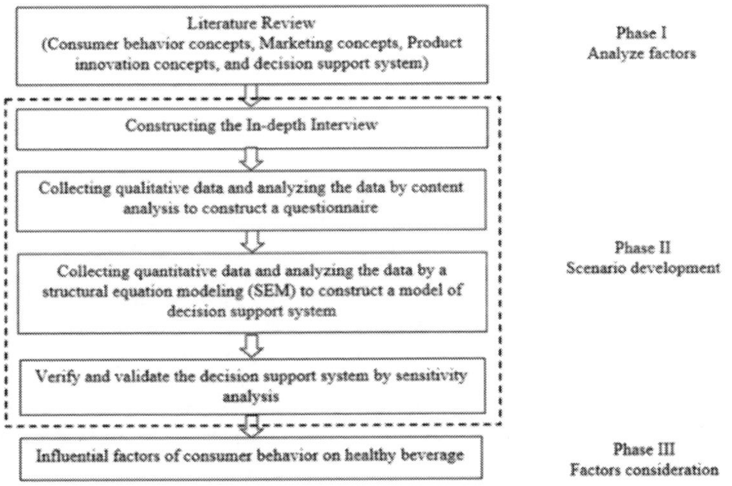

Figure 1. Research methodology for consumer behavior on healthy beverage

a DSS. The model is verified by using sensitivity analysis tests in the last step. Finally, the influential factors of CB on healthy beverages are expressed in Phase III: Factor consideration.

In this research, an in-depth interview is applied to collect qualitative data from 120 experts in the marketing field, DSS field, F&B industry, and entrepreneurs in F&B. The questions focused on an overview of the F&B industry, tourism industry, problems and obstacles, factors influencing CB, type of healthy beverages for elderly tourists, packaging, and marketing strategies. The questionnaire is divided into three sections as follows. Section 1 acquires general information about each respondent (e.g., gender, age, and health condition). Section 2 obtains feedback whether which factors possibly influence CB on healthy beverages. Then, a short comment for suggestion in section 3. The Cronbach's alpha coefficient of the whole questionnaire for 410 samples is 0.921.

3 Results and Discussion

3.1 Qualitative analysis from In-depth interview

The In-depth interview of 120 samplings is composed of both academic and industrial experts and entrepreneurs in the F&B sector, especially SMEs entrepreneurs from 6 parts of Thailand. There are North, South, East, West, Central, and Bangkok. In Appendix Table A.

3.2 Collecting quantitative data via marketing surveys

The respondents are female (56.60%) and aged between 60-65 years old (52.70%). Most of them are married (80.00%). They do exercise daily (70.00%), have a portion of healthy food at least 1 meal a day (45.20%), and have good health (70.00%). Construct's composite reliabilities (CR), Cronbach's alpha and average variance extracted (AVE) are employed to evaluate reliability, internal consistency, construct validity and the reflective measurement model. The result shows in Table 1 that the highest Cronbach's alpha coefficient is CB (0.93). Marketing stimuli (MS) and Product Innovation (PI) are 0.91 and 0.89, respectively. The AVE coefficients are ranging from 0.71 to 0.78 with which the highest coefficient is MS (0.78). CB and PI are 0.74 and 0.71 respectively. The highest CR coefficient is MS (0.79). CB and PI are 0.76 and 0.73, respectively. Ranging of the CR coefficients is from 0.73 to 0.79. Furthermore, the correlation coefficient between CB and MS is the highest (0.81), CB and PI are 0.72, and MS and PI are 0.78. Summarily, these data are shown that all figures have passed the acceptance criteria as shown in Table 2.

3.3 Data analysis by structural equation modeling (SEM)

Three hypotheses are verified which are H1, H2, and H3. H1 is a marketing stimulus that has a positive effect on CB, H2 is product innovation that has a positive effect on CB, and H3 is a marketing stimulus that has a positive effect on product innovation.

Structural Equation Modeling (SEM) is applied to examine these three hypotheses. The overall fit model is inspected by diagnostic indices which are shown in Table 3 [23]. If the construct validity is congruent with empirical data meant a good fit model.

Table 1: Means, SD, Cronbach's alpha, AVE, CR, and Correlation

Var.	$\bar{x}$	S.D.	AVE	CR	CB	MS	PI
CB	3.53	0.72	0.74	0.76	(.93)		
MS	3.78	0.58	0.78	0.79	.81**	(.91)	
PI	3.35	0.67	0.71	0.73	.72**	.78**	(.89)

Cronbach's alpha is presented in parenthesis.
**Correlation is significant at the .01 level (2-tailed)
CB = consumer behavior, MS = Marketing stimuli, PI = Product innovation

Table 2: Standard statistic criterion

Statistics	Criterion	Reference
AVE	> 0.50	[23]
CR	> 0.70	[23]
Cronbach's alpha	> 0.70	[24]
Correlation coefficient	< 0.85	[24]

Table 3: Statistics criterion for model fit

Statistics	Criterion	Results
Relative Chi-Square ($\chi 2$/df)	< 2.00	1.055
Comparative Fit Index (CFI)	> 0.90	0.998
Goodness of Fit Index (GFI)	> 0.90	0.978
Adjusted Goodness of Fit Index (AGFI)	> 0.90	0.955
Root Mean Square Residual (RMR)	< 0.05	0.007
Root Mean Square Error of Approximation (RMSEA)	< 0.05	0.012

Figure 2 expresses the SEM fit model for CB on healthy beverages for elderly Chinese tourists. The model is composed of MS and PI. MS has a direct effect on CB with a strong relationship at the significant level of .05. While MS affects CB indirectly through PI with a slight relationship at a significant level of .05.

The result shows that H1, H2, and H3 are accepted at a significant level of .05. In H1, MS has a positive effect on CB with a coefficient of 0.66. This implied that price, product, promotion, place, attitude, reference group, and consumer culture affect consumer behavior on brand, trust, and product expectation. [5] explains the price, quality of product, promotion package, and distribution channel and product expectation. Accordingly, the reference group and consumer culture directly affect brand and purchasing decisions [19]. A study [25] found that advertisements, reference groups, and products influenced healthy food and beverage purchasing decisions at a significant level of .05, especially brand consciousness. (place) are directly influential factors for the brand and trust.

https://doi.org/10.1051/e3sconf/202235502012

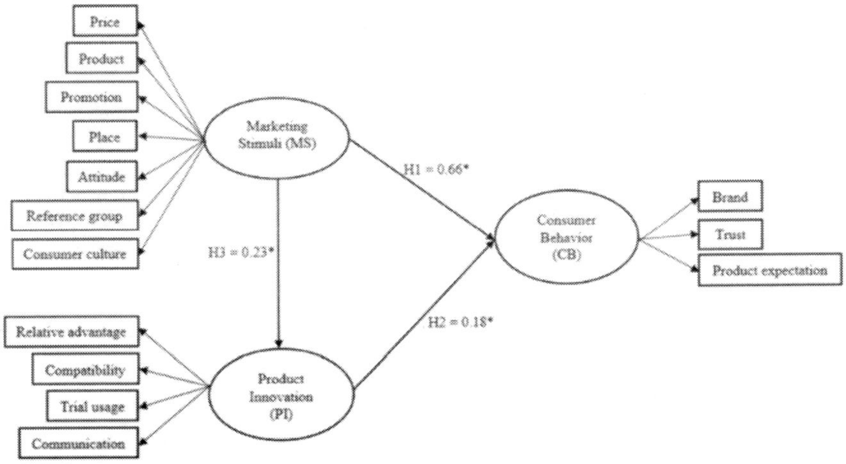

Figure 2. The SEM model for CB on healthy beverages for elderly Chinese tourist

In H2, PI has a slightly positive effect on CB with a coefficient of 0.18. This inferred that relative advantage, compatibility, trial usage, and communication have little effect on consumer behavior. [5] explains that consumers always compare product advantages to brands and their expectations. [25] confirmed that relative advantage and information influenced healthy food and beverage purchasing decisions at a significant level of .05, especially when considering the brand and product expectations. Then, product innovation has to be compatible with lifestyle and needs. Moreover, product innovation is necessary to compare with substitutes or former products.

In H3, MS has a slightly positive effect on the PI with a coefficient of 0.23. This means that MS has a little positive relationship with PI. The price, product, place, promotion, attitude, reference group, and consumer culture are slightly influential factors for product innovation. [5] explains that consumers have to be informed about the product advantage and product quality before making a purchasing decision. Reference groups and consumer culture are affected by communication together with trial usage which can increase sales volume. [19] explained MS is a key factor in the DSS of CB. Most consumers make a purchasing decision for a healthy beverage with a reasonable price, product benefit, and a promotion program. Normally, a consumer chooses types of healthy beverages based upon reference group, product trial, and attitude. Consumers purchase the beverages depending on brand reliability, quality expectation, and health benefit.

3.4 Creating a decision support system

The DSS model for CB of elderly Chinese tourists on healthy beverages is divided into 3 parts: Input, Output, and Description of the computer program. The input part provides blocks to fill in a username and a user number. During step 1, the number of their opinion for each independent variable is selected. Details of the numbers are as shown in Table 4. Once the save button is clicked, the output and description of the data appear.

In step 2, the output shows the statistical values of variables such as price to brand, trust, and product expectation of which the beta is 0.310, 0.192, and 0.290, respectively; and the t-value is 5.246, 2.963, and 3.535 respectively.

Table 4: Interpretation of number

No.	Definition
1	Least influence on consumer behavior
2	Less influence on consumer behavior
3	Average influence on consumer behavior
4	More influence on consumer behavior
5	Most influence on consumer behavior

In step 3, the description appears to explain the effect of the independent variable on the dependent variable. According to the above sample, price is the most influential of which the coefficient is 0.310. In contrast, it has the least influence on trust with a coefficient of 0.192.

3.5 Verifying by sensitivity analysis

Uncertainty of the model is studied using sensitivity analysis by apportioning to distinct data sets of uncertainty in the input blocks [24]. The sensitivity analysis forecasts the effects of outputs. A robust model is a focus when comparing the output. The model is robust if the output has no significant change. Conversely, the model is not robust if the output has a significant change.

The sensitivity model is done by estimating new independent variable data sets, which are increased by 20% and, on the other hand, are decreased by 20% from the mean value. Afterwards, the new data sets are input into the SEM and observed for the robustness of the model.

There are eleven independent variables tested in the sensitivity analysis. The result shows price (X1+20%), product (X2+20%, X2–20%), promotion (X3+20%, X3-20%), place (X4+20%, X4–20%), reference group (X6+20%, X6–20%), consumer culture (X7+20%, X7-20%), trial usage (X10+20%), and communication (X11+20%, X11-20%) are robust. Price (X1–20%), attitude (X5+20%, X5-20%), relative advantage (X8+20%, X8-20%), compatibility (X9+20%, X9-20%), and trial usage (X10–20%) are non-robust. The result is shown in Table 5.

Table 5: The result of sensitivity analysis

Variables	Mean Value	
	+20%	–20%
Price (X1)	√	X
Product (X2)	√	√
Promotion (X3)	√	√
Place (X4)	√	√
Attitude (X5)	X	X
Reference group (X6)	√	√
Consumer culture (X7)	√	√
Relative advantage (X8)	X	X
Similarity (X9)	X	X
Trial usage (X10)	√	X
Communication (X11)	√	√

√ = Model is robust
X = Model is non robust

The result expresses that marketing stimuli, price (X1–20%) and attitude (X5+20%, X5-20%), are not robust because they change when the mean values are adjusted at both plus 20% and minus 20%. Accordingly, product innovation: relative advantage (X8+20%, X8-20%), compatibility (X9+20%, X9-20%), and trial usage (X10-20%) are not robust as well. This means that CB changed when the price of products decreased. Similarly, attitudes changed for both increase and decrease due to its subjective norm and market trend. While, the CB is changed when relative advantage and compatibility increase and decrease altogether with decreasing trial usage. Normally, the consumer always compares existing products or substitutes. The compatibility variable concerns lifestyle and needs. When an entrepreneur launches a new product to market, it is necessary to prepare a trial sample for a consumer. If not, then, the consumer would not trust the new product. The consumer who intends to purchase a healthy beverage chooses the beverage based on preferences, product quality, promotion, reference group, consumer culture, and communication. Therefore, an entrepreneur needs to disseminate more information about product advantages to the public and raise consumer perception of healthy beverages. Finally, attitude, relative advantage, and compatibility are of key importance to CB outcomes. The study of [10] and [11] informed that marketing stimuli influence the decision-making of consumers. Moreover, the entrepreneur also has to conduct product innovation, remarkably for relative advantage and compatibility when producing the beverage.

4 Conclusions

The COVID-19 pandemic affected the world economy and every industrial sector, especially the hospitality industry. The tourism and Food & Beverage sectors generate the majority of revenue for this industry. Therefore, entrepreneurs in this sector have to improve their capabilities to gain a competitive advantage. In the post-COVID era, consumer behavior has changed to a new normal, new lifestyle, and new trend. Thus, the study of consumer behavior in the F&B market is unavoidable. SMEs still lack know-how for product development and there is no efficient tool to support them in decision-making. The research objectives are (1) to examine major influential factors for consumer behavior of elderly Chinese tourists on healthy beverages; and (2) to structure a decision support system of consumer behavior toward healthy beverages. The result is a DSS of CB that is applicable to elderly Chinese tourists. The obtained tool helps entrepreneurs become aware of the influential factors of CB on healthy beverages whenever they need to launch a new product to the market.

The research result expressed that marketing stimuli: price, product, promotion, place, attitude, reference group, and consumer culture affect CB, directly and indirectly, affect CB through product innovation. Then, product innovation: relative advantage, compatibility, trail usage, and communication have a slightly direct effect on CB. The CB is composed of the brand, trust, and product expectations. The model is appropriate at a significant level of .05. When sensitivity analysis is applied, the result proves that most variables of the model are robust. An increasing or decreasing attitude, relative advantage, and compatibility, and decreasing price and trial usage of healthy beverages can influence CB, which means changing, increasing and decreasing attitude (norms or trend), information of product advantage, right products for lifestyles or preferences, and decreasing price of product potentially alter CB.

The DSS is an effective tool to assist Chinese tourists' satisfaction, which means that entrepreneurs gain a competitive advantage while increasing profit, promoting sales volume growth rate, and expanding product innovation to meet customer requirements. Afterwards, sensitivity analysis shows that price, attitude, relative advantage, compatibility, and trial usage are influential factors for CB of elderly Chinese tourists on a healthy beverage. When launching a new product, the entrepreneur has to ensure that the consumer's attitude is classified. The relative advantage is a fully informed consumer and a product that resonates with their lifestyle and needs. When launching a product to the market, trial samples have to be provided to the consumers.

On account of the highly competitive market for a healthy beverage, the DSS of healthy beverages for elderly Chinese tourists is an advantageous tool for entrepreneurs especially SMEs business because of its easy accessibility, user-friendliness, and applicability. Thus, it is sustainable development for the business if a DSS model can be successfully developed as it can save time and costs for entrepreneurs in developing marketing strategies and products in response to customer needs.

This research is still limited as the sampling size is too small because what is focused on here are only the healthy beverages produced by SMEs. It is suggested to conduct further study on other groups of consumers, e.g., teenagers or health-conscious persons. Furthermore, the DSS is also applicable to other industries to study influential factors such as energy drinks, healthy food, etc.

Appendix

Table A: Results of In-Depth Interview

Topic	Results
1. Overview of the F&B industry	1) Due to the COVID-19 pandemic, the F&B industry has been impacted by the decreasing global demands, strong competition from newcomers in the market, and strengthening of the Thai baht currency. The F&B sector is sensitive to global trends and changes in consumers preferences. Nowadays, health consciousness becomes more concerned because of the crisis. Consumers are oriented to healthy food and beverages, which brings a good opportunity to the entrepreneurs in this sector. 2) Most experts said that the F&B industry had greatly grown compared to the last two years prior to the COVID-19 crisis. 3) A large number of product innovations with respect to healthy beverages were launched to the market upon consumer requirements. 4) The most important factors of the sector's growth are dynamic urbanization, advanced use of technology, growing e-marketplaces, export orientation, and increased interest in healthy food. 5) Thai F&B products drew attention from foreign countries and became a target market while also enjoying an advantage. 6) Critical points of the Thai F&B industry include dependence on tourism and export markets, food security issues, and organic farming
2. Overview of the tourism industry	1) The tourism industry brings widespread opportunities for employment and business expansions. 2) The tourism industry is a majority sector that contributes to a large amount of Thailand's GDP. 3) Thailand is known as one of the most visited destinations in the world. 4) In the post-COVID era, global tourism trends are more oriented to health and hygiene. Tourists' preferences shift from mass to niche segments; for example, medical and wellness, community-based tourism, and responsible tourism. 5) The technology adoption has forced out the tourism supply chain to enable contactless payment. 6) Tourism demands in the new normal era consist of niche travel, safety-first travel, and digitalization trends.
3. Problems and obstacles	1) Logistics system needs to develop effectively to bring products to foreign markets. 2) Labeling is required by the Food and Drug Administration (FDA), the process of which is time consuming and requires a lot of supporting documents. 3) Consumer trends are changing, and entrepreneurs need to be aware of the movements and innovations of their competitors. 4) Infrastructure has to be improved, both transportation and utilities.
4. Factors influencing consumer behavior	Most experts said that there were several factors e.g., psychological (including motivation, perception, learning, beliefs, and attitudes), personal (including age and life-cycle stage, occupation, economic circumstances, lifestyle, and personality), social (including reference groups, family, roles and status), and cultural factors.
5. Types of healthy beverages for elderly tourists	1) Types of healthy beverages are divided based on consumer needs. There are (1) beverages for therapeutic nutrition with higher calories; (2) beverages that are refreshing and alternative to creamy shakes; (3) beverages to aid recovery from illness, injury, or surgery (4) beverages to minimize blood sugar spikes; (5) beverages for medication programs; and (6) plant-based or vegan beverages. 2) Normally, the elderly are advised to drink still water and to minimize drinking sweetened, carbonated drinks, and/or drinks containing caffeine. 3) Most experts said that healthy beverages consisted of fruit juices, vegetable juices, low-calorie/low-fat/non-fat milk, green tea, mix-berries juice, etc.
8. Packaging	1) Packaging is compact, lightweight, not slippery, and can be comfortably carried on. 2) Label is clearly understandable. 3) Packaging is impressive.
9. Marketing strategies on healthy beverages for elderly Chinese tourists	1) Contact a travel agency. 2) Present choices of healthy beverages at tourist attractions. 3) Present interesting promotions and launch an interesting advertisement. 4) Prepare some samples for testing. 5) Communicate in the Chinese language. 6) Acquire reference groups.

Acknowledgements

This research has received funding support from The National Science, Research and Innovation Fund (NSRF) (Grant No. KMUTNB-FF-65-05).

References

[1] A. Kengpol, T. Pichitkarnkar, K. Elfvengren. A decision support system for consumer behavior of Chinese in-bound tourists on functional beverage: An Empirical Study during COVID-19 with Thailand Sandbox, Applied Science and Engineering Progress, 15, 1 (2022): 5420

[2] J. Ofori, Y. Peggy. Novel technologies for the production of functional foods in Bio-nanotechnology: A Revolution in Food, *Biomedical and Health Sciences*. Chichester, UK: John Wiley & Sons (2013)

[3] W. Yonwikai, Business Development Guidelines to Support Tourism Behaviours of Elders Travelling in Thailand, Dusit Thani College Journal, 13, 2 (2019)

[4] Thai Websites. Tourism Statistics Thailand 2000–2021, [Online]. Available: https://www.thaiwebsites.com/tourism.asp (2021)

[5] P. Kotler, G. Armstrong, *Principles of Marketing*. 16th ed., Essex, UK: Pearson Education Limited, (2016)

[6] B. Chua, H. Kim, S. Lee, H. Han, The role of brand personality, self-congruity, and sensory experience in elucidation sky lounge users' behaviour, Journal of Travel & Tourism Marketing, 36, 3 (2018): 1–14

[7] N. Rubio, J. Oubiña, N. Villaseñor, Brand awareness–Brand quality inference and consumer's risk perception in store brands of food products, Food Quality and Preference, 32 (2014): 289–298

[8] W. Efanny, J. Haryanto, M. Kashif, H. Widyanto, The relationship between marketing mix and retailer-perceived brand equity, IMP Journal, 12, 1 (2018): 192–208

[9] A. Nayeem, S. Raja, R. Shojib, The impact of marketing mix in fascinating customers perception: Case study on online banks in Bangladesh, IEEE-SEM, 8, 1 (2020): 231–241

[10] A. Feil. C. Cyrne, F. Sindelar, J. Barden, M. Dalmoro, Profiles of sustainable food consumption: Consumer behavior toward organic food in southern region of Brazil, Journal of Cleaner Production, 258 (2020): 120690

[11] J. Singh, L. Quamina, T. Xue, Ten million followers and counting: How digital brand alliances between online influencers and brands impact consumer value: An abstract perceptions, in *Proceedings of the Academy of Marketing Science Conference of 2017 (AMSAC 2017)*, (2017): 24– 26

[12] L. Schiffman, L. Kanuk, *Tourist behaviour*. 10th ed. Prentice Hall, (2009)

[13] P. Drucker, *Management challenges for the 21st century*. Harper Collins., (1999)

[14] N. Kanagal, Innovation and product innovation in marketing strategy, Journal of Management and Marketing Research, 18 (2015)

[15] Q. Luo, M. Suacamram, Product Innovation and National Image of Chinese Products in the Eyes of Thai People, SAGE Open, (2022): 1-13

[16] A. Nataya, J. E. Sutanto, The Effect of Product Innovation and Service Innovation towards Marketing Performance (Case Study on Plastic Producer in Surabaya), International Journal of Business and Management Invention, 7, 8 (2018): 61-66

[17] T. Müller, F. Schuberth, J. Henseler, PLS path modeling – A confirmatory approach to study tourism technology and tourist behaviour, Journal of Hospitality and Tourism Technology, 9, 3 (2018): 249–266

[18] L. Sie, K. Phelan, S. Pegg, The interrelationships between self-determined motivations, memorable experiences and overall satisfaction: A case of older Australian educational tourists, Journal of Hospitality and Tourism Technology, 9, 3 (2018): 354–379

[19] W. Wang, The influence of perceived technological congruence of smartphone application and air travel experience on consumers' attitudes toward price change and adoption, Journal of Hospitality and Tourism Technology, 10, 2 (2019): 122–135

[20] R. Pereira, Factors influencing consumer perceptions of Web-based decision support systems, Logistics Information Management, 12, 1/2 (1999): 157–181

[21] A. Massari, F. Manca, F. Girone, Multiple correspondence analysis for customer segmentation in large retail groups, Electronic Journal of Applied Statistical Analysis, 9, 4 (2016): 637– 654

[22] J. F. Hair, *Multivariate Data Analysis,* 7th ed. New Jersey: Pearson Prentice-Hall, (2010)

[23] R.B. Kline, *Principles and Practice of Structural Equation Modeling,* 2nd ed. New York: Gilford Press, (2005)

[24] A. Saltelli, M. Ratto, T. Andres, F. Campolongo, J. Cariboni, D. Gatelli, M. Saisana, S. Tarantola, *Global Sensitivity Analysis: The Primer*. West Sussex, England: John Wiley & Sons Ltd., (2008)

[25] K. Ahalya, B. Sreeya, Effect of Brand Image On Beverages: A Research on Health Drinks In Chennai, International Journal of Innovative Technology and Exploring Engineering (IJITEE), 8,11 (2019): 3167-316

 https://doi.org/10.1051/e3sconf/202235502013

Effect of pepsin on antioxidant and antibacterial activity of protein hydrolysate from salted jellyfish (*Lobonema smithii* and *Rhopilema hispidum*) by-products

Pratchaya Muangrod[1], *Wiriya* Charoenchokpanich[1], *Sittiruk* Roytrakul[2], *Vilai* Rungsardthong[1,3], *Savitri* Vatanyoopaisarn[1,3], *Sawanya* Charoenlappanit[2], *Benjamaporn* Wonganu[4] and *Benjawan* Thumthanaruk[1,3,*]

[1]Department of Agro-Industrial Food and Environmental Technology, Faculty of Applied Science, King Mongkut's University of Technology North Bangkok, Bangkok, Thailand
[2]National Center for Genetic Engineering and Biotechnology, National Science and Technology Development Agency (NSTDA), Pathum Thani, Thailand
[3]Food and Argo-Industry Research Center, King Mongkut's University of Technology North Bangkok, Bangkok 10800, Thailand
[4]Department of Biotechnology, Faculty of Applied Science, King Mongkut's University of Technology North Bangkok, Bangkok, Thailand

Abstract. Protein hydrolysates are products of protein degradation that provide various sizes of peptides and free amino acids. Protein hydrolysate from the different types of enzymes and raw materials provides different bioactivity, such as antioxidant and antibacterial activity. Salted jellyfish by-products have the potential to be a source for protein hydrolysate production because of their low price and having collagen protein. This research aimed to evaluate the antioxidant and antibacterial activity of protein hydrolysates from jellyfish by-products. The dried salted jellyfish by-products from the umbrella and oral arm part of white-type (*Lobonema smithii*) and sand-type (*Rhopilema hispidum*) were desalted and enzymatically hydrolyzed by 5% (w/w) pepsin for 24 h at 37°C. Bioactivity assays showed that the hydrolysate of the oral arms part of white-type jellyfish exhibited the highest antioxidant activity (13.27%). While protein hydrolysate of umbrella part of sand-type jellyfish showed the highest antibacterial activity against *Vibrio parahaemolyticus* up to 13.61%. The results demonstrated that peptic hydrolysate of different types and parts of jellyfish by-products provided different antioxidant or antibacterial activity, thereby increasing the potential uses of jellyfish protein hydrolysate as a functional food.

Keyword. Protein hydrolysate, Salted jellyfish by-products, Antioxidant activity, Antibacterial activity

1 Introduction

Protein hydrolysates are protein degradation products that contain peptides and free amino acids of varying sizes [1, 2]. Protein hydrolysate can be produced by various methods, including acid or alkaline hydrolysis and enzymatic hydrolysis [1, 3]. The amino acid sequence and composition are indicators of the bioactivity of protein. Different bioactivity of protein hydrolysate was received from the different methods of protein hydrolysate production. [1, 4-6].

Enzymatic hydrolysis of proteins is a promising method to produce protein hydrolysate due to the protease enzymes will cleave specific peptide bonds and yield a smaller molecular weight and size of peptides than native proteins [1]. A low molecular weight peptide with around 2 to 20 amino acid residues is highly bioactive [6]. Previous studies reported that protein hydrolysate obtained from enzymatic hydrolysis has higher bioactivity than protein hydrolysate obtained from other methods [7]. Many commercial proteases were used in protein hydrolysate production, such as bromelain, papain, alcalase, collagenase, trypsin, and pepsin [8, 9]. Different protease enzyme types also provided protein hydrolysate with different bioactivity [8].

However, different types of raw material used for protein hydrolysate production also provide different amino acid sequences and amino acid compositions [10-13], which provide different bioactivities [3, 14, 15]. To date, several studies have reported the protein hydrolysate from various food proteins, such as cocoa bean [16], nuts [17], milk [18-20], eggs [21, 22], meat [23, 24], turkey [25], chicken [25], pork [25], beef [25], fish sauce [26], and fish [27]. Different sources of proteins show different bioactivity such as anticancer, antihypertensive, antibacterial, and antioxidant [3, 14, 15]. Antibacterial activity is the ability to reduce or inhibit the growth of bacteria. Some bacteria can cause disease. For example, Gram-negative halophilic bacteria called *Vibrio parahaemolyticus* is the most common cause of food poisoning [28, 29]. Antioxidant activity is considered one of the major biological activities of most protein hydrolysates [3]. The antioxidant activity is the ability to scavenge or reduce free radicals. Free radicals can cause diseases, such as diabetes, arthritis, aging,

* Corresponding author: benjawan.t@sci.kmutnb.ac.th

cancer, and atherosclerosis when the quantity of free radicals exceeds normal [9, 26]. Currently, protein hydrolysate from novel protein sources is of interest for lowering the risk of several diseases due to its antibacterial and antioxidant properties [26]. The use of a marine source has grown in popularity because the protein hydrolysates produced from marine species have been reported as a source of antibacterial and antioxidant peptides [30].

Jellyfish are also a marine species that can be consumed after pickling with salt [31, 32]. However, many salted jellyfish by-products having irregular shapes or broken pieces occur during the various processing steps, such as cleaning, trimming, or sizing. The by-products of salted jellyfish are selected to produce protein hydrolysate because of valuable collagen that can be turned into a functional food peptide ingredient [31, 33]. The benefit of this work is not only for the reduction of unused waste released to environment, it can increase the value of these by products by converting to high-valued food additives. To date, protein hydrolysates from few jellyfish species have been reported to possess antioxidant activities [5, 9, 14], but the antioxidant and antibacterial activity of protein hydrolysate from sand-type (*Rhopilema hispidum*) and white-type jellyfish hydrolyzed with pepsin has not been well discovered. Therefore, this research aimed to evaluate the antibacterial and antioxidant activity of protein hydrolysates from pepsin hydrolysis of jellyfish by-products (*Lobonema smithii* and *Rhopilema hispidum*).

2 Materials and methods

2.1 Preparation of jellyfish protein powder (JPP)

Salted jellyfish by-products, the abnormal shapes or broken pieces (*Lobonema smithii* and *Rhopilema hispidum*) were obtained from Mahachai Food and Trading Co., Ltd., Samut Sakhon, Thailand. The salted jellyfish by-products were stored in polyethylene bags and kept at 10°C until used. The salted jellyfish by-products preparation was prepared as previously described [33-35]. First, salted jellyfish by-products were separated into four samples that are umbrella part of white-type jellyfish (UW), oral arms part of white-type jellyfish (OW), and umbrella part of sand-type jellyfish (US), and oral arms part of sand-type jellyfish (OS). Then, each part of salted jellyfish by-products was washed with tap water at a ratio of 1:40 (w/v), 2 cycles, and 15 min per cycle with a jellyfish washing machine to remove salt and then drained for 30 min. After that, the samples were dried for 24 h at 60°C using a tray dryer (ED 400, Binder, USA). The dried samples were then broken down into tiny pieces and passed through a sieve with a mesh size of 100. The chemical compositions of the jellyfish protein powder, including its moisture, protein, fat, and ash content, were examined. The jellyfish protein powder was stored at room temperature in sealed PE bags until use.

2.2 Preparation of jellyfish protein hydrolysate (JPH)

JPH was prepared according to Chi *et al.* [36] and Lima *et al.* [30] with a slight modification (Figure 1). First, to inactivate the endogenous enzymes, JPP from the umbrella part of white-type jellyfish (UW-JPP), JPP from the oral arms part of white type jellyfish (OW-JPP), JPP from the umbrella part of sand-type jellyfish (US-JPP), and JPP from the oral arms part of sand-type jellyfish (OS-JPP) were added in 0.05 M sodium acetate buffer at 95°C for 10 min (the JPP: buffer ratio of 1: 25 w/v) and then equilibrated to the optimum temperature conditions. Next, the hydrolysis was started by adding pepsin enzyme into the mixture at the enzyme: substrate ratio of 1: 20 g (w/w). The reaction was continued in a shaking incubator (WIS-20R, WiseCube, Korea) at 150 rpm for 24 h at 37°C. After that, the enzyme was inactivated by shaking the mixture at 100 rpm in a temperature-controlled water bath shaker (Memmert, Schwabach, Germany) for 10 min at 95°C. The mixture was centrifuged at 9,500 × g for 30 min. The supernatant of JPH was filtered by Whatman filter paper No 1. Finally, the sample after filtration were stored at −18°C until analysis.

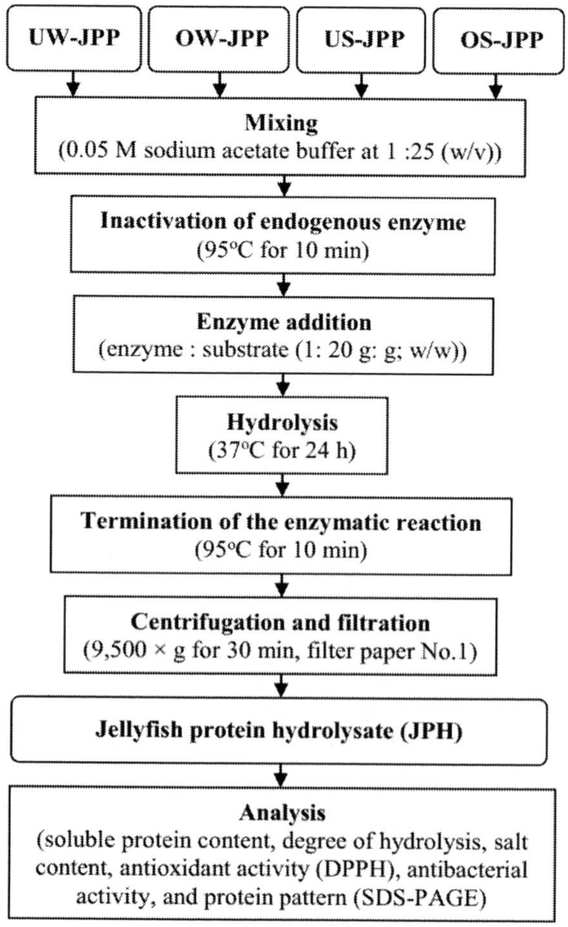

Figure 1. Preparation of jellyfish protein hydrolysate (JPH).

2.3 Analysis

2.3.1 Chemical composition

The chemical composition of JPP was measured according to the AOAC standard methods [37], which included protein, moisture, fat, and ash content. The Kjeldahl method was used to determine the protein content. The nitrogen value was converted to protein using a factor of 5.55 [38].

2.3.2 Soluble protein content

The Lowry method [39] was used to evaluate the soluble protein content of JPH. Bovine serum albumin was chosen as the standard.

2.3.3 Degree of hydrolysis (DH)

The degree of hydrolysis (DH) was measured according to the method of Sathivel *et al.* [40] with a slight modification. Briefly, 5 mL of jellyfish protein hydrolysate (JPH) was combined with 5 mL of 20% trichloroacetic acid (TCA) to obtain 10% TCA-soluble nitrogen and then centrifuged at 5000 × g at 25°C for 15 min. The Kjeldahl method was used to calculate the total nitrogen in JPH as well as the soluble nitrogen in the supernatant of 10% TCA-soluble nitrogen. The DH was calculated according to the following equation (1):

$$DH\,(\%) = [10\%\ TCA\text{-}soluble\ N\ /\ Total\ N] \times 100 \qquad (1)$$

where 10% TCA soluble N = the amount of nitrogen in 10% TCA-soluble nitrogen, and Total N = the amount of nitrogen in the protein hydrolysate solution.

2.3.4 Sodium chloride content

The sodium chloride (NaCl) content of jellyfish protein hydrolysate was measured using a conductivity meter (TDS Meter 308, Systronics, India) [35]. The percentage of NaCl in jellyfish protein hydrolysate was extrapolated from the standard curve of NaCl.

2.3.5 Antioxidant activity (DPPH radical scavenging assay)

DPPH radical scavenging activity was measured according to the method of Khositanon *et al.* [26] and Molyneux [41] with slight modifications. Briefly, 10 µL of jellyfish protein hydrolysate was combined with 190 µL of 0.1 mmol/L DPPH in 70% ethanol. Then, the reaction mixture was left in the dark at room temperature for 30 min. After that, the absorbance of the reaction was analyzed at 492 nm using a microplate reader (M 965 plus, Metertech, Taiwan). Ascorbic acid was used as a positive control in this experiment. The %inhibition of DPPH radical scavenging activity was calculated by the following equation (2):

$$\%inhibition = [(A_{control} - A_{sample}) / A_{control}] \times 100 \qquad (2)$$

where $A_{control}$ = the absorbance of the control that contained all reagents except the test samples. A_{sample} = the absorbance of the JPH with reagents added (0.1 mmol/L DPPH in 70% ethanol).

2.3.6 Antibacterial activity

Antibacterial activity of JPH samples was determined against pathogenic bacteria (*Vibrio parahaemolyticus*) according to the method of Muhialdin *et al.* [42] with a slight modification. Briefly, *Vibrio parahaemolyticus* was grown on tryptic soy agar for 24 h at 37°C. After that, the colony of *Vibrio parahaemolyticus* was transferred to the tryptic soy broth (TSB) and incubated at 37°C for 5 h at 200 rpm. A 20 µL of JPH (0.5 mg/mL) was placed in 96 well microtiter plates, and 80 µL of TSB containing *Vibrio parahaemolyticus* was transferred to the well. Then, the plate was agitated at 37°C for 6 h at 200 rpm. The microplate reader (M 965 plus, Metertech, Taiwan) was used to determine the absorbance of the sample. The following equation (3) was used to compute the growth inhibition percentage.

$$\%inhibition = [(OD_{control} - OD_{sample}) / OD_{control}] \times 100 \quad (3)$$

where $OD_{control}$ = the absorbance of the control that contained all reagents except the test samples. OD_{sample} = the absorbance of the JPH with reagents added.

2.3.7 Sodium dodecyl sulfate-polyacrylamide gel electrophoresis (SDS-PAGE)

The protein pattern of JPH was determined using the sodium dodecyl sulfate-polyacrylamide gel electrophoresis (SDS-PAGE), which was slightly modified from Laemmli method [43]. The separating gel (12%) and stacking gel (5%) were used to perform the polyacrylamide gel. The JPH sample was dissolved in a sample buffer (SDS 0.9 g, glycerol 3 mL, bromophenol blue 1 mg, 0.5 M Tris-hydroxymethyl-aminomethane 3.9 mL, β-Mercaptoethanol 1.5 mL, distilled water 1.6 mL) and then boiled for 3 min. The protein molecular weight marker (PageRulerTM Plus Prestained protein ladder, Thermo Fisher Scientific) was used to evaluate the molecular weight of the JPH. Electrophoresis was conducted using the polyacrylamide gel electrophoresis machine (Mini-slab size electrophoresis system, ATTO, Japan) at 50 V. After electrophoresis, gels were stained with silver stained.

2.4 Statistical analysis

All experiments were carried out in triplicate. The data was displayed as mean and standard deviation. The statistical program for social science (SPSS) 22.0 was used to conduct an analysis of variance (ANOVA) on the data and to perform Duncan's multiple range test to ascertain whether there was a significant difference (p 0.05).

3 Results and discussions

3.1 Chemical composition of JPP

Proximate composition of jellyfish protein powder from the umbrella part of white-type jellyfish (UW), oral arms part of white-type jellyfish (OW), umbrella part of sand-type jellyfish (US), and oral arms part of sand-type jellyfish (OS) are shown in Table 1. The UW-JPP and OW-JPP had 74.31-74.80% protein, 8.97-9.45% moisture, 1.31-1.32% fat, and 15.11-15.28% ash. There was a slightly different between the oral arm and umbrella part of white-type jellyfish powder. These results were similar to other studies, which reported that the JPP had protein, moisture, fat, and ash content of 76.41-76.59%, 7.42-7.69%, 1.32-1.35%, and 15.54-15.76% [33, 35]. The US-JPP and OS-JPP had 75.80-75.86% protein, 8.01-8.42% moisture, 1.27-1.29% fat, and 16.44-16.72% ash. However, the proximate composition of the oral arm and umbrella part of sand-type jellyfish powder showed the difference between the proximate composition of the oral arm and the umbrella part of white-type jellyfish powder. The JPP from each part of two salted jellyfish species, including white-type jellyfish and sand-type jellyfish, had high protein content with low fat, revealing the efficiency of producing jellyfish protein hydrolysate.

Table 1. Chemical composition of jellyfish protein powder.

Sample	Chemical composition (%)			
	Protein*	Moisture*	Fat*	Ash*
UW-JPP	74.31±0.50^{b}	8.97±0.14^{b}	1.32±0.00^{a}	15.28±0.11^{b}
OW-JPP	74.80±0.15^{b}	9.45±0.12^{a}	1.31±0.00^{ab}	15.11±0.09^{b}
US-JPP	75.80±0.19^{a}	8.01±0.13^{c}	1.29±0.01^{b}	16.44±0.33^{a}
OS-JPP	75.86±0.25^{a}	8.42±0.17^{d}	1.27±0.01^{c}	16.72±0.36^{a}

*Different superscripts (a, b, c, and d) in the same column mean a significant difference in value ($p<0.05$)

3.2 Soluble protein content of JPH

The soluble protein content of jellyfish protein hydrolysate is shown in Table 2. During the hydrolysis, the pepsin enzyme digested jellyfish collagen, resulting in hydrolyzed jellyfish proteins, peptides and amino acids. The soluble protein contents of the umbrella part of white-type jellyfish protein hydrolysate (UWJPH), oral arms part of white-type jellyfish protein hydrolysate (OWJPH), umbrella part of sand-type jellyfish protein hydrolysate (USJPH), and oral arms part of sand-type jellyfish protein hydrolysate (USJPH) samples were 100.69-115.64 μg/μL. Higher soluble protein content was recorded for enzymatic pepsin hydrolyzed jellyfish from sand-type jellyfish than white-type jellyfish. These results were lower than the previous study, which reported the soluble protein of enzymatic pepsin hydrolyzed jellyfish by-product at 37°C for 24 h was 1,589.30 μg/μL [35] due to raw material, collagen proteins content, and the activity of the enzyme pepsin used in hydrolysis.

Table 2. Soluble protein content and degree of hydrolysis of jellyfish protein hydrolysate.

Sample	Soluble protein content* (μg/μL)	Degree of hydrolysis* (%)
UWJPH	100.69±0.59^{d}	47.96±0.47^{a}
OWJPH	103.74±0.85^{c}	46.98±0.55^{b}
USJPH	143.45±0.28^{a}	40.47±0.21^{d}
OSJPH	115.64±0.59^{b}	41.78±0.25^{c}

*Different superscripts (a, b, c, and d) in the same column mean a significant difference in value ($p<0.05$)

3.3 Degree of hydrolysis (DH) of JPH

The DH value is used to determine how completely pepsin hydrolyzed jellyfish protein. In this research, the degrees of hydrolysis of UWJPH, OWJPH, USJPH, and OSJPH samples were 47.96%, 46.98%, 40.47%, and 41.78%, respectively (Table 2). The peptide bonds of a jellyfish protein were hydrolyzed during pepsin hydrolysis, resulting in a free amino acid and a short-chain peptide [44], depending on hydrolysis parameters, such as temperature, hydrolysis time, pepsin concentration, and even source of protein. In this study, all of the jellyfish protein powder samples were hydrolyzed with 5% pepsin at 37°C for 24 h. Thus, the DH values of all JPH samples were similar. This result was quite different from the previous studies, DHs of JPH samples hydrolyzed with pepsin for 6, 12, 18, and 24 h at 37°C were 52.20-55.41% [35].

3.4 Sodium chloride content of JPH

The sodium chloride contents and conductivity values of UWJPH, OWJPH, USJPH, and OSJPH samples are presented in Table 3. The sodium chloride content and conductivity values of UWJPH, OWJPH, USJPH, and OSJPH samples were 0.05-0.10% and 1.25-2.18 ms/cm, respectively. These results were not different from previous study, which reported that desalted jellyfish by-products hydrolyzed with pepsin for 6, 12, 18, and 24 h at 37°C, showed conductivity values in the range of 1.85-1.99 ms/cm, and all jellyfish protein hydrolysate contained about 0.09% sodium chloride. [35]. The conductivity value and salt content of UWJPH, OWJPH, USJPH, and OSJPH were significantly different. The difference in salting procedures used by salted jellyfish producers might yield difference in conductivity values and salt content in each type of jellyfish and each part of jellyfish. Thus, desalted jellyfish protein hydrolysates had a different salt content.

3.5 Antioxidant activity of JPH

DPPH is a radical that is often used to indicate the antioxidant activity of the sample [7, 45]. The hydrolysis conditions have a significant impact on the protein hydrolysate's antioxidant properties (hydrolysis time and temperature), amino acid composition and sequence, and protease enzyme (types, activity, and concentration) [7]. In this study, the values of % DPPH radical scavenging activity of UWJPH, OWJPH, USJPH, and OSJPH were 8.13±0.01%, 13.27±0.46%, 8.40±0.46%, and 10.02±0.46%, respectively (Figure 2). These results were slightly different from the previous study that hydrolyzed jellyfish protein with trypsin for 9 h at 50°C [9] and pepsin for 6-24 h at 37°C [35] because of different raw materials and enzymes (types, concentrations, and activity) provided protein hydrolysate with different amino acid sequences and compositions [7]. These differences result in different bioactivities of JPH. However, the results showed that all JPH were capable of scavenging DPPH radicals.

Table 3. Conductivity value and salt content of jellyfish protein hydrolysate.

Sample	Conductivity* (ms/cm)	Salt content* (%)
UWJPH	1.41±0.01^d	0.06±0.00^c
OWJPH	1.25±0.01^c	0.05±0.00^d
USJPH	1.81±0.01^b	0.08±0.00^b
OSJPH	2.18±0.01^a	0.10±0.00^a

*Different superscripts (a, b, c, and d) in the same column mean a significant difference in value (p<0.05)

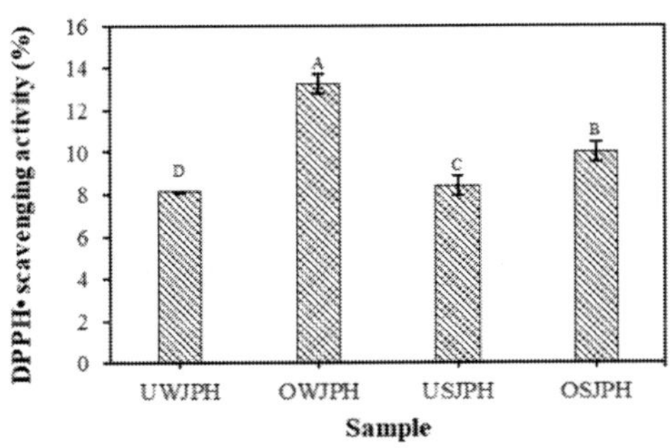

Figure 2. DPPH radical scavenging activity (%) of jellyfish protein hydrolysate (JPH); bars with the different superscripts were significantly different (p<0.05).

3.6 Antibacterial activity of JPH

The antibacterial activity of the jellyfish protein hydrolysate samples was evaluated through the growth inhibition activity against pathogenic bacteria (*Vibrio parahaemolyticus*). The results showed that the antibacterial activity of UWJPH, OWJPH, USJPH, and OSJPH samples were 10.53±0.80%, 8.31±0.27%, 13.61±0.25%, and 7.28±0.32%, respectively (Figure 3). These results were quite different from the previous research, which found that the hydrolysates possessing a higher DH provided higher antibacterial activity [30]. However, all JPH samples were able to inhibit the growth of pathogenic bacteria.

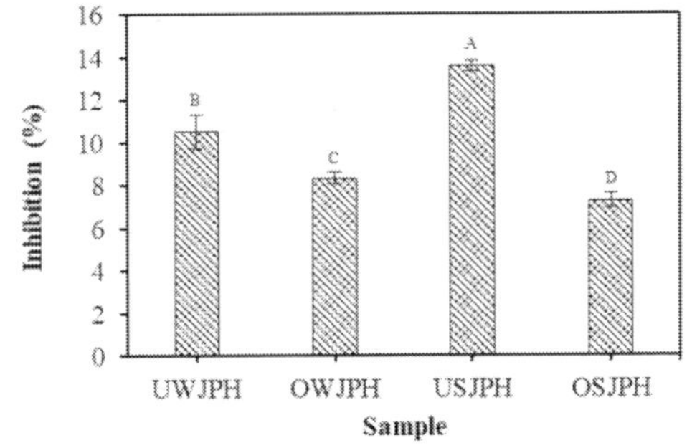

Figure 3. Antibacterial activity (% inhibition) of jellyfish protein hydrolysate (JPH); bars with the different superscripts were significantly different (p<0.05).

3.7 Protein pattern of JPH

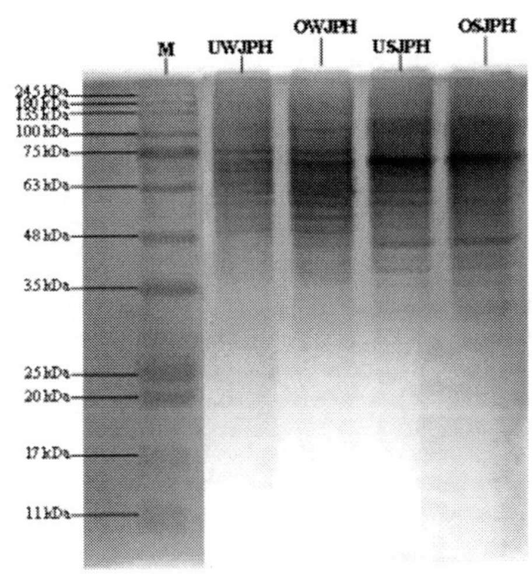

Figure 4. Protein pattern of jellyfish protein hydrolysate (JPH); molecular weight standard (M), umbrella part of white-type jellyfish protein hydrolysate (UWJPH), oral arms part of white-type jellyfish protein hydrolysate (OWJPH), umbrella part of sand-type jellyfish protein hydrolysate (USJPH), and oral arms part of sand-type jellyfish protein hydrolysate (OSJPH)

The protein pattern of pepsin hydrolyzed jellyfish protein is presented in Figure 4. The jellyfish protein hydrolysate from the same species of jellyfish showed a similar protein pattern, while the protein pattern of hydrolysates from white-type jellyfish and sand-type jellyfish were different. The jellyfish protein hydrolysate from sand-type jellyfish contained smaller peptides than that of hydrolysate from white-type jellyfish. Low-molecular-weight protein fragments affect enzymatic pepsin hydrolyzed jellyfish's antioxidant and antibacterial activities. The low-molecular-weight protein or small peptides from protein hydrolysate possessed more potent bioactivity than the high-molecular-weight protein [26]. Collagen's interchain cross-links were hydrolyzed during extraction. In this study, all JPHs had a molecular weight higher than 11 kDa. Thus, the JPH samples exhibited weak antioxidant and antibacterial activity. Protein size was influenced by raw material, hydrolysis process, enzyme activity, and extraction time. Therefore, factors of increased hydrolysis time, increased enzyme concentration, and protein hydrolysate purification, are needed to improve the bioactivity of jellyfish protein hydrolysate.

4 Conclusions

All jellyfish protein hydrolysate samples had the ability to scavenge DPPH radicals and to inhibit the growth of *Vibrio parahaemolyticus*. These findings increase the value of salted jellyfish by-products while reducing waste in the salted jellyfish industry. The oral arms part of white-type jellyfish protein hydrolysate (OWJPH) exhibited the highest antioxidant activity, while the umbrella part of sand-type jellyfish protein hydrolysate (USJPH) provided the highest antibacterial activity. The results demonstrated that different types and parts of jellyfish by-products provided different bioactivities.

Acknowledgment

This research was funded by King Mongkut's University of Technology North Bangkok and National Science and Technology Development Agency, Thailand (Contract no. B.020-2563), and the National Research Council of Thailand through the NRCT Senior Research Scholar Program (Contract no. 814-2020). The authors would like to thank Mahachai Food and Trading Co., Ltd., Samut Sakhon, Thailand, for supporting salted jellyfish by-products.

References

1. S. Severin, W.S. Xia, Enzymatic hydrolysis of whey proteins by two different proteases and their effect on the functional properties of resulting protein hydrolysates, Journal of Food Biochemistry, 30(1) (2006): 77-97
2. T. Julmanlik, S. Kongruang, Functional properties and applications of egg white protein hydrolysates, Journal of Food Technology of Siam University, 14(1) (2019): 69-87
3. K.M.I. Bashir, J.H. Sohn, J.S. Kim, J.S. Choi, Identification and characterization of novel antioxidant peptides from mackerel (Scomber japonicus) muscle protein hydrolysates, Food Chemistry, 323 (2020): 126809
4. P. Ambigaipalan, F. Shahidi, Bioactive peptides from shrimp shell processing discards: Antioxidant and biological activities, Journal of Functional Foods, 34 (2017): 7-17
5. X. Guo, X. Sun, Y. Zhang, R. Wang, X. Yan, Interactions between soy protein hydrolyzates and wheat proteins in noodle making dough, Food Chemistry, 245 (2018): 500-507
6. N. Ktari, R.B.S.B. Salem, I. Bkhairia, S.B. Slima, R. Nasri, R.B. Salah, M. Nasri, Functional properties and biological activities of peptides from zebra blenny protein hydrolysates fractionated using ultrafiltration, Food Bioscience, 34 (2020): 100539
7. S. Silaprueng, B. Thumthanaruk, P. Wongsa-ngasri, Comparative functional properties of jellyfish (Lobonema smithii) protein hydrolysate as influenced by bromelain and hydrochloric acid, Journal of Food Science and Agricultural Technology, 1 (2015): 171-176
8. G. Guillén, M.E. López Caballero, A. Alemán, A.L.D. Lacey, B. Giménez, P. Montero García, in Sea by-Products as Real Material: New Ways of Application, edited by E.L. Bihan (Transworld Research Network, Trivandrum, (2010): 89-115
9. Z. Barzideh, A.A. Latiff, C.Y. Gan, M. Abedin, A.K. Alias, ACE inhibitory and antioxidant activities of collagen hydrolysates from the ribbon jellyfish (Chrysaora sp.). Food Technology and Biotechnology, 52(4) (2014): 495-504
10. H. Korhonen, A. Pihlanto, Bioactive peptides: production and functionality, International Dairy Journal, 16(9) (2006): 945-960
11. K. Fields, T.J. Falla, K. Rodan, L. Bush, Bioactive peptides: signaling the future, Journal of Cosmetic Dermatology, 8(1) (2009): 8-13
12. E. Maestri, M. Marmiroli, N. Marmiroli, Bioactive peptides in plant-derived foodstuffs, Journal of Proteomics, 147 (2016): 140-155
13. H.M. Rawel, G. Huschek, S.T. Sagu, T. Homann, Cocoa bean proteins-characterization, changes and modifications due to ripening and post-harvest processing, Nutrients, 11(2) (2019): 428
14. Y.L. Zhuang, X. Zhao, B.F. Li, Optimization of antioxidant activity by response surface methodology in hydrolysates of jellyfish (Rhopilema esculentum) umbrella collagen, Journal of Zhejiang University Science B, 10(8) (2009): 572-579
15. B. Li, F. Chen, X. Wang, B. Ji, Y. Wu, Isolation and identification of antioxidative peptides from porcine collagen hydrolysate by consecutive chromatography and electrospray ionization–mass spectrometry, Food Chemistry, 102(4) (2007): 1135-1143
16. L.A. Domínguez-Pérez, L.M. Beltrán-Barrientos, A.F. González-Córdova, A. Hernández-Mendoza, B. Vallejo-Cordoba, Artisanal cocoa bean fermentation: From cocoa bean proteins to bioactive peptides with potential health benefits, Journal of Functional Foods, 73 (2020): 104134
17. J. Sealey-Voyksner, J. Zweigenbaum, R. Voyksner, Discovery of highly conserved unique peanut and tree nut peptides by LC–MS/MS for multi-allergen detection, Food Chemistry, 194 (2016): 201-211
18. F. Tonolo, A. Folda, L. Cesaro, V. Scalcon, O. Marin, S. Ferro, A. Bindoli, M.P. Rigobello, Milk-

derived bioactive peptides exhibit antioxidant activity through the Keap1-Nrf2 signaling pathway, Journal of Functional Foods, 64 (2020): 103696

19. M. Amorim, J.O. Pereira, L.B. Silva, R.C.S.C. Ormenese, M.T.B. Pacheco, M. Pintado, M., Use of whey peptide fraction in coated cashew nut as functional ingredient and salt replacer, Lebensmittel-Wissenschaft & Technologie, 92 (2018): 204-211
20. C. Zhao, T.J. Ashaolu, Bioactivity and safety of whey peptides, Lebensmittel-Wissenschaft & Technologie, 134 (2020): 109935
21. J. Zheng, T. Bu, L. Liu, G. He, S. Li, J. Wu, Naturally occurring low molecular peptides identified in egg white show antioxidant activity, Food Research International, 138 (2020): 109766
22. S. Moreno-Fernández, M. Garcés-Rimón, M. Miguel, Egg-derived peptides and hydrolysates: a new bioactive treasure for cardiometabolic diseases, Trends in Food Science & Technology, 104 (2020): 208-218
23. L. Mora, M. Reig, F. Toldrá, Bioactive peptides generated from meat industry by-products, Food Research International, 65 (2014): 344-349
24. T. Sayd, C. Dufour, C. Chambon, C. Buffière, D. Remond, V. Sante-Lhoutellier, Combined in vivo and in silico approaches for predicting the release of bioactive peptides from meat digestion, Food Chemistry, 249 (2018): 111-118
25. S. Martini, A. Conte, D. Tagliazucchi, Comparative peptidomic profile and bioactivities of cooked beef, pork, chicken and turkey meat after In vitro gastro-intestinal digestion, Journal of Proteomics, 208 (2019): 1-10
26. P. Khositanon, N. Panya, S. Roytrakul, S. Krobthong, S. Chanroj, W. Choksawangkarn, Effects of fermentation periods on antioxidant and angiotensin I-converting enzyme inhibitory activities of peptides from fish sauce by-products, Lebensmittel-Wissenschaft & Technologie, 135 (2021): 110122
27. A. Cerrato, A.L. Capriotti, F. Capuano, C. Cavaliere, A.M.I. Montone, C.M. Montone, S. Piovesana, Z. Chiozzi, A. Laganà, Identification and antimicrobial activity of medium-sized and short peptides from yellowfin tuna (Thunnus albacares) simulated gastrointestinal digestion, Foods, 9 (2020): 1-12
28. Y. Geng, K. Tan, L. Liu, X.X. Sun, B. Zhao, J. Wang, Development and evaluation of a rapid and sensitive RPA assay for specific detection of Vibrio parahaemolyticus in seafood, BMC Microbiology, 19(1) (2019): 1-9
29. Q. Fan, Y. Yuan, T. Zhang, W. Song, Q. Sheng, T. Yue, Inhibitory effects of lactobionic acid on Vibrio parahaemolyticus planktonic cells and biofilms, Food Microbiology, 103 (2022): 103963
30. K.O. Lima, C.D.C.D. Quadros, M.D. Rocha, J.T.J.G.D. Lacerda, M.A. Juliano, M. Dias, M.A. Mendes, C. Prentice, Bioactivity and bioaccessibility of protein hydrolyzates from industrial byproducts of Stripped weakfish (Cynoscion guatucupa), Lebensmittel-Wissenschaft & Technologie, 111 (2019): 408-413
31. Y.H.P. Hsieh, F.M. Leong, J. Rudloe, Jellyfish as food, Hydrobiologia, 451 (2001): 11-17
32. M.T. Pedersen, J.R. Brewer, L. Duelund, P.L. Hansen, On the gastrophysics of jellyfish preparation, International Journal of Gastronomy and Food Science, 9 (2017): 34-38
33. P. Muangrod, V. Rungsardthong, S. Vatanyoopaisarn, Y. Tamaki, E. Kuraya, B. Thumthanaruk, Effect of wash cycle on physical and chemical properties of rehydrated jellyfish by-products and jellyfish protein powder, Science, Engineering and Health Studies, 15 (2021): 21030004-21030004
34. W. Charoenchokpanich, V. Rungsardthong, S. Vatanyoopaisarn, B. Thumthanaruk, Y. Tamaki, Salt reduction in salted jellyfish (Lobonema smithii) using a mechanical washing machine, Science, Engineering and Health Studies, 14 (2020): 184-192
35. P. Muangrod, W. Charoenchokpanich, V. Rungsardthong, S. Vatanyoopaisarn, B. Wonganu, S. Roytrakul, B. Thumthanaruk, Effect of pepsin hydrolysis on antioxidant activity of jellyfish protein hydrolysate, E3S Web of Conferences, 302 (2021): 02010
36. C.F. Chi, F.Y. Hu, B. Wang, X.J. Ren, S.G. Deng, C.W. Wu, Purification and characterization of three antioxidant peptides from protein hydrolyzate of croceine croaker (Pseudosciaena crocea) muscle, Food Chemistry, 168 (2015): 662-667
37. AOAC, Official Methods of Analysis of AOAC International, AOAC International, USA (2000)
38. C.S. James, Analytical Chemistry of Foods, Springer, London (1995)
39. O.H. Lowry, N.J. Rosebrough, A.L. Farr, R.J. Randall, Protein measurement with the Folin phenol reagent, Journal of Biological Chemistry, 193 (1951): 265-275
40. S. Sathivel, P.J. Bechtel, J. Babbitt, S. Smiley, C. Crapo, K.D. Reppond, W. Prinyawiwatkul, Biochemical and functional properties of herring (Clupea harengus) byproduct hydrolysates. Journal of Food Science, 68(7) (2003): 2196-2200
41. P. Molyneux, The use of stable free radical diphenylpicrylhydrazyl (DPPH) for estimating antioxidant activity, Songklanakarin Journal of Science and Technology, 26 (2004): 211–219
42. B.J. Muhialdin, N.F.A. Rani, A.S.M. Hussin, Identification of antioxidant and antibacterial activities for the bioactive peptides generated from bitter beans (Parkia speciosa) via boiling and fermentation processes, Lebensmittel-Wissenschaft & Technologie, 131 (2020): 109776

43. U.K. Laemmli, Cleavage of structural proteins during the assembly of the head of bacteriophage T4, Nature, 227(5259) (1970): 680-685
44. I. Batish, D. Brits, P. Valencia, C. Miyai, S. Rafeeq, Y. Xu, M. Galanopoulos, E. Sismour, R. Ovissipour, Effects of enzymatic hydrolysis on the functional properties, antioxidant activity and protein structure of black soldier fly (Hermetia illucens) protein, Insects, 11(876) (2020): 1-12
45. K. Shimada, K. Fujikawa, K. Yahara, T. Nakamura, Antioxidative properties of xanthan on the autoxidation of soybean oil in cyclodextrin emulsion, Journal of Agricultural and Food Chemistry, 40(6) (1992): 945-948

Influence of extraction times on physical and functional properties of gelatin from salted jellyfish by-products

Wiriya Charoenchokpanich[1], *Pratchaya* Muangrod[1], *Sittiruk* Roytrakul[2], *Vilai* Rungsardthong[1,3], *Savitri* Vatanyoopaisarn[1,3], *Benjamaporn* Wonganu[4] and *Benjawan* Thumthanaruk[1,3,*]

[1]Department of Agro-Industrial Food and Environmental Technology, Faculty of Applied Science, King Mongkut's University of Technology North Bangkok, Bangkok, Thailand
[2]National Center for Genetic Engineering and Biotechnology, National Science and Technology Development Agency (NSTDA), Pathum Thani, Thailand
[3]Food and Argo-Industry Research Center, King Mongkut's University of Technology North Bangkok, Bangkok, Thailand
[4]Department of Biotechnology, Faculty of Applied Science, King Mongkut's University of Technology North Bangkok, Bangkok, Thailand

Abstract. By-products of the marine industry have gained attention for producing valuable food ingredients like gelatin, which might benefit food applications and decrease food waste. Gelatin is the only protein-based food hydrocolloid, mainly used for gelling, viscosity, or emulsifying in the food industry. So far, a number of researchers have reported that by-products of salted jellyfish can produce jellyfish gelatin. The quality of jellyfish gelatin gel depends on several factors including hydrochloric acid pretreatment, extraction temperature, and extraction time. However, the functional properties such as foaming and emulsifying of jellyfish gelatin are not well understood. This research was aimed at investigating the hydrochloric acid pretreatment effect of extraction times (12, 24, and 48 h) at 60 °C on the resulting gelatin's yield, physical, and functional properties. Results showed that jellyfish gelatin's yield, gel strength, and viscosity significantly increased with increasing extraction times. Jellyfish gelatin yields were 2.74-14.07%. The gel strength of jellyfish gelatin extracted for 48 h (325.97±2.84 g) was higher than that of jellyfish gelatins extracted for 12 h (210.46±3.97 g) and 24 h (261.60±3.25 g). All jellyfish gelatins can form gels at 4 °C. Viscosity values of jellyfish gelatin were 23.00-24.50 centipoise. The foaming capacity and foaming stability of jellyfish gelatin were 12.28-17.54% and 10.52-15.78%, respectively. The emulsification activity index of jellyfish gelatin was 13.11-13.30 m^2/g, and the emulsification stability index was 39.19-56.42%. As a result, varied gelatin extraction periods influenced jellyfish gelatin's physical and functional properties, indicating that the extended extraction time of 48 h delivered the jellyfish gelatin that can be used as a foaming and emulsifying agent. Therefore, turning the jellyfish by-products into food ingredients like gelatin would increase product values and potential uses in the food and medical applications.

Keyword. Salted jellyfish by-products, Gelatin, Extraction times, Physical properties, Functional properties

1 Introduction

Gelatin is a natural biopolymer derived from the hydrolysis of collagen-producing from acid-base or enzymatic methods. Due to its functional properties, gelatin has been extensively applied in the food, pharmaceutical, medical, and cosmetic industries. Grand View Research [1] reported that by 2027, the global gelatin market would be expected to reach 6.67 billion US dollars. Around 90% of gelatin production on the market is mainly produced from the bone, ligaments, and skin of bovine and porcine, which is higher than gelatin extraction from marine sources [2-3]. However, gelatin production from these mammalian animals has faced certain religious constraints [4]. Therefore, alternative gelatin, especially marine gelatin, has gradually attracted attention in food and pharmaceutical applications. Furthermore, there are no biological contaminants in marine gelatin, enough to approve product safety.

Gel strength, viscosity, and setting time are the three important functional properties of gelatin [5]. However, typical marine gelatins have lower gel characteristics. Thus, improves marine gelatins comparable to mammalian gelatin sources are needed. Marine gelatin is mainly produced from by-products to decrease food waste and the functional qualities of marine gelatin. Several researchers have reported gelatin extracted from marine by-products such as cold-water fish (chum salmon [6]), warm-water fish (cobia [7], pangasius catfish [8], and tilapia [9-10]), and jellyfish [11-12].

Thailand is one of the top exporters of edible jellyfish, especially jellyfish (*Lobonema smithii*), generally exported in salted jellyfish form to China, Japan, and South Korea. The Ministry of Commerce [13] report presented exports of salted jellyfish in 2021 with a

* Corresponding author: benjawan.t@sci.kmutnb.ac.th

https://doi.org/10.1051/e3sconf/202235502014

value of greater than 559 million baht. Many salted jellyfish by-products occur at processing steps, with irregularly shaped and broken pieces, which have gained increasing attention for gelatin extraction research. The low price of salted jellyfish by-products but rich in collagen protein with minimal fat content [14-18] have potentially been a source for gelatin extraction. The attempts to improve jellyfish gelatin by varying hydrochloric acid (HCl) pretreatment conditions (concentration and time), extraction temperature, and extraction time [11-12, 15-16] have been reported. Charoenchokpanich *et al.* [12] reported that the condition of 2 h HCl pretreatment at 25 °C and 3 h water extraction at 60 °C was insufficient for the jellyfish gelatin process and delivered the low gel strength of 214.47 g. Research on the functional properties of jellyfish gelatin produced from salted jellyfish by-products has been limited. Therefore, this research aimed to investigate the duration of hot water extraction (at 60 °C) varying from 12, 24, and 48 h on jellyfish gelatin's yield, gel properties, and functional properties. The benefit of this work not only provides alternative raw material for gelatin, but also reduce the food waste production as well.

2 Materials and methods

2.1 Preparation of jellyfish by-products gelatin

The process diagram for producing jellyfish by-products gelatin is shown in Fig. 1. The salted jellyfish (*Lobonema smithii*) by-products were received from Mahachai Food and Trading Co., Ltd., Samut Sakhon, Thailand. The salted jellyfish by-products were kept at 10 °C and stored in polyethylene (PE) bags until use. A mechanically washing machine briefly performed the wash process at 2 cycles for 15 min per cycle and then drained for 10 min [14, 18].

For gelatin extraction, jellyfish by-products gelatin was prepared according to Lueyot *et al.* [11] and Charoenchokpanich *et al.* [12] with a slight modification. Firstly, the desalted by-products (1 kg) were agitated at 150 rpm and 4 °C for 2 h with 0.5 M sodium hydroxide (NaOH) solution (1:3, w/v) in a shaking incubator (WIS-20R, WiseCube, Korea). The alkali-treated sample was washed with a mechanically washing machine at 3 cycles for 15 min per cycle and then shaken in 0.2 M HCl (1:2, w/v) at a speed of 150 rpm for 1 h at 25 °C. The acid-treated sample was thoroughly washed with a mechanically washing machine at 3 cycles for 15 min per cycle or until neutral pH. The sample was then extracted with distilled water (1:2, w/v) at 60 °C for 12, 24, and 48 h at a speed of 80 rpm in a temperature-controlled water bath shaker (Memmert, Schwabach, Germany) and filtered using a Buchner funnel with Whatman No. 4 filter paper. Finally, the sample was dried with a freeze dryer (GFD5L, Grisrianthong, Thailand). The freeze-dried jellyfish by-products gelatins of JFG12, JFG24, and JFG48 were kept in an aluminum foil bag and a desiccator.

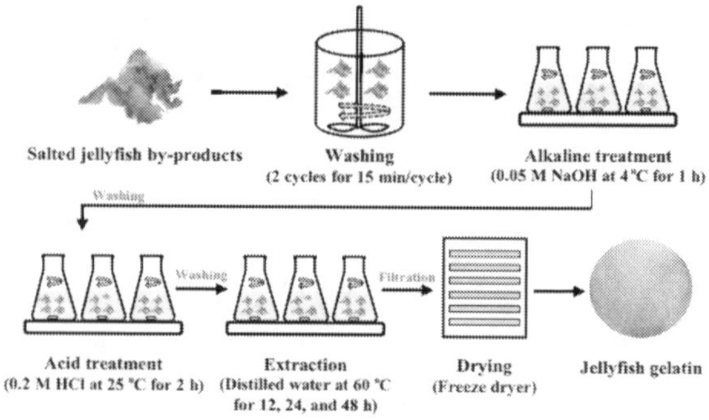

Fig. 1. Process diagram for the preparation of gelatin from salted jellyfish by-products.

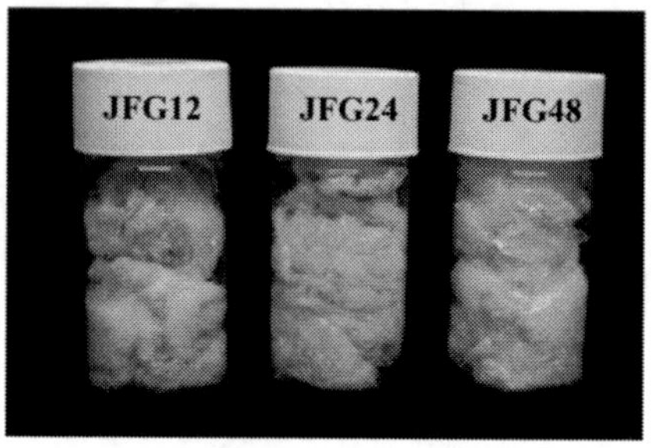

Fig. 2. Freeze-dried jellyfish by-products gelatins at various extraction times of 12 (JFG12), 24 (JFG24), and 48 h (JFG48).

2.2 Analysis

2.2.1 Yield

Each freeze-dried jellyfish by-product gelatin sample was weighed and calculated for the yield. The yield was derived from the ratio of the freeze-dried gelatin weight (M_2) to the freeze-dried, salted jellyfish by-products weight (M_1) and calculated using the following equation (1):

$$Yield\,(\%)\,(\text{dry weight basis}) = (M_2 / M_1) \times 100 \quad (1)$$

2.2.2 Salt (sodium chloride) content

The salt content of the gelatin solution was measured using a conductivity meter (TDS Meter 308, Systronics, India), and the calculated value was extrapolated from the standard sodium chloride 0-1,000 ppm graph [12].

2.2.3 Sodium dodecyl sulfate-polyacrylamide gel electrophoresis (SDS-PAGE)

The preparation of jellyfish by-products gelatin was desalted using a dialysis technique overnight before performing SDS-PAGE. The comparison of protein pattern of jellyfish by-products gelatin was performed using SDS-PAGE according to Laemmli [19] method with a slight modification. The polyacrylamide gel was made up of 5% stacking gel and 12% separating gel. A high molecular weight protein marker (11-250 kDa) was used to estimate the molecular weight of the jellyfish gelatin protein (PageRuler™ Plus Prestained protein ladder, Thermo Fisher Scientific). All jellyfish by-products gelatin (200 μg protein) samples were heated at 95 °C for 3 min. Then, each cooled sample was loaded onto the gel and subjected to an electrophoresis unit (Mini-slab size electrophoresis system, ATTO, Japan) with a constant current of 50 V in vertical slab gels.

2.2.4 Gel preparation

According to the standard method of GMIA [20], the preparation of gelatin gel (6.67%, w/v) was slightly modified. The freeze-dried jellyfish by-products gelatin was dissolved in deionized water and stirred at 60 °C until completely dissolved. After that, 10 mL of gelatin solution (6.67%) was poured into a vial of 27 mm in diameter and 58 mm in height. The gel sample was kept at 4 °C for 18 before the analysis.

2.2.5 Setting time

The setting time of gelatin was performed as previously described [21] with a slight modification. The jellyfish by-product gelatin solution (6.67%) was prepared using the above method in section 2.2.4. The jellyfish's by-product gelatin in the vial was incubated at 4 °C. During the incubation process, the aluminum needle of 1 mm in diameter and 60 mm in height was inserted into the jellyfish by-products gelatin solution and lifted every 10 s. The setting time of jellyfish by-products gelatin is the amount of time (min) that the needle cannot be separated from the jellyfish by-products gelatin sample.

2.2.6 Determination of gel strength

The gelatin gel sample (6.67%) in a vial from section 2.2.4 was used to determine gel strength using a texture analyzer (TA-XT2i, Stable Micro Systems, UK). The gel strength was performed as previously described [12].

2.2.7 Viscosity

The viscosity of gelatin solution (6.67%) in section 2.2.4 after heating at 60 °C was measured using a Brookfield viscometer (DV - II +, BROOKFIELD, USA) at a speed of 90 rpm and room temperature.

2.2.8 Color measurement

The gelatin gel sample in section 2.2.4 (10 mL) was filled in the glass receptacle and measured using a colorimeter (ColorQUEST 45/0, Hunter Lab, USA). The measured color was expressed as L* (lightness), a* (red-green), and b* (yellow-blue). The hue angle (h*) of gelatin was calculated according to the following equation (2):

$$Hue\ angle = \tan^{-1} (b^* / a^*) \qquad (2)$$

2.2.9 Foaming properties

The foaming properties of gelatin were estimated as previously described [22] with a slight modification. 20 mL of gelatin solution (1%) was homogenized using a homogenizer (T-25 Ultra-Turrax, IKA Labortechnik, Germany) at speed of 13,000 rpm for 1 min. After that, the mixture was carefully transferred into a measuring cylinder (100 mL) without delay. The sample was then kept at room temperature for 30 min. Finally, the volume was measured. The foaming capacity (FC) and foaming stability (FS) of gelatin were calculated according to the following equation (3) and (4):

$$FC\ (\%) = [(V_1 - V_0) / V_0] \times 100 \qquad (3)$$

$$FS\ (\%) = [(V_{30} - V_0) / V_0] \times 100 \qquad (4)$$

where V_0 = volume before homogenization (mL), V_1 = volume after homogenization (mL), and V_{30} = volume after 30 min (mL).

2.2.10 Emulsifying properties

The emulsion properties of gelatin were determined as previously described [23] with a slight modification. 10 mL of soybean oil was mixed with 6 mL of gelatin solution (1%) and then homogenized using a homogenizer (T-25 Ultra-Turrax, IKA Labortechnik, Germany) at speed of 20,000 rpm for 2 min. The mixed sample was taken out at 0 and 10 min and then mixed with 0.1% sodium dodecyl sulfate (SDS) at a ratio of 1:100 (v/v). The absorbance was measured at 500 nm using a spectrophotometer (SP 830 plus, Metertech, Taiwan). The emulsion activity index (EAI) and emulsion stability index (ESI) of gelatin were calculated according to the following equation (5) and (6):

$$EAI\ (m^2/g) = [(2 \times 2.303 \times A_0 \times N) / (\emptyset \times C \times 10{,}000)] \qquad (5)$$

$$ESI\ (\%) = [A_0 / (A_0 - A_{10})] \times \Delta t \qquad (6)$$

where A_0 = absorbance at 500 nm of the sample was kept for 0 min, N = dilution factor, Ø = the volume of oil portion, C = gelatin concentration, A_{10} = absorbance at 500 nm of the sample was kept for 10 min, and Δt = 10 min.

2.3 Statistical analysis

All experiments were performed in triplicate. The results were presented as mean±standard deviations. Significant differences ($p<0.05$) between the means values were analyzed by analysis of variance (ANOVA) using Duncan's multiple in the statistical package for social science (SPSS) 22.0.

3 Results and discussions

3.1 Gelatin yield

The gelatin yield of freeze-dried jellyfish by-products gelatins derived from hot water extraction for 12, 24, and 48 h were 2.74%, 4.8 5%, and 14.07% (a dry weight basis), respectively (Table 1). The gelatin yield of jellyfish by-products gelatin was generally increased as extraction time increased. The jellyfish by-products gelatin yield extracted at 60 °C for 12, 24, and 48 h differed from the previous study. The jellyfish gelatin (type A) pretreated with sulfuric acid (pH 2) solution and extracted at 75 °C for 12 h showed the highest gelatin yield of 39.47% (on a dry weight basis) [15]. The highest gelatin yield of type A gelatin from jellyfish (*Lobonema smithii*) was 40.54% (on a dry weight basis) when pretreated with hydrochloric acid (pH 1) solution and extracted at 75 °C for 12 h [16]. Compared to the same jellyfish by-products, the jellyfish gelatin yield reported in this study was higher than the previous work that showed a gelatin yield of 6.34% (on a dry weight basis) [12], which applied an extraction time of 3 h. It was noted that the longer extraction time than 3 h did increase the gelatin yield. Compared to other marine samples, the highest gelatin yield of blacktip shark and the brown-banded bamboo shark gelatin was 21.17-24.76% and 19.06–22.81% (based on wet weight), respectively, when extracted at 45, 60, and 75 °C for 6 and 12 h [24]. The increased extraction time could de-stabilize cross-linked proteins at the covalent bonds in the collagen network, resulting in free α-chains or β-chains being liberated from the protein [25]. Consequently, the extraction of gelatin was enhanced, as shown by higher gelatin yield.

Table 1. Physicochemical properties of jellyfish by-products gelatin.

Properties		Sample		
		JFG12	JFG24	JFG48
Gelatin yield* (%)		2.74±0.36^{c}	4.85±0.58^{b}	14.07±0.53^{a}
Conductivity^{ns} (ms/cm)		0.32±0.01	0.32±0.01	0.32±0.01
NaCl^{ns} (ppm)		113.01±0.01	113.01±0.01	113.01±0.01
Setting time* (min)		43.57±1.29^{a}	23.67±1.15^{b}	20.03±0.06^{c}
Gel strength* (g)		210.46±3.97^{c}	261.60±3.25^{b}	325.97±2.84^{a}
Viscosity^{ns} (cP)		23.00±1.00	24.00±1.73	24.5±0.87
Color*	*L**	29.43±0.01^{c}	31.29±0.03^{b}	37.16±0.01^{a}
	*a**	1.32±0.02^{c}	1.57±0.01^{a}	1.428±0.03^{b}
	*b**	1.23±0.03^{c}	5.10±0.04^{b}	6.92±0.16^{a}
	*h**	46.88±0.99^{a}	17.09±0.03^{b}	11.66±0.18^{c}

*Different superscript letters in the same row indicate a significant difference ($p<0.05$).
^{ns}The samples were not significantly different ($p\geq0.05$).

3.2 Conductivity and salt content

The conductivity of JFG12, JFG24, and JFG48 samples was 0.32 ms/cm, and sodium chloride (NaCl) was 113.01 ppm (Table 1). The different extraction times did not significantly affect jellyfish by-products gelatin's conductivity and salt content. The salt content of jellyfish by-products gelatin in this study showed lower than in other research. Hydrolyzed salted, dried cannonball jellyfish (*Stomolophus meleagris*) gelatin (H-SDJ) showed a salt content of 56.61% remaining in the sample [26]. This might be due to the unsuccessful removal of salt from the washing and soaking step. The presence of electrolytes has directly influenced low gelatin gel quality. Sow and Yang [27] reported that the addition of 1.5% NaCl into fish gelatin effect reduced gel strength because of loss in molecular order in secondary protein structure. As a result, excess salt content must be first removed prior before continuing further extraction steps.

3.3 Protein pattern

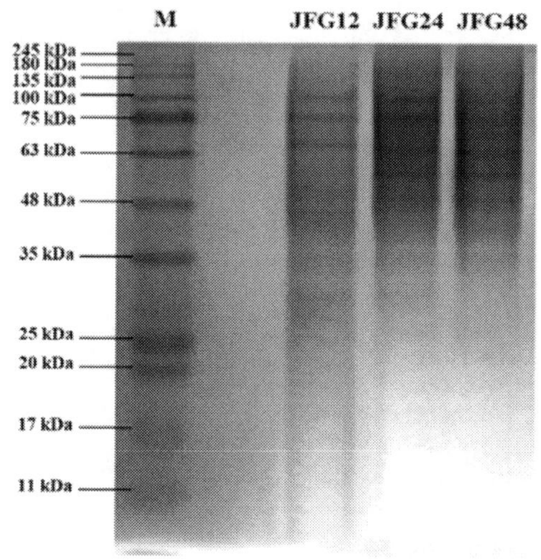

Fig. 3. Protein pattern of jellyfish by-products gelatins (M) molecular weight standard and jellyfish by-products gelatin extracted at different times of 12 h (JFG12), 24 h (JFG24), and 48 h (JFG48).

The protein pattern of jellyfish by-products gelatin is presented in Fig. 3. The protein patterns of JFG12, JFG24, and JFG48 samples found protein having a molecular weight of 65-125 kDa, 56-125 kDa, and 52-125 kDa, respectively. Moreover, the α-chain was observed in all samples, which had MW of 80-125 kDa [28]. The physical and functional properties of gelatins are influenced by their molecular weight distribution, structure, and subunit makeup, which includes α-chain and low-molecular-weight protein fragments. The recovered proteins' molecular weight distribution differed slightly from gelatins extracted with a longer extraction time. The degradation of gelatin's major protein components (α-, β- and γ-chain) became more prominent as the extraction time increased, thereby increasing low molecular weight peptides. The hydrolysis process may affect the molecular weight of the extracted jellyfish by-products gelatin, which contributes to the splitting of the peptide chains as the extraction temperature increases [29]. The inter-chain cross-links of collagen were hydrolyzed during gelatin extraction, resulting in gelatin with varying molecular weight peptides [30]. The low content of α-chain in gelatin may affect the physical and functional properties of gelatin [31].

3.4 Physical properties

3.4.1 Setting time

The setting time required to develop jellyfish by-products gelatin gel at 4 °C is shown in Table 1. All jellyfish by-products gelatin (6.67%) solution samples can form a gel at 4 °C. The setting time decreased as the extraction time increased. The shortest setting time was for the JFG48 sample, while the longest time was for the JFG12 sample. However, the setting time of jellyfish by-products gelatin was longer than commercial bovine gelatin, commercial fish gelatin, and other work. The setting time of commercial bovine and fish gelatin was 10.37±0.05 and 12.00±1.73 min, respectively. The shortest setting time of tilapia gelatin solution was 0.87 min when extracted at 45 °C for 6 h at pH 5 [32]. For jellyfish by-products gelatin, using a more extended time during the extraction may cause the triple helix structure of jellyfish collagen to denature [10], resulting in increased helix-to-coil transition and converted to soluble gelatin and high gelling ability. On the other hand, the shortened extraction time may be caused by crude gelatin, resulting in taking a long time to form a reasonable gel network arrangement.

3.4.2 Gel strength

The gel strength of gelatin is one of the essential physical properties that influence the textural quality of food, pharmaceutical, and cosmetic products. The gel strength of JFG12, JFG24, and JFG48 samples is shown in Table 1. The gel strength of jellyfish by-products gelatin was in the range of 210.46-325.97 g after being extracted at 60 °C for 12, 24, and 48 h. The gel strength of jellyfish by-products gelatin was increased as extraction time increased. This result is consistent with the trend corresponding to the setting time of gelatin. The JFG48 sample showed the highest gel strength of 325.97 g. The gel strength of all jellyfish by-products gelatin gel in this study was lower than that of commercial bovine gelatin (358.62 g) [12]. However, the gel strength of commercial fish gelatin (306.11 g) was lower than that of the JFG48 sample. Previous research has found that jellyfish by-products gelatin has lower gel strength due to differences in collagen type, the alpha and beta chain ratio in the gelatin, and the gelatin peptides of the alpha chain [11]. The lowest gel strength of jellyfish (*Lobonema smithii*) gelatin was 108.36 g [15]. The bloom of hydrolyzed salted, dried cannonball jellyfish gelatin (H-SDJ) (6.67%) was 3.4 g [26], which is lower than that of other research because H-SDJ had a high salt content. The salt may affect the gel strength because Na^+ and Cl^- may interfere with hydrogen bonds with the carbonyl oxygen of an adjacent peptide group, resulting in lower gelling properties. Compared to the same jellyfish by-products (*Lobonema smithii*), the gel strength of the JFG48 sample was higher than previously studied. The jellyfish gelatin produced from by-products had the gel strength of 223.61-323.74 g [11-12] but the value was still lower than that of marine gelatin from the previous work. The highest gel strength value of tilapia gelatin was 382.3 g when extracted at 45 °C for 6 h [32]. Therefore, compared to the gel strength from other studies, factors of raw material, salt content, pretreatment conditions, extraction temperature, and extraction times influenced collagen conversion into gelatin, thereby affecting its quality.

3.4.3 Viscosity

Viscosity is another significant factor in gel formation. The viscosity of jellyfish by-products gelatin is shown in Table 1. The viscosity of jellyfish by-products gelatin was in a range of 23.00-24.50 cP. The viscosity of all jellyfish by-products gelatin was higher than that of those previously studied. Lueyot *et al.* [11] reported that the highest viscosity of jellyfish by-products gelatin was 7.73 cP when extracted at 60 °C for 12 h. The different extraction times did not significantly affect viscosity. The viscosity of commercial bovine and fish gelatin was 27.00±1.73 cP and 24.00±1.00 cP, respectively. These results were slightly different from other studies, which reported that the viscosity of commercial bovine gelatin was 9.8 cP [2]. However, the viscosity of all jellyfish by-products gelatin samples was lower than that of commercial bovine gelatin. In contrast, the viscosity of commercial fish gelatin showed no significant differences from all jellyfish by-products gelatin samples. Compared to other gelatins, the viscosity of all jellyfish by-products gelatin samples is within an acceptable range of 2.0-13.0 cP [2, 33]. The viscosity is influenced by several parameters, including raw material sources, extraction conditions, gelatin gel preparation, and the test instrument's precision and repeatability [11].

3.4.4 Color

The color of gelatin is one of the factors that affect food product preference. The color of jellyfish by-products gelatin is presented in Table 1. JFG12, JFG24, and JFG48 gel (6.67%) samples showed that the lightness (L*) increased as extraction time increased. The hue value of JFG12, JFG24, and JFG48 gel samples was 46.88±0.99, 17.09±0.03, and 11.66±0.18, respectively. The hue value of jellyfish by-products gelatin gel decreased significantly with increasing extraction time. The color of jellyfish by-products gelatin showed light to dark brown, as shown in Fig. 4. These results were different from the previous study, which reported a hue value of jellyfish gelatin in the range of 32.43-33.05 after being extracted at 60 °C for 3 h; the sample was then dried using a tray dryer at 60 °C for 3 days [12]. The intense brown color of jellyfish by-products gelatin gel could be due to the color of the raw material, pretreatment condition, extraction temperature, extraction time, drying, and Maillard reaction, in which reducing sugars react with a free amino acid at the appropriate temperature during jellyfish by-products gelatin drying [30].

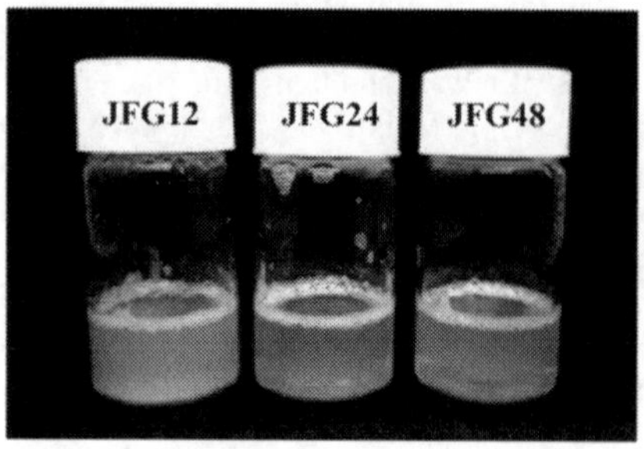

Fig. 4. Jellyfish gelatins gel (6.67%) extracted at different times of 12 h (JFG12), 24 h (JFG24), and 48 h (JFG48).

3.5 Functional properties

3.5.1 Foaming properties

The foaming properties of gelatin are widely used in dessert products, especially soft, light, and porous products, such as bakery products, meringue, and marshmallows [16]. The foaming capacity (FC) and foaming stability (FS) of jellyfish by-products gelatin are presented in Table 2.

Table 2. Functional properties of jellyfish by-products gelatin.

Properties	Sample		
	JFG12	JFG24	JFG48
FC* (%)	12.28 ± 1.52^{b}	16.67 ± 1.52^{a}	17.54 ± 1.52^{a}
FS* (%)	10.53 ± 0.00^{b}	12.28 ± 1.52^{b}	15.79 ± 2.63^{a}
EAI* (m^2/g)	13.30 ± 0.02^{a}	13.12 ± 0.01^{b}	13.11 ± 0.02^{b}
ESI* (%)	56.42 ± 0.39^{a}	44.32 ± 0.14^{b}	39.19 ± 2.50^{c}

*Different superscript letters in the same row indicate a significant difference ($p < 0.05$).

The foaming capacity of JFG12, JFG24, and JFG48 samples were in the range of 12.28-17.58%. The foaming capacity of jellyfish by-products gelatin was generally increased as extraction time increased. A similar result was reported in type A gelatin from jellyfish (*Lobonema smithii*) [16]. The foaming capacity of commercial bovine and fish gelatin samples was 20.17±4.02% and 19.29±1.52%, respectively. All jellyfish by-products gelatin samples had slightly lower foaming capacity than commercial bovine, and fish gelatin due to the migration of shorter peptides to the air-water interface [34]. During foam formation, the higher foaming capacity is associated with readily the absorbed molecule of gelatin that can migrate rapidly in the gas-liquid interface. The foaming stability of JFG12, JFG24, and JFG48 samples was 10.52-15.78% which was lowered than those of commercial bovine and fish gelatin samples exhibiting 17.54±4.02% and 14.91±1.52%, respectively. The foaming stability of jellyfish by-products gelatin decreased as extraction time increased. Foaming qualities are affected by the degree of interaction between protein molecules and the existence of hydrophobic regions [35]. The results suggested that extended extraction time increased the foaming properties of jellyfish by-products gelatin. The jellyfish by-products gelatin could function as a foaming agent in food applications.

3.5.2 Emulsifying properties

Gelatin can function as an emulsifier that plays an essential role in the stabilization of emulsion in the food and pharmaceutical industry. In this investigation, the emulsion was made using soybean oil, water and jellyfish by-products gelatin as an emulsifier. The emulsion activity index (EAI) and emulsion stability index (ESI) of jellyfish by-products gelatin are shown in Table 2. The emulsion activity index of JFG12, JFG24, and JFG48 samples were 13.11-13.30 m^2/g, and the emulsification stability index was 39.18-56.42%. The emulsion activity index of commercial bovine and fish gelatin was 13.91±0.22 m^2/g and 13.72±0.13 m^2/g, respectively, and the emulsification stability index was 58.34±0.64% and 52.63±1.22%, respectively. The EAI and ESI of the JFG12 sample were higher than that of any jellyfish by-products gelatin (JFG24 and JFG48 samples), but lower than commercial bovine and fish gelatin. The EAI and ESI of jellyfish by-products gelatin were decreased as extraction time increased. The degradation of jellyfish collagen might be due to the generation of short-chain peptides during long extraction times. The increased short-chain peptides yield more charged groups, which are more hydrophilic and tend to adsorb in the water phase, resulting in a decrease in encapsulated oil droplets. The longer peptide chains might have a stronger film surrounding the oil, resulting in enhanced emulsion stability [32]. Emulsification stability depends on the rate of reaching equilibrium surface tension, steric stabilization, volume and surface viscosity, and electrical repulsion between droplets [36]. Thus, the effect of extraction times plays a vital role in producing jellyfish gelatin with good emulsifying properties.

4 Conclusions

Jellyfish by-products may be a promising source for gelatin production, which reduces food waste from the salted jellyfish industry. The factor of extraction time directly impacts jellyfish gelatin quality. The resulting molecular weight distribution and the triple helix content of gelatin influence the physical and functional properties of jellyfish by-products gelatin. All jellyfish gelatins derived from different extraction times showed lower gel strength values than commercial bovine gelatin. In conclusion, the proper condition for producing jellyfish gelatin with the highest gel strength of 325.97±2.84 g was 60 °C for 48 h. The foaming and emulsifying properties of jellyfish gelatin were similar to commercial bovine and fish gelatin. Thus, jellyfish by-products gelatin may be used in food applications due to the gelation gel, foam formation, and emulsification qualities.

Acknowledgment

This research was funded by King Mongkut's University of Technology North Bangkok and National Science and Technology Development Agency, Thailand (Contract no. B.021-2563), and the National Research Council of Thailand through the NRCT Senior Research Scholar Program (Contract no. 814-2020). The authors would like to thank Mahachai Food and Trading Co., Ltd., Samut Sakhon, Thailand, for their kind support of salted jellyfish.

References

1. Grand View Research, https://www.grandviewresearch.com/press-release/global-gelatin-market (2020)
2. L. Lin, J. M. Regenstein, S. Lv, J. Lu, S. Jiang, An overview of gelatin derived from aquatic animals: Properties and modification, Trends in Food Science & Technology, 68 (2017): 102-112
3. J. Tkaczewska, M. Morawska, P. Kulawik, M. Zając a, Characterization of carp (*Cyprinus carpio*) skin gelatin extracted using different pre-treatments method, Food Hydrocolloids, 81 (2018): 169-179
4. P. Kittiphattanabawon, S. Benjakul, S. Sinthusamran, H. Kishimura, Gelatin from clown featherback skin: extraction conditions, LWT, 66 (2016): 186-192
5. T. Huang, Z. Tu, X. Shangguan, X. Sha, H. Wang, L. Zhang, N. Bansal, Fish gelatin modifications: A comprehensive review, Trends in Food Science & Technology, 86 (2019): 260-269
6. Y. Liu, L. Xia, H. Jia, Q. Li, W. Jin, X. Dong, J. Pan, Physiochemical and functional properties of chum salmon (*Oncorhynchus keta*) skin gelatin extracted at different temperatures, Journal of the Science of Food and Agriculture, 97, 15 (2017): 5406-5413
7. R. S. G. Silva, S. F. Bandeira, L. A. A. Pinto, Characteristics and chemical composition of skins gelatin from cobia (*Rachycentron canadum*), LWT, 57, 2 (2014): 580-585
8. F. Mahmoodani, V. Sanaei Ardekani, S. F. See, S. M. Yusop, A. S. Babji, Optimization and physical properties of gelatin extracted from pangasius catfish (*Pangasius sutchi*) bone, Journal of Food Science and Technology, 51, 11 (2014): 3104-3113
9. T. Zhang, R. Sun, M. Ding, L. Tao, L. Liu, N. Tao, X. Wang, J. Zhong, Effect of extraction methods on the structural characteristics, functional properties, and emulsion stabilization ability of *Tilapia* skin gelatins, Food Chemistry, 328 (2020): 127114
10. B. Wonganu, Application of gelatin derived from waste tilapia scales to an antibiotic hydrogel pad, E3S Web of Conferences, 141 (2020): 03004
11. A. Lueyot, V. Rungsardthong, S. Vatanyoopaisarn, P. Hutangura, B. Wonganu, P. Wongsa-Ngasri, S. Charoenlappani, S. Roytrakul, B. Thumthanaruk, Influence of collagen and some proteins on gel properties of jellyfish gelatin, PLOS One, 16, 6 (2021): e0253254
12. W. Charoenchokpanich, P. Muangrod, V. Rungsardthong, S. Vatanyoopaisarn, B. Wonganu, S. Roytrakul, B. Thumthanaruk, Effect of hydrochloric acid extraction on yield and gel properties of gelatine from salted jellyfish by-products, E3S Web of Conferences, 302 (2021): 02009
13. Ministry of Commerce, http://www.ops3.moc.go.th/infor/MenuComTH/stru1_export/export_topn_re/report.asp (2021)
14. W. Charoenchokpanich, V. Rungsardthong, S. Vatanyoopaisarn, B. Thumthanaruk, Y. Tamaki, Salt reduction in salted jellyfish (*Lobonema smithii*) using a mechanical washing machine, Science, Engineering and Health Studies, 14, 3 (2020): 184-192
15. P. Chancharern, N. Laohakunjit, O. Kerdchoechuen, B. Thumthanaruk, Extraction of type A and type B gelatin from jellyfish (*Lobonema smithii*), International Food Research Journal, 23, 1 (2016): 419-424
16. U. Rodsuwan, B. Thumthanaruk, O. Kerdchoechuen, N. Laohakunjit, Functional properties of type A gelatin from jellyfish (*Lobonema smithii*), International Food Research Journal, 23, 2 (2016): 507-514
17. S. Cho, J.R. Ahn, J.S. Koo, S.B. Kim, Physicochemical properties of gelatin from jellyfish *Rhopilema hispidum*, Fisheries and Aquatic Sciences, 17, 3 (2014): 299-304
18. P. Muangrod, V. Rungsardthong, S. Vatanyoopaisarn, Y. Tamaki, E. Kuraya, B. Thumthanaruk, Effect of wash cycle on physical and chemical properties of rehydrated jellyfish by-products and jellyfish protein powder, Science, Engineering and Health Studies, 15 (2021): 21030004
19. U.K. Laemmli, Cleavage of structural proteins during the assembly of the head of bacteriophage T4, Nature, 227, 5259 (1970): 680-685
20. GMIA, *Standard methods for the sampling and testing of gelatins* (Gelatin Manufacturers of America, Inc., New York, 1986)
21. J.H. Muyonga, C.G.B. Cole, K.G. Duodu, Fourier transform infrared (FTIR) spectroscopic study of acid soluble collagen and gelatin from skins and bones of young and adult Nile perch (*Lates niloticus*), Food Chemistry, 86, 3 (2004): 325-332
22. S.K. Sathe, D.K. Salunkhe, Functional properties of the great northern bean (*Phaseolus vulgaris* L.) proteins: Emulsion, foaming, viscosity, and gelation properties, Journal of Food Science, 46, 1 (1981): 71-81

23. K.N. Pearce, J.E. Kinsella, Emulsifying properties of proteins: Evaluation of a turbidimetric technique, Journal of Agricultural and Food Chemistry, 26, 3 (1978): 716-723
24. P. Kittiphattanabawon, S. Benjakul, W. Visessanguan, F. Shahidi, Comparative study on characteristics of gelatin from the skins of brownbanded bamboo shark and blacktip shark as affected by extraction conditions, Food Hydrocolloids, 24, 2-3 (2010): 164-171
25. S. Sinthusamran, S. Benjakul, H. Kishimura, Characteristics and gel properties of gelatin from skin of seabass (*Lates calcarifer*) as influenced by extraction conditions, Food Chemistry, 152 (2014): 276-284
26. P.G. Chiarelli, R.B. Pegg, G.D. Kumar, K.M. Solval, Exploring the feasibility of developing novel gelatin powders from salted, dried cannonball jellyfish (*Stomolophus meleagris*), Food Bioscience, 44 (2021): 101397
27. L.C. Sow, H. Yang, Effects of salt and sugar addition on the physicochemical properties and nanostructure of fish gelatin, Food Hydrocolloids, 45 (2015): 72-82
28. V. Gudipati, *Chapter 3. Fish Gelatin: A Versatile Ingredient for The Food and Pharmaceutical Industries in Marine Proteins and Peptides: Biological Activities and Applications*, edited by S.K. Kim (Wiley-Blackwell, Oxford, UK, 2013), pp. 271–295
29. N.M. Sarbon, F. Badii, N.K. Howell, Preparation and characterisation of chicken skin gelatin as an alternative to mammalian gelatin, Food Hydrocolloids, 30, 1 (2013): 143-151
30. P. Kaewruang, S. Benjakul, T. Prodpran, S. Nalinanon, Physicochemical and functional properties of gelatin from the skin of unicorn leatherjacket (*Aluterus monoceros*) as affected by extraction conditions, Food Bioscience, 2 (2013): 1-9
31. A. Erge, Ö. Zorba, Optimization of gelatin extraction from chicken mechanically deboned meat residue using alkaline pre-treatment, LWT, 97 (2018): 205-212
32. W. Liao, Y. Zhu, Y. Lu, Y. Wang, X. Dong, G. Xia, X. Shen, Effect of extraction variables on the physical and functional properties of tilapia gelatin, LWT, 146 (2021): 111514
33. E.T.A. Tümerkan, Ü. Cansu, G. Boran, J.M. Regenstein, F. Özoğul, Physiochemical and functional properties of gelatin obtained from tuna, frog and chicken skins, Food Chemistry, 287 (2019): 273-279
34. H. Kchaou, N. Benbettaied, M. Jridi, M. Nasri, F. Debeaufort, Influence of Maillard reaction and temperature on functional, structure and bioactive properties of fish gelatin films, Food Hydrocolloids, 97 (2019): 105196
35. K. Elavarasan, V.N. Kumar, B.A. Shamasundar, Antioxidant and functional properties of fish protein hydrolysates from fresh water carp (*Catla catla*) as influenced by the nature of enzyme, Journal of Food Processing and Preservation, 38, 3 (2014): 1207–1214
36. F. Ravera, K. Dziza, E. Santini, L. Cristofolini, L. Liggieri, Emulsification and emulsion stability: The role of the interfacial properties, Advances in Colloid and Interface Science, 288 (2021): 102344

* Corresponding author: benjawan.t@sci.kmutnb.ac.th

A comparative study of conventional and supercritical carbon dioxide extraction methods for the recovery of bioactive compound from Lion's Mane mushroom (*Hericium erinaceus*)

Pinida Joradon[1], *Vilai* Rungsardthong[1,*], *Uracha* Ruktanonchai[2], *Khomson* Suttisintong[2], *Tawin* Iempridee[2], *Benjawan* Thumthanaruk[1], *Savitri* Vatanyoopaisarn[1], and *Dutsadee* Uttapap[3]

[1]Department of Agro-Industrial, Food and Environmental Technology, Faculty of Applied Science, Food and Agro-industrial Research Center, King Mongkut's University of Technology North Bangkok, Bangkok, Thailand
[2]Nanotechnology Center, National Science and Technology Development Agency, Thailand Science Park, Pathum Thani, Thailand
[3]Division of Biochemical Technology, School of Bioresources and Technology, King Mongkut's University of Technology Thonburi, Bangkok, Thailand

Abstract. Lion's Mane mushroom (*Hericium erinaceus*), LM, is a medicinal mushroom which has various bioactive compounds within its fruiting bodies. However, during the cultivation, a large amount of the irregular-shape LM (Ir-LM) was generated. This mushroom type was considered a mushroom by-product. In this study, conventional solvent extraction (Maceration) and supercritical fluid extraction using carbon dioxide (SCFE-CO_2) were performed to compare the recovery of ergosterol, hericenone C, and hericene A content from Reg-LM and Ir-LM fruiting bodies. Furthermore, two extraction conditions (40°C at 200 bar and 70°C at 350 bar) were conducted for the SCFE-CO_2 technique. The results showed that SCFE-CO_2 methods produced a higher recovery of ergosterol and hericenone C as compared to the Maceration techniques. SCFE-CO_2 extracts were determined for their antioxidant activities. The DPPH radical scavenging activity of the extract from 70°C at 350 bar was significantly higher ($p < 0.05$) than the extract obtained from 40°C at 200 bar. The results revealed the use of green technology supercritical fluid extraction using carbon dioxide to recover bioactive compounds from mushroom by-products and apply for high-value added products.

Keyword. Lion's Mane mushroom (*Hericium erinaceus*), Ergosterol, Conventional solvent extractions, Supercritical fluid extraction using carbon dioxide, Antioxidant activity

1 Introduction

Lion's Mane mushroom (*Hericium erinaceus*), LM, is an important medicinal mushroom containing many bioactive compounds which have been used for medicines and food supplements [1]. Several bioactive compounds belonging to the genus *Hericium* have been isolated from LM mushroom [2]. The compounds indicate responses on antitumor, antibacterial, hypoglycemic and neuroprotective [3]. Gasecka *et al.* 2020 [4] reported that more than 4.5 mg/g dry weight of ergosterol was found in the fruiting bodies of fresh LM mushroom. The anti-inflammatory activity of ergosterol is correlated to the suppression of the proinflammatory cytokines including TNF-α, IL-6 and IL-1β [5]. Hericenone L isolated LM mushroom fruiting bodies show cytotoxic activity against the EC109 cell line by the MTT assay [6]. Kawagishia *et al.* 1991 [7] reported that hericenones C, D and E can stimulate the synthesis of nerve growth factor (NGF). Hericene A, one of the bioactive compounds in LM mushroom, strongly inhibited α-glucosidase activity with an IC_{50} of 6.7 μM [8].

Nowadays, mushrooms are cultivated for sustainable production and consumption, by sometimes using agricultutal waste as substrate, which fit to the concept of circular economy [9]. However, irregular-shape LM (Ir-LM), considered a by-product, was found during mushroom cultivation. These by-products are sold at low prices in the market. Aguayo *et al.* 2017 [10] reported that high amounts of button mushroom (*Agaricus Bisporus*) fruiting bodies by-product were generated during mushroom production (representing around 20% of total production). These button mushroom by-products were discarded due to their misshapen caps or stalks that do not meet the specifications set by the retailers. However, amounts of β-glucan 1.01 g/100 g dry mass can be obtained from these button mushroom by-products. β-glucan is a soluble dietary fiber that can enhance immune function and lower blood cholesterol [10]. Therefore, an investigation for alternative and profitable solutions is required. Mushroom by-products could be used for the preparation of extracts enriched in bioactive compounds that could be applied by the pharmaceutical and food industry.

Several extraction methods have been developed with the goal of producing extracts with higher yields and reduced costs. Organic solvents including methanol, ethanol, and acetone are used in extraction. The usage of

* Corresponding author: vilai.r@sci.kmutnb.ac.th

https://doi.org/10.1051/e3sconf/202235502015

such solvents, on the other hand, is hazardous to human health as well as the environment [11]. Supercritical fluid extraction using carbon dioxide (SCFE-CO_2) is a promising method for extracting natural oils and oleoresins. SCFE-CO_2 uses carbon dioxide as the extraction solvent under supercritical conditions. The critical temperature and pressure for carbon dioxide are 31.1 °C and 72.8 bar, which enables it to preserve the sensitive heat compounds and also their quality. Carbon dioxide is inert, non-toxic, non-corrosive, non-flammable and recyclable [12].

The optimal condition for the extraction of ergosterol from *Lentinula edodes* by SCFE-CO_2 was reported as 350 bar and 70 °C, providing the highest ergosterol at 18% (w/w) or 180 mg/g dw [13]. Prasad *et al.* (2017) [11] investigated the effect of SCFE-CO_2 on *L. edodes* extract production and antioxidant activity. They found supercritical extracts gave low yields (0.56 % when pure CO_2 was used as the solvent and 1.02 % when 5% methanol was added as a co-solvent) but produced high quality extracts [strong antioxidant activity, such as 1,1-diphenyl-2-picrylhydrazyl (DPPH) radical scavenging activity, polyphenol content, and ferric reducing antioxidant power (FRAP)] with a greater number of organic compounds than extracts obtained by the solvent extraction. However, there are very few studies on the extraction of bioactive compounds from LM mushroom by SCFE-CO_2. Parada et al. 2015 [14] revealed that the temperature and pressure at 40°C, 200 bar could provide the LM extract with high antioxidant properties, while the use of 70°C, 350 bar by Morales *et al.* 2017 [13] exhibited the highest ergosterol content from shiitake mushrooms. The recovery of extracts enriched in bioactive compounds from the mushroom by-product and used as additional ingredients in foods or beverages would help value-adding the produces, reduce production and waste management costs and also environmental impact.

In this study, conventional solvent extraction (Maceration) and supercritical fluid extraction using carbon dioxide (SCFE-CO_2) were performed to compare the recovery of ergosterol content and other major compounds from Ir-LM fruiting bodies. Furthermore, two extraction conditions (40°C at 200 bar and 70°C at 350 bar) were conducted for the SCFE-CO_2 technique by varying pressure and temperature. The antioxidant capacities of the extracts from each SCFE-CO_2 condition were determined.

2 Materials and methods

2.1 Materials

Regular-shape (Reg-LM) and irregular-shape (Ir-LM) fruiting bodies of LM were purchased from Fresh and Friendly Farm Co., Ltd. at Thanyaburi district in Pathum Thani province, Thailand, and stored at -20°C in dark condition until further use. Ergosterol (95%) was obtained from Sigma-Aldrich Química (Madrid, Spain). Gallic acid (98%) was procured from Sigma-Aldrich, USA. Folin-Ciocalteu reagent and 2,2-diphenyl-1-picrylhydrazyl (DPPH) were purchased from Sisco Research Laboratories Pvt. Ltd., India. Sodium carbonate (99.5% pure) was purchased from Merck, India. Trolox reagent was obtained from M Tedia, USA. Solvents such as water (HPLC grade) and methanol (HPLC grade) were obtained from LAB-SCAN (Gliwice, Polland) and absolute ethanol was from PanReac (Barcelona, Spain). Carbon dioxide gas was supplied by the Bombay Carbon Dioxide Gas Company, Mumbai, India. The reagents and solvents were of analytical quality.

2.2 Mushroom preparation

Mushroom fruiting bodies were lyophilized by a vacuum freeze drier (GFD30S, Grisrianthong, Thailand). The freeze-dried samples were milled and sieved to the particle size of around 75-150 µm and stored in a container at -20 °C until analysis. Overview of the experiments and analyses carried out in this study for the conventional method (Maceration) and supercritical fluid extraction using carbon dioxide (SCFE-CO_2) of LM mushroom is shown in Figure 1.

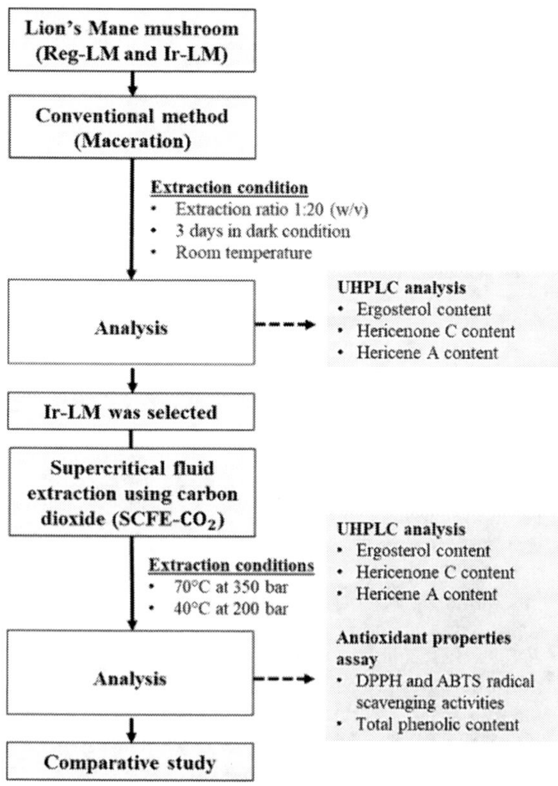

Figure 1. Overview of the experiments and analyses carried out in this study for the conventional method (Maceration) and supercritical fluid extraction using carbon dioxide (SCFE-CO_2) of Lion's Mane mushroom.

2.3 Extraction of bioactive compounds by conventional method (Maceration)

Dried powder of Reg-LM and Ir-LM, 10 g, were extracted using 200 mL absolute ethanol at room temperature (25 ±

3 °C) for 3 days in a dark condition. The sample was stored in the dark to prevent the transformation of ergosterol into vitamin D_2 or a variety of photoirradiation products such as previtamin D_2, tachysterol, and lumisterol [15]. The crude extracts were then filtered and concentrated to dryness using a rotary evaporator (R114, Buchi, Switzerland) at a controlled temperature (50 ± 1 °C) and pressure (150 bar). Ergosterol, hericenone C, and hericene A in the extract were determined and calculated for their content in the mushroom samples.

2.4 Extraction of bioactive compounds by supercritical fluid extraction using carbon dioxide (SCFE-CO_2)

Supercritical fluid extraction using carbon dioxide was used to extract the mushroom (Ir-LM). Carbon dioxide with a purity of > 99.9% was used. The extraction was carried out on Applied Separations, laboratory-scale supercritical equipment, (Allentown, PA, USA). The extractions were performed at two conditions: 40°C and 200 bar, and 70°C and 350 bar. The conditions for the extraction were from Parada *et al.* 2015 [14] and Morales *et al.* 2017 [13]. The extraction time was 40 min with a flow rate of CO_2 at 4 mL/min. Polypropylene wool and the sample were put in the vessel before being placed in the oven. The extracts were collected and stored at -20°C until further analysis. Storage at -20°C appeared to be appropriate for ergosterol extracts from mushrooms because it could reduce ergosterol oxidation [16]. The SCFE-CO_2 extracts and mushroom powder before and after the extraction were shown in Figure 2.

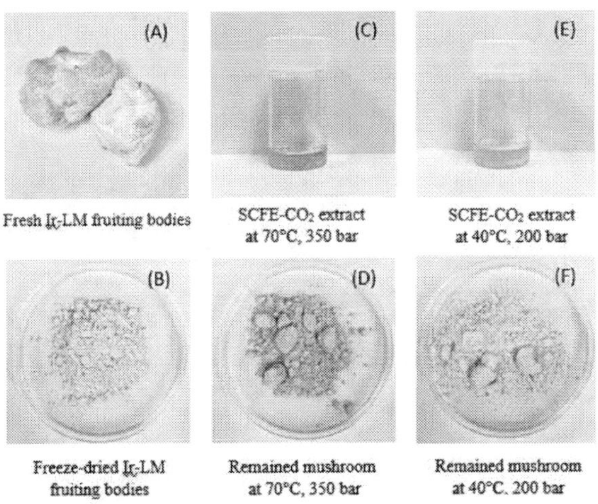

Figure 2. Fresh (A) and freeze-dried Ir-LM fruiting bodies (B); their extracts by SCFE-CO2 at 70°C, 350 bar (C), 40°C, 200 bar (E); and mushroom remained after extraction (D, F).

2.5 Analysis

2.5.1 UHPLC analysis

Determination of ergosterol, hericenone C and hericene A in the extract were performed using ultra high-performance liquid chromatography, UHPLC (Acquity Arc HPLC, waters, UK). The UV-VIS LC detector was connected to the UHPLC system, which was equipped with an Aquity-C18 (100 x 3 mm) analytical column. The mobile phase consisted of a mixture of methanol and water at a ratio of 98:2. The chromatograms were detected at 282 nm. The biologically active compounds content in the LMs as mg/g extract and mg/g dried LM calculated as mg of each bioactive compound (calculated from the calibration curve of standard ergosterol) in 1 g of the extract or 1 g of dried LM.

2.5.2 DPPH scavenging activity

To determine the antioxidant capacity of SCFE-CO_2 extracts, DPPH radical scavenging assay was adopted from a previous study with minor modification [17]. Briefly, DPPH radical solution was prepared by dissolving DPPH in absolute methanol at a concentration of 0.5 mM. The experiments were conducted in a 96-well plate. In each well, an aliquot (150 µL) of the DPPH solution was added to 50 µL of each extract. The plate was vigorously shaken and incubated in the dark for 30 min at room temperature. Trolox was used to constructing a calibration curve. The absorbance of the solution was measured on a microplate reader at 540 nm. All the experiments were performed in triplicates and calculated as mg TE /g dried LM: The scavenging capacity was calculated using Eq. 1.

$$\% \, Inhibition = [(A_{\text{control}} - A_{\text{sample}}) / A_{control}] \times 100 \quad (1)$$

Where A_{control} is the absorbance of the control that contained all reagents except the samples. A_{sample} is the absorbance of the mushroom extracts with reagents added.

2.5.3 ABTS scavenging activity

The ABTS radical scavenging assay was modified slightly from Ahmed et al. 2012 [18]. ABTS was prepared by mixing ABTS with $K_2S_2O_8$ solution and allowed the mixture to react at room temperature for 16 h. The ABTS solution was diluted with methanol to obtain the absorbance of 1.1-1.2 at 750 nm. One hundred microliters of ABTS solution were mixed with 50 µL of the extract in 96-well microtiter plates. The absorbance was monitored for 30 min at room temperature. Methanol was used as a blank. All measurements were performed in triplicate. The antioxidant activity was determined from the calibration curve of Trolox and calculated as mg TE/g dried LM followed Eq. 1.

2.5.4 Total phenolic content (TPC)

Total phenolic content (TPC) of SCFE-CO_2 extracts (40°C, 200 bar and 70°C, 350 bar) was evaluated using the Folin-Ciocalteu reagent. A 96-well microplate reader (Bio-Rad, iMark, USA) was used to determine the TPC value by measuring the absorbance at 750 nm. Gallic acid was used as a standard, and the results were calculated as gallic acid equivalents (GAE)/g of dry weight sample [19]. All experiments were performed in triplicates.

2.6 Statistical analysis

Data were analysed using the SPSS 11.6 for Windows (SPSS Inc.) by one-way analysis of variance (ANOVA). An independent t-test was used to compare the means of each treatment and the Scheffe test at $p < 0.05$ significance level was applied to compare the differences between each sample.

3 Results and discussion

3.1. Extraction of bioactive compounds by a conventional method

The content of ergosterol, hericenone C, and hericene A content in the extract from LM mushrooms are presented in Table 1. The Maceration of the LM mushroom revealed that Ir-LM contained significantly higher ($p < 0.05$) ergosterol content (2.09 ± 0.03 mg/g dried LM) than that of Reg-LM (1.74 ± 0.02 mg/g dried LM). Therefore, Ir-LM was selected for further study using SCFE-CO_2 at varying extraction times and temperatures. Ir-LM is considered as by-product because their inferior morphology (including misshapen caps) does not meet the specifications set by retailers, and is sold at low prices in the mushroom market. Therefore, alternative solutions like the SCFE-CO_2 methodology are required to increase the value of these mushroom by-products from the mushroom farm. However, the Reg-LM was not selected for SCFE-CO_2 in this research because the regular-shape LM can be sold fresh at a high price.

3.2 Extraction of bioactive compounds by supercritical fluid extraction using carbon dioxide

The extraction of bioactive compounds in Ir-LM obtained from Maceration (Table 1) and SCFE-CO_2 extraction (Table 2) methods were compared. The comparison study was performed based on the content of each specific bioactive compound including ergosterol, hericenone C, and hericene A which appeared in the extracts from both maceration and SCFE-CO_2 methods. The information is presented as mg/g of dried LM and mg/g of extract.

When expressing the results in mg/g dried LM, the extract from Maceration presented higher ergosterol content (2.09 ± 0.03 mg/g dried LM) than those of SCFE-CO_2 extracts, 1.11 ± 0.01 mg/g dried LM and 1.10 ± 0.01 mg/g dried LM, for 70°C at 350 bar and 40°C at 200 bar, respectively. This result might be caused by the extraction time of SCFE-CO_2 used in this study (40 min). However, further investigation on the effects of extraction time is required. Parada *et al.* 2015 [14] revealed that the extraction yield of Ir-LM extracted by SCFE-CO_2 (20 MPa, 40 °C) with pure CO_2 was increased with an increase in the extraction time (during 2 h). Longer extraction times allow supercritical fluid carbon dioxide (SCF-CO_2) to penetrate deeper into the sample matrix, resulting in greater solute dissolution and interaction with SCF-CO_2 over time [20].

Interestingly, ergosterol content in the Ir-LM extract obtained from SCFE-CO_2 (Table 2) at 70°C, 350 bar was the highest (61.15 ± 0.11 mg/g extract) when expressing the results as mg/g extract, followed by SCFE-CO_2 at 40°C, 200 bar (60.19 ± 0.31 mg/g extract) and Maceration (22.17 ± 0.31 mg/g extract). The contents of hericenone C and hericene A in the extracts obtained from both Maceration and SCFE-CO_2 also showed the same trend with the ergosterol content. Besides, Ir-LM extract obtained from SCFE-CO_2 for 70°C at 350 bar exhibited the highest hericenone C (43.35 ± 0.06 mg/g extract) and hericene A (6.53 ± 0.04 mg/g extract) content, followed by SCFE-CO_2 for 40°C at 200 bar (Table 2). The results represented that SCFE-CO_2 yielded a higher purity of ergosterol, hericenone C, and hericene A than the use of Maceration. The possible reason for the lower purity of bioactive compounds in the Ir-LM extract obtained by Maceration might be because the polar solvent (ethanol) used for the extraction can also be polyphenols or other compounds [21]. HPLC chromatograms of ergosterol standard and the extracts of Ir-LM by both extraction techniques are shown in Figure 3.

Table 1. Ergosterol, hericenone C, and hericene A concentrations in regular-shape (Reg-LM) and irregular-shape (Ir-LM) Lion's Mane mushroom extracted by conventional method (Mar).

Dried sample	Extract (g)	Extraction yield (%)	Bioactive compound	mg/g dried LM	mg/g extract
Reg-LM-Mar 10 g	1.04	10.40	Ergosterol	1.74 ± 0.02^b	15.72 ± 0.17^b
			Hericenone C	0.44 ± 0.01^c	3.99 ± 0.06^c
			Hericene A	0.03 ± 0.01^f	0.26 ± 0.01^e
Ir-LM-Mar 10 g	0.90	9.41	Ergosterol	2.09 ± 0.03^a	22.17 ± 0.31^a
			Hericenone C	0.35 ± 0.01^d	3.74 ± 0.14^c
			Hericene A	0.07 ± 0.01^e	0.74 ± 0.01^d

Different superscripts in the same column of each unit mean a significant difference at $p < 0.05$

Table 2. Ergosterol, hericenone C, and hericene A concentrations in irregular-shape Lion's Mane (Ir-LM) mushroom extracted by supercritical fluid extraction using carbon dioxide (SCFE-CO_2).

Dried sample	Extract (mg)	Extraction yield (%)	Bioactive compound	mg/g dried LM	mg/g extract
Ir-LM-SCFE-CO_2 (70°C, 350 bar) 3 g	52.20	1.74	Ergosterol	1.11 ± 0.01^a	61.15 ± 0.11^a
			Hericenone C	0.79 ± 0.01^b	43.35 ± 0.06^b
			Hericene A	0.12 ± 0.01^d	6.53 ± 0.04^d
Ir-LM-SCFE-CO_2 (40°C, 200 bar) 3 g	52.46	1.75	Ergosterol	1.10 ± 0.01^a	60.19 ± 0.31^a
			Hericenone C	0.32 ± 0.01^c	25.81 ± 0.08^c
			Hericene A	0.09 ± 0.01^e	4.75 ± 0.01^e

Different superscripts in the same column of each unit mean a significant difference at $p < 0.05$

The extraction of non-polar molecules is ensured by SCFE-CO_2 because supercritical fluid carbon dioxide (SCF-CO_2) is non-polar [12]. However, the CO_2 molecule

 https://doi.org/10.1051/e3sconf/202235502015

at the critical state exhibits low polarity due to the presence of a quadrupole moment. Thus, it can dissolve some polar compounds at strong extraction conditions [22]. Ergosterol has four ring structure with a hydrophilic hydroxyl, and an acyl side chain [23]. Song et al. 2009 [18] reported that SCFE-CO_2 exhibits dominance in adjustable polarity, and has the ability to extract ergosterol from *Ganoderma lucidum*.

There are very few reports on the extraction of polar compounds using carbon dioxide in a supercritical state as in SCFE-CO_2, due to the fact that SCF-CO_2 is a non-polar solvent. However, SCF-CO_2 might dissolve the slight polar compound like ergosterol under specific conditions of temperature and pressure. These are two possible reasons to support the result, firstly, the presence of a quadrupole moment in the CO_2 molecule at the supercritical state. Secondly, under appropriate conditions, the high vapor pressure of polar solute may significantly affect the SCFE-CO_2 and increase the extractive efficiency and the extract properties [18,22-24]. In addition, Wrona et al. 2019 [25] postulated that the extraction of polar compounds by non-polar solvent may be induced by the diffusion mechanism of bioactive chemicals from the internal sections of the plant matrix due to the high pressure and temperature used.

Another possible reason for higher bioactive compound contents in SCFE-CO_2 from 70°C at 350 bar is an increase in pressure increases the density of SCF-CO_2 which in turn increase the solvation capacity of SCF-CO_2 and leads to solubilization of more solute. Higher pressure extraction reduces the intermolecular distances between the solute and SCF-CO_2, allowing a higher mass transfer between them [26].

The effect of temperature and pressure on the color of SCFE-CO_2 extracts and remaining mushrooms after extraction were obvious as presented in Figure 2. Lower temperature and pressure (40°C, 200 bar) gave a lighter yellow-color extract (Figure 2E) and remaining mushroom (Figure 2F), while higher temperature and pressures (70°C, 350 bar) gave deep yellow-color extract (Figure 2C) and remaining mushroom (Figure 2D). Chang *et al.* 2021 [27] reported that stir-fried LM mushroom had significantly higher ($P < 0.05$) L* and +a* values than microwaved, steamed, and boiled cooking techniques. This could be due to the formation of Maillard reaction products. Besides, heat treatment might also induce non-enzymatic oxidative changes in polyphenol, resulting in the development of darker compounds [28]. This might be the possible reason for the darker color shown in the mushroom cake extracted by SCFE-CO_2 at 70°C, 350 bar.

Nevertheless, one of the major chromatogram peaks (Rt = 5.368) observed in Ir-LM extract from Maceration did not appear in both SCFE-CO_2 chromatograms as presented in Figure 3. This peak appeared at a shorter retention time (Rt) than hericenone C, ergosterol, and hericene A peak that represented a high polar compound. Reversed phase-ultra-high-performance liquid chromatography exhibits a strong attraction between the polar solvent and polar molecule in the mixture being passed through the column. Therefore, the polar molecules move through the column more quickly than non-polar compounds [29]. To extract this compound (Rt = 5.368) by SCFE-CO_2, the extraction temperature and pressure should be optimized or the modifiers such as ethanol could be used to increase the polarity of the solvent mixture to extract higher polar compounds including ergosterol.

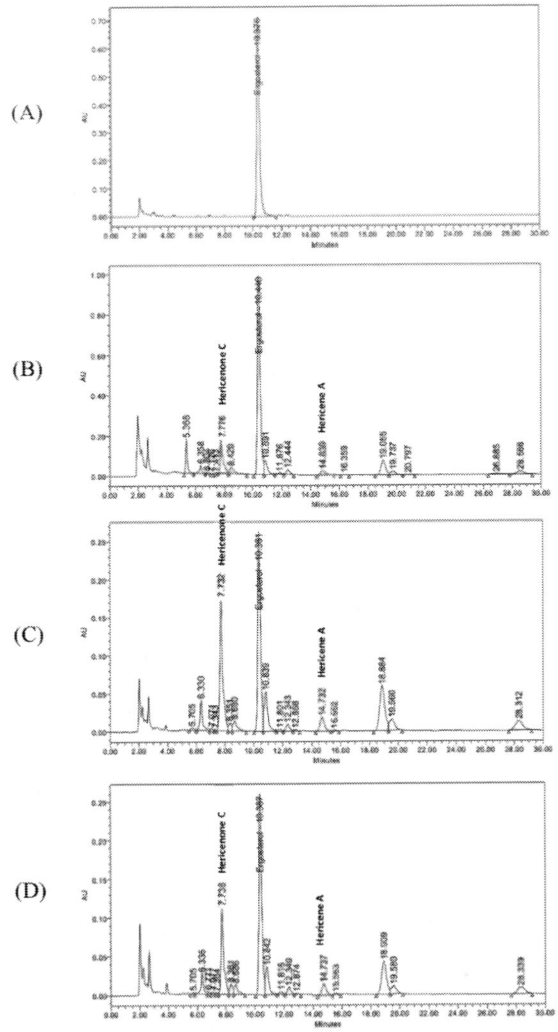

Figure 3. HPLC chromatogram of ergosterol standard (A), bioactive compounds extracted from Ir-LM by Maceration (B), SCFE-CO_2 at 70°C, 350 bar (C), and SCFE-CO_2 at 40°C, 200 bar (D).

Modifiers cause the sample matrix to swell, resulting in the increased mass transfer rate between the solute and solvent. Modifiers can also enhance the extract flow better by lowering its viscosity. All these factors work together to increase yield [30]. Selectivity of the supercritical extraction fluid is the main factor in determining the functional qualities of the final extract. Changing the extraction conditions, such as temperature, pressure, and type and concentration of co-solvent could allow a selective extraction of the desired compounds from biomass [31].

Mazzutti *et al.* 2012 [32] used supercritical fluid extraction with pure CO_2 and CO_2 plus 2.5 %, 5.0 %, and 10.0 % (w/w) ethanol as a co-solvent to extract *Agaricus brasiliensis* mushrooms. They discovered that the optimal conditions for obtaining high yields (1.19 %) using pure CO_2 were 30 MPa and 323.15 K. The extraction yields were much higher, reaching 4.2 % when 10% ethanol was used as a co-solvent to recover polar molecules. Ergosterol extraction from *Pleurotus pulmonarius* mushrooms was investigated using the SCFE-CO_2 [33]. Compared to ultrasound-assisted extraction with water as solvent (UAEW) (0.9 mg/g) and SCFE-CO_2 without co-solvent, the results showed that SCFE-CO_2 with co-solvent (5.0 % EtOH) led to the best acceptable approach to generate the extracts rich in ergosterol content, reaching a value of 40.1 mg/g.

3.3 Antioxidant activity and TPC of Ir-LM extract obtained from SCFE-CO_2

DPPH, ABTS scavenging ability, and total phenolic contents of SCFE-CO_2 extracts are revealed in Table 3. The mushroom extract by SCFE-CO_2 was diluted with absolute methanol to determine their antioxidant properties. The Ir-LM extract obtained from SCFE-CO_2 at 70°C at 350 bar (0.09 ± 0.01 mg TE/g dried LM) was found significantly higher in DPPH radical scavenging activity than the extract obtained at 40°C, 200 bar (0.07 ± 0.01 mg TE/g dried LM). The IC_{50} by DPPH of both SCFE-CO_2 extracts were 10.26 mg/ml and 12.99 mg/ml of SCFE-CO_2 with 70°C at 350 bar and 40°C at 200 bar, respectively. However, no significant difference in ABTS scavenging ability and total phenolic contents between both SCFE-CO_2 extracts was observed (Table 3).

Table 3. The antioxidant activities and total phenolic content of Ir-LM were extracted by supercritical fluid extraction using carbon dioxide (SCFE-CO_2).

Extract	DPPH (mg TE/g dried LM)	ABTS (mg TE/g dried LM)	Total phenolic content (mg GAE/g dried LM)
Ir-LM-SCFE-CO_2 (70°C, 350 bar)	0.09 ± 0.01^a	0.11 ± 0.01^a	0.20 ± 0.01^a
Ir-LM-SCFE-CO_2 (40°C, 250 bar)	0.07 ± 0.01^b	0.11 ± 0.01^a	0.19 ± 0.01^a

Different superscripts in the same column mean significant difference at $p < 0.05$
DPPH: 2,2-diphenyl-1-picrylhydrazyl
ABTS: 3-ethylbenzthiazoline-6-sulphonic acid

Dupont *et al.* 2021 [34] described that the presence of two double bonds in the B-ring of ergosterol could confer antioxidant properties. The antioxidant activity of the lipophilic fraction of button mushrooms was shown to be dominated by ergosterol [35]. Recently, there is an increasing interest in bioactive compounds from natural sources including antioxidants from Sacha Inchi oil [36-37], bamboo mushroom [38] and LM obtained from maceration [39-40]. The antioxidant activities of both SCFE-CO_2 extracts might depend on the ability of other compounds in the crude extracts including phenolic compounds. Total phenolic contents were found in both Ir-LM extract obtained from SCFE-CO_2 with 70°C, 350 bar (0.20 ± 0.01 mg GAE/g dried LM) and the extract obtained from SCFE-CO_2 at 40°C, 200 bar (0.19 ± 0.01 mg GAE/g dried LM).

4 Conclusion

Irregular-shape Lion's Mane mushroom extracts obtained by supercritical extractions from 70°C at 350 bar presented the highest purity of the ergosterol, hericenone C, and hericene A contents when compared to the extract obtained by SCFE-CO_2 from 40°C at 200 bar and the conventional solvent extraction or Maceration technique. The extracts obtained by SCFE-CO_2 from 70°C at 350 bar also showed a higher antioxidant activity determined by the DPPH method than the extract obtained from 40 °C at 200 bar. However, the SCFE-CO_2 method with two extraction conditions used for the extraction of bioactive compounds in this research proved to be more promising than conventional methods used in the purification of interesting compounds in Ir-LM. This research enables the extraction of bioactive compounds from mushroom by-products by SCFE-CO_2, a green extraction technology, for industrial scale applications.

Acknowledgements

This research was funded by King Mongkut's University of Technology North Bangkok and National Science and Technology Development Agency, Thailand (Grant number 024/2563). The research was also partially supported by NRCT Senior Research Scholar Program (Contract No.814-2020). In addition, the authors would like to thank Dr. Thongchai Koobkokkruad for his kind support on the use of supercritical carbon dioxide extractor, and Dr. Aphiwat Pankaew for his recommendation on HPLC analysis.

References

[1] B. Thongbai, S. Rapior, K.D. Hyde, K. Wittstein, M. Stadler, Hericium erinaceus, an amazing medicinal mushroom, Mycological Progress, 14(91) (2015): 1-23

[2] Z.G. Chen, P. Buchanan, S.Y. Quek, Identification and determination of compounds unique to Hericium in an edible New Zealand mushroom Hericium novae-zealandiae, Food Analytical Methods, 15 (2022): 67-74

[3] J. Chen, X. Zeng, L.Y. Yang, X.Q. Zhang, J.M. Li, K. Ma, H.W. Liu, S.X. Guo, Genomic and transcriptomic analyses reveal differential regulation of diverse terpenoid and polyketides secondary metabolites in Hericium erinaceus, Scientific Reports, 7 (2017): 10151

[4] M. Gasecka, M. Siwulski, Z. Magdziak, S. Budzyn´ska, K. Stuper-Szablewska, P. Niedzielski, M. Mleczek, The effect of drying temperature on

bioactive compounds and antioxidant activity of Leccinum scabrum (Bull.) Gray and Hericium erinaceus (Bull.) Pers, Journal of Food Science and Technology, 57(2) (2020): 513-525

[5] W. Huan, Z. Tianzhu, L. Yu, W. Shumin, Effects of ergosterol on COPD in mice via JAK3/STAT3/NF-kappa-B pathway, Inflammation, 40(3) (2017): 884-893

[6] M.A. Bing-Ji, M.A. Jin-Chuan, R.U.A.N. Yuan, Hericenone L, a new aromatic compound from the fruiting bodies of Hericium erinaceums, Chinese Journal of Natural Medicines, 10(5) (2012): 0363-0365

[7] H. Kawagishi, M. Ando, H. Sakamoto, S. Yoshida, F. Ojima, Y. Ishiguro, N. Ukai, S. Furukawa, Hericenones C, D and E, stimulators of nerve growth factor (NGF)- synthesis, from the mushroom Hericium erinaceum, Tetrahedron letters, 32(35) (1991): 4561-4564

[8] S.K. Lee, S.H. Ryu, A. Turk, S.W. Yeon, Y.H. Jo, Y.K. Han, B.Y. Hwang, K.Y. Lee, M.K. Lee, Characterization of α-glucosidase inhibitory constituents of the fruiting body of Lion's Mane mushroom (Hericium erinaceus), Journal of Ethnopharmacology, 262 (2020): 113197

[9] A. Turk, S.W. Yeon, S.H. Ryu, S.M. Ko, B.S. Kim, B.Y. Hwang, M.K. Lee, Effect of culture conditions on the content of hericene A, an α-glucosidase inhibitory constituent of Hericium erinaceus, Scientia Horticulturae, 288 (2021): 110407

[10] I. Aguiló-Aguayo, J. Walton, I. Viñas, B.K. Tiwari, Ultrasound assisted extraction of polysaccharides from mushroom by-products, Lebensmittel-Wissenschaft & Technologie Food Science and Technology, 77 (2017): 92-99

[11] S. Prasad, H. Rathore, S. Sharma, A.S. Yadav, S.N. Naik, Studies on effects of supercritical CO_2 extraction on yield and antioxidant activity of L.edodes extract, Research Journal of Pharmaceutical, Biological and Chemical Sciences, 8(4) (2017): 1144-1154

[12] N. Nagavekar, R.S. Singhal, Supercritical fluid extraction of Curcuma longa and Curcuma amada oleoresin: Optimization of extraction conditions, extract profiling, and comparison of bioactivities, Industrial Crops and Products, 134 (2019): 134-145

[13] D. Morales, A. Gil-Ramirez, F.R. Smiderle, A.J. Piris, A. Ruiz-Rodriguez, C. Soler-Rivasa, Vitamin D-enriched extracts obtained from shiitake mushrooms (Lentinula edodes) by supercritical fluid extraction and UV-irradiation, Innovative Food Science and Emerging Technologies, 41 (2017): 330-336.

[14] M. Parada, A. Rodríguez-Blanco, F.F.A. Magan, H. Domínguez, Sequential extraction of Hericium erinaceus using green solvents, Lebensmittel-Wissenschaft & Technologie Food Science and Technology, 64 (2015): 397-404

[15] P. Mattila, A.M. Lampi, R. Ronkainen, J. Toivo, V. Piironen, Sterol and vitamin D_2 contents in some wild and cultivated mushrooms, Food Chemistry, 76, 3 (2002), 293-298

[16] A. Villares, L. Mateo-Vivaracho, A. García-Lafuente, E. Guillamón, Storage temperature and UV-irradiation influence on the ergosterol content in edible mushrooms, Food Chemistry, 147 (2014): 252-256

[17] A.S. Ahmed, E.E. Elgorashi, N. Moodley, L.J. McGaw, V. Naidoo, J.N. Eloff, The antimicrobial, antioxidative, anti-inflammatory activity and cytotoxicity of different fractions of four South African Bauhinia Species used traditionally to treat diarrhoea, Journal of Ethnopharmacology, 143 (2012): 826-839

[18] S. Song, X. Jia, Y. Chen, L. Wang, Li. Feng, Conditions for extraction of ergosterol from fruits of Ganoderma lucidum using supercritical fluid extraction, China Journal of Chinese Materia Medica, 34(14) (2009): 1783-1785

[19] A. Rosa, A. Maxia, D. Putzu, A. Atzeri, B. Era A. Fais, C. Sanna, A. Piras, Chemical composition of Lycium europaeum fruit oil obtained by supercritical CO_2 extraction and evaluation of its antioxidant activity, cytotoxicity and cell absorption, Food Chemistry, 230 (2017): 82-90

[20] G.N. Sapkale, S.M. Patil, U.S. Surwase, P.K. Bhatbhage, Supercritical fluid extraction, International Journal of Chemical Science, 8 (2010): 729-743

[21] S.A. Heleno, P. Diz, M.A. Prieto, L. Barros, A. Rodrigues, M.F. Barreiro, I.C.F.R. Ferreira, Optimization of ultrasound-assisted extraction to obtain mycosterols from Agaricus bisporus L. by response surface methodology and comparison with conventional Soxhlet extraction, Food Chemistry, 197 (2016): 1054-1063

[22] M. Mukhopadhyay, Natural Extract Using Supercritical Carbon Dioxide, CRC Press, USA (2000)

[23] G.G. Piva, E. Casalta, J.L. Legras, C. Tesnière, J.M. Sablayrolles, D. Ferreira, A. Ortiz-Julien, V. Galeote, J.R. Mouret, Characterization and role of sterols in Saccharomyces cerevisiae during white wine alcoholic fermentation, Fermentation, 8(2) (2022): 90

[24] O. Wrona, K. Rafińska, C. Możeński, B. Buszewski, Supercritical Fluid Extraction of Bioactive Compounds from Plant Materials, Journal of AOAC International, 100(6) (2017): 1624-1635

[25] O. Wrona, K. Rafi´nska, J. Walczak-Skierska, C. Mo˙ze´nski, B. Buszewski, Extraction and determination of polar bioactive compounds from alfalfa (Medicago sativa L.) using supercritical techniques, Molecules, 24 (2019): 4608

[26] T.P.K. Murthy, B. Manohar, Modelling solubility of phenolics of mango ginger extract in supercritical carbon dioxide using equation of state and empirical

models, Journal of Food Science and Technology, 52 (2015): 5557-5567

[27] K.A. Chang, H.N. Kow, T.E. Tan, K.L. Tan, L.Y. Chew, Y.P. Neo, V. Sabaratnam, Effect of domestic cooking methods on total phenolic content, antioxidant activity and sensory characteristics of Hericium erinaceus, International Journal of Food Science and Technology, 56 (2021): 5639-5646

[28] S. ivanovic, R. Buescher, S.K. Kim, Mushroom texture, cell wall composition, color and ultrastructure as affected by pH and temperature, Journal of Food Science, 68 (2003): 1860-1865

[29] S.D. Sarker, L. Nahar, Applications of highperformance liquid chromatography in the analysis of herbal products, Elsevier, Netherlands, (2015)

[30] M.D.A. Saldaña, F. Temelli, S. E. Guigard, B. Tomberli, C.G. Gray, Apparent solubility of lycopene and β-carotene in supercritical CO_2, CO_2^+ ethanol and CO_2^+canola oil using dynamic extraction of tomatoes, Journal of Food Engineering, 99 (2010): 1-8

[31] T. Arumugham, K. Rambabu, S.W. Hasan, P.L. Show, J. Rinklebe, F. Banat, Supercritical carbon dioxide extraction of plant phytochemicals for biological and environmental applications, A review Chemosphere, 271 (2021): 129525

[32] S. Mazzutti, S.R.S. Ferreiraa, C.A.S. Riehl, S.J. Artur, F.A. Smaniac, J. Martínez, Supercritical fluid extraction of Agaricus brasiliensis: Antioxidant and antimicrobial activities, Journal of Supercritical Fluids, 70 (2012): 48-56

[33] I. Milovanovic, G. Zengin, S. Maksimovicc, V. Tadicd, Supercritical and ultrasound-assisted extracts from Pleurotus pulmonarius mushroom: chemical profiles, antioxidative, and enzyme inhibitory properties, Journal of the Science of Food and Agriculture, 101(6) (2021): 2284-2293

[34] S. Dupont, P. Fleurat-Lessard, R.G. Cruz, C. Lafarge, C. Grangeteau, F. Yahou, P. Gerbeau-Pissot, O.A. Júnior, P. Gervais, F. Simon-Plas, P. Cayot, L. Beney, Antioxidant properties of ergosterol and its role in yeast resistance to oxidation, Antioxidants, 10 (2021): 1024

[35] S. Shao, M. Hernandez, J.K. Kramer, D.L. Rinker, R. Tsao, Ergosterol profiles, fatty acid composition, and antioxidant activities of button mushrooms as affected by tissue part and developmental stage, Journal of Agricultural and Food Chemistry, 58 (2010): 11616-11625

[36] S. Suwannasang, B. Thumthanaruk, Q. Zhong, D. Uttapap, C. Puttanlek, S. Vatanyoopaisarn, V. Rungsardthong, The improved properties of zein encapsulating and stabilizing Sacha Inchi oil by surfactant combination of lecithin and Tween 80, Food and Bioprocess Technology, 14 (2021): 2078-2090.

[37] S. Suwannasang, Q. Zhong, B. Thumthanaruk, S. Vatanyoopaisarn, D. Uttapap, C. Puttanlek, V. Rungsardthong, Physicochemical properties of yogurt fortified with microencapsulated Sacha Inchi oil, Lebensmittel-Wissenschaft & Technologie, 61 (2022): 113375

[38] F. Binheam, B. Thumthanaruk, S. Vatanyoopaisarn, V. Rungsardthong, S. Poodchakarn, Antioxidant capacity of the bamboo mushroom (Dictyophora indusiata) and the effects of temperature and relative humidity on its growth, In proceeding of the 3rd International Conference on Informatics, Agriculture, Management, Business administration, Engineering, Science and Technology - IAMBEST 2022, Chumphon (2022)

[39] P. Joradon, V. Rungsardthong, U. Ruktanonchai, K. Suttisintong, T. Iempridee, B. Thumthanaruk, S. Vatanyoopaisarn, N. Sumonsiri, D. Uttapap, Ergosterol content and antioxidant activity of Lion's Mane mushroom (Hericium erinaceus) and its induction to vitamin D_2 by UVC-irradiation, in proceeding of the 8th International Conference on Agricultural and Biological Sciences - ABS 2022, Shenzhen (2022)

[40] N. Tachabenjarong, V. Rungsardthong, U. Ruktanonchi, K. Suttisintong, T. lempridee, B. Thumthanaruk, S. Vatanyoopaisam, S. Poodchakarn, D. Uttapap, Bioactive compound and antioxidant activity of Lion's mane mushroom (Hericium erinaceus) from different growth period, In proceeding of the International Conference on Research, Invention, and Innovation Congress - RI²C 2022, Bangkok (2022)

Bioactive compounds and antioxidant activity of Lion's Mane mushroom (*Hericium erinaceus*) from different growth periods

Nattapat Tachabenjarong[1], *Vilai* Rungsardthong[1,*], *Uracha* Ruktanonchi[2], *Somrerk* Poodchakarn[3], *Benjawan* Thumthanaruk[1], *Savitri* Vatanyoopaisarn[1], *Khomson* Suttisintong[2], *Tawin* Iempridee[2] *and Dudsadee* Uttapap[4]

[1]Department of Agro-Industrial, Food and Environment, Faculty of Applied Science, Food and Agro-industrial Research Center, King Mongkut's University of Technology North Bangkok, Bangkok 10800, Thailand
[2]National Nanotechnology Center, National Science and Technology Development agency, Thailand Science Park, Pathumthani 12120, Thailand
[3]Department of Production and Robotics Engineering, Faculty of Engineering, King Mongkut's University of Technology North Bangkok, Bangkok 10800, Thailand
[4]Division of Biochemical Technology, School of Bioresources and Technology, King Mongkut's University of Technology Thonburi, Bangkok 10150, Thailand

Abstract. Lion's Mane mushroom (*Hericium erinaceus*) is one of the most popularly edible and medicinal mushrooms. However, there is still a lack of knowledge on the relationship between growth period and bioactive content in the mushroom. The objectives of this research were to study bioactive compounds and antioxidant activity of Lion's Mane mushroom at various growth periods. The mushroom was cultivated and harvested at growth periods of 14, 21, and 28-days. The samples were dried by lyophilization and extracted with ethanol. Bioactive compounds (ergosterol, hericenone C, and hericene A), total phenolic content, total flavonoid content, and antioxidant activity of the samples were analyzed. HPLC analysis demonstrated the highest concentration of ergosterol, hericenone C and hericene A in the 14-days, 21-days, and 28-days samples, respectively. Total phenolic content and total flavonoid content of the dried sample were not statistically significant different ($p>0.05$). The 21-days sample showed higher activity than the 14-days and 28-days samples for both DPPH and ABTS radical scavenging assays. In conclusion, the 14-days sample showed the highest concentrations of bioactive compounds, while the 21-days sample showed the highest yield and antioxidant activity. The 28-days sample exhibited a change in morphology and color. This study demonstrates that the growth periods of Lion's Mane mushroom play a role in their bioactive compounds and antioxidant activities. Finally, the correlation of growth periods to the content of other bioactive compounds can be used for mushroom cultivation and applications in foods, food supplements and cosmetics.

Keyword. Hericenone C, Ergosterol, Hericene A, ABTS, DPPH

1 Introduction

Mushrooms are good sources of nutrients and more than 2000 species of edible and/or medicinal mushrooms were identified [1]. Mushroom intake receives very high attention due to its flavour, beneficial health properties, and high nutritional value especially protein [2, 3]. Some examples of edible mushrooms are the White button mushroom (*Agaricus bisporus*), Maitake mushroom (*Grifola frondosa*) and Enoki mushroom (*Flammulina velutipes*). Several mushrooms are sources of health-promoting compounds, and some are medicinal mushrooms such as Lion's Mane mushroom (*Hericium erinaceus*), Reishi mushroom (*Ganoderma lucidum*) and Chaga mushroom (*Inonotus obliquus*), which indicate high potentials for many health benefits, such as hyperglycemia effect, improved lipid metabolism, anti-obesity, anti-inflammatory, anti-proliferative, and antioxidant capacity [4-7].

Lion's Mane mushroom (*Hericium erinaceus*), LM, also known as Pompom mushroom, Monkey Head mushroom, Yamabushitake mushroom (Japanese) and Houtou mushroom (Chinese), is one of the medicinal mushrooms that contains many bioactive compounds [8]. This mushroom looks like a pack of white silk thread (5-20 cm) and grows in oak, walnut, and perennials. Generally, Lion's Mane mushroom are found in the Northern Hemisphere, such as Europe and East Asia, especially in Japan [9]. Bioactive compounds from Lion's Mane mushroom have many health benefits, such as neuroprotective effects, anti-obesity, antioxidant, anti-microbial, anti-inflammatory, and improved immunology [10-16]. Antioxidant activity from several compounds, such as phenolic compounds, flavonoid compounds, and unique bioactive compounds has been demonstrated. It can reduce brain degeneration in people with mild cognitive impairment in both mice and the

* Corresponding author: vilai.r@sci.kmutnb.ac.th

elderly and exhibits more stimulating effects on the nervous system than other mushrooms [17, 18].

Mushroom cultivation is an important profession in the Thai economy. In many cases for commercialization, agricultural wastes, e.g. rice straw, are utilized as substrates for mushroom cultivation, which is well agreed to the circular economy. Nowadays, the development of mushroom cultivation is important for control and preferable production. Farm environments and growth substrates significantly affect the quantity and quality of the mushroom obtained [19-22]. There are many studies on the effects of humidity, light, temperature, and growth substrate on bioactive compounds and the activity of the mushroom but only a few studies on the effect of growth period were reported. Antioxidant activity, total phenolic content, and total flavonoid content in each growth period of the giant Leucopax mushroom (*Leucopaxillus giganteus*) mycelia were demonstrated only at a laboratory scale [23]. In this study, Lion's Mane mushroom were cultivated in a farm with controlled environments, in terms of temperature and humidity, and the bioactive compounds in the mushroom from different growth periods were investigated. The information obtained will help the mushroom farm or processor to choose the optimal growth period of the mushroom for sale or further product development. The control of cultivation conditions would relate to the sustainable development of natural resource management.

2 Material and methods

2.1 Raw materials and chemicals

Lion's Mane mushrooms were cultivated at Fresh & Friendly Farm Co., Ltd., Pathumthani province, Thailand. Ethanol (AR grade) was obtained from PanReac AppliChem ITW Reagents (Barcelona, Spain), while ergosterol was supplied by Sigma-Aldrich (Madrid, Spain). Gallic acid was bought from Sigma-Aldrich (USA). Folin-Ciocalteu reagent, 2,2'-azino-bis (ABTS) and 2,2-diphenyl-1-picrylhydrazyl (DPPH) were obtained from Sisco Research Laboratory Pvt. Ltd. Sodium carbonate (Na_2CO_3), sodium nitrite ($NaNO_2$), sodium hydroxide (NaOH), aluminium chloride ($AlCl_3$) and potassium persulfate ($K_2S_2O_8$) were purchased from Merch (India). Trolox (6-hydroxy-2,5,7,8-tetramethylchroman-2-carboxylic acid) was obtained from Copens Scientific (M) Sdn Bhd (Tedia, USA), while water (HPLC grade) and methanol (HPLC grade) were supplied from LAB-SCAN (Gliwice, Poland).

2.2 Lion's Mane mushroom cultivation

In this study, Lion's Mane mushroom was cultivated under controlled conditions including temperature and humidity. The room temperature was controlled at 15-16°C throughout the cultivation. The cultivation parameters including carbon dioxide concentration were monitored continuously for 28 days by corresponding sensors in the inoculation room via the online system. The growth substrates were composed of sawdust, rice bran, and corn cob at the ratio of 40:30:30, and their acidity was adjusted to pH 9-10 using calcium carbonate ($CaCO_3$). After sterilization, the growth substrate in the plastic bottles was inoculated with Lion's Mane mushroom spawn and inoculated for their mycelium growth for 18-20 days. The growth substrate composts were opened to let the fruiting bodies grow in the inoculation room. Lion's Mane mushrooms were harvested and determined for their morphology, yield, bioactive compounds, and antioxidant activity at 14-days, 21-days, and 28-days of cultivation.

2.3 Preparation of Lion's Mane mushroom

Lion's Mane mushrooms harvested from different growth periods were lyophilized using a freeze drier (GFD30S, Grisrianthong, Thailand) at -30 °C for 36 h. The dried samples were ground and separated into small sizes (particle size < 150 μm). One hundred mL of ethanol was added to 5 g of the dried sample and extracted for 72 h in the dark. The ethanol extracts were filtered by filter papers (Whatman® qualitative filter paper, Grade 1) and dried by a rotary vacuum evaporator (Rotavapor® R114, Buchi, Switzerland). The crude extract was stored at -4 °C until use. The yield percentage of the dried mushroom powder was calculated as g dried LM/100g fresh LM while % crude extract was calculated from the crude extract (g) obtained from 100 g of dried LM. The moisture content of the dried Lion's Mane mushroom was analyzed by followed AOAC 1999 [24].

2.4 Analysis

2.4.1 Determination of bioactive compounds

The samples were dissolved in methanol for the analysis of total phenolic content, total flavonoid content, bioactive compounds, and antioxidant activity. The crude extracts in methanol were filtered by filter membranes (PES 0.22 μm 13 mm Chromplus) before the analysis by high performance liquid chromatography (HPLC, Agilent 1200, Germany). Column C18 column (250 x 4.6 mm, 5 μm particle size, Zorbax Eclipse Plus C18, Agilent) with a guard column (12.5 x 4.6 mm, 5 μm particle size, Zorbax Eclipse Plus, Agilent) were used. The analysis was conducted at 25 °C, with a UV detector at a wavelength of 282 nm. The mobile phase was methanol: water of 98:2. Ergosterol standard was used as a reference for quantification. The content of ergosterol, hericenone C, and hericene A were calculated from their peak area compared with the area of ergosterol standard.

2.4.2 Total phenolic content

Each extract was measured for its total phenolic content by Folin-Ciocalteu (FC) method followed by Cheung, Cheung, & Ooi 2003 [25] with slight modification. Ten μL of crude extract and 8 μL of Folin-Ciocalteu reagent were loaded onto a 96-well microplate. After 5 min, 20 μL of 20% sodium carbonate was added and made up to 200 μL with methanol. The reaction was measured for absorbance at 750 nm by a microplate reader (iMark™ Microplate Absorbance Reader, BIORAD, Japan). The results were calculated as gallic acid equivalent (GAE)/g dried LM.

2.4.3 Total flavonoid content

Total flavonoid content in the samples was determined by aluminium chloride assay followed by Moon *et al.* 2018 [26] with slight modification. Ten μL of crude extract and 10 μL of 5% sodium nitrite were loaded onto 96-well microplates containing 170 μL of distilled water. After 6 min, 10 μL of aluminium chloride and 50 μL of 1 M sodium hydroxide were added, after 5 min incubation. The absorbance at 540 nm was measured using a microplate reader (iMark™ Microplate Absorbance Reader, BIO RAD, Japan). The results were calculated as catechin equivalent (CE)/g dried LM.

2.4.4 Antioxidant activity

Antioxidant activity of the samples was analyzed by 2,2-diphenyl-1-picrylhydrazyl (DPPH) radical scavenging assay and 2,2'-azino-bis (ABTS) radical scavenging assay followed by Gasecka *et al.* 2020 [13] with slight modification. DPPH radical scavenging activity of the samples was determined by adding 100 μL DPPH reagent and 50 μL of the extracts onto a 96-well microplate. After 30 min, the reaction was measured for absorbance at 540 nm by a microplate reader (iMark™ Microplate Absorbance Reader, BIO RAD, Japan). For ABTS radical scavenging assay, 7mM ABTS was prepared in 5 mL distilled water and 175 μL of 140 mM potassium persulfate was added to the flask and made up to 10 mL with distilled water. After 16 h, 50 μL of crude extract and 100 μL of ABTS solution were loaded onto a 96-well microplate. After 30 min, the reaction was measured for absorbance at 750 nm by a microplate reader (iMark™ Microplate Absorbance Reader, BIORAD, Japan). The DPPH and ABTS radical scavenging activity were calculated as follows (1):

$$\%\ Inhibition = [(A_{control} - A_{sample}) / A_{\mathrm{control}}] \times 100 \quad (1)$$

where $A_{control}$ is the absorbance of the control that contained all reagents except the samples. A_{sample} is the absorbance of the samples with reagents added.

The radical scavenging activity from DPPH and ABTS radical scavenging assay were expressed as IC_{50} value (mg extract/mL), which is the inhibitory concentration at which 50% of radicals were scavenged. A low IC_{50} value indicates the strong activity of the sample.

2.5 Statistical analysis

The results were expressed as the mean value of three replicates (n=3) and standard deviation. Differences between mean values were evaluated by one-way analysis of variance (ANOVA) followed by a post hoc Tukey's test.

3 Results and discussions

3.1 Cultivation and preparation of dried Lion's Mane mushroom

Environmental conditions during the cultivation of Lion's Mane mushroom were monitored using corresponding sensors and recorded (Table 1). The growth periods in this study were shorter than 35-days in the study of Imtiaj *et al.* 2008 [21], and similar to Hassan 2007 [27], who reported the cultivation for 14–20 days. Different cultivation conditions and growth substrates significantly affect the growth periods and morphology of the mushrooms. According to the standard and mushroom specification of Fresh & Friendly Farm, Lion's Mane mushrooms should present a white exterior, short white silk thread, and firm texture. The optimal morphology of the cultivated mushroom was observed at 14-days, while 21-days and 28-days samples showed more yellow colour, irregular shape, and long yellow silk (Table 2). The yield of the dried Lion's Mane mushroom at 14-days, 21-days, and 28-days cultivation were in a range of 13.5-16.17% (g/100g dried LM) as presented in Table 3. The 28-days sample indicated the lowest yield of dried Lion's Mane mushroom which might be due to the accumulation of water within their cells. However, crude extract content from all growth periods was not significantly different ($p > 0.05$).

Table 1. Environmental conditions during the cultivation of the Lion's Mane mushroom farm.

Cultivation conditions	Cultivation time (day)	
	0-7	8-28
Temperature (°C)	16.17±0.18	15.30±0.61
Humidity (%)	91.37±2.32	88.64±3.33
Light (lx)	4.27±0.16	0.82±0.37
Carbon dioxide (ppm)	538±197	542±116

3.2 Bioactive compounds

HPLC chromatograms of 14-days, 21-days, and 28-days LM extracts show similar peak and retention times of bioactive compounds (Figure 1). Hericenone C, ergosterol, and hericene A were found at 14.9 min, 17 min, and 28.54 min of retention time, respectively. We have previously isolated and purified hericenone C and hericene A from Lion's Mane mushroom and confirmed the compounds by NMR (Nuclear magnetic resonance). Both purified compounds were identified for their retention times by HPLC previously (data not shown). The results showed that the highest content of hericenone C, ergosterol, and hericene A were detected

https://doi.org/10.1051/e3sconf/202235502016

at 14-days as 1.23±0.10 mg/g dried LM, 3.54±0.30 mg/g dried LM, 0.09±0.01 mg/g dried LM and decreased to 0.73±0.09 mg/g dried LM, 2.62±0.31 mg/g dried LM, 0.03±0.01 mg/g dried LM and 0.48±0.01 mg/g dried LM, 1.87±0.06 mg/g dried LM at 21-days and 28-days, respectively, as presented in Table 4. Hericenone C and Hericene A are bioactive compounds specifically found in Lion's Mane mushroom. Hericene A were not detected at 28-days of cultivation. Bioactive compounds were produced in the early stages of the growth period 14-days and decreased after that. Ergosterol content found in this study, 3.54 mg/g dried LM, was much higher than ergosterol content reported by Gasecka *et al.* 2020 [13], 4.5 mg/kg dried LM.

Table 2. Fresh mushroom, dried powder, and suspension in ethanol of Lion's Mane mushroom from different growth periods.

Lion's Mane mushroom	Growth periods		
	14-days	21-days	28-days
Fresh mushroom		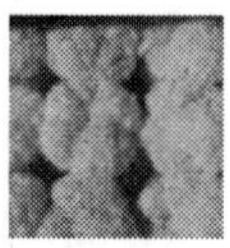	
Dried LM powder			
Suspension in ethanol	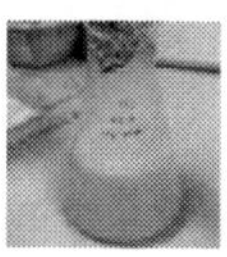	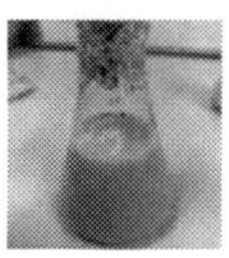	

Table 3. Yield, moisture, total phenolic contents, and total flavonoid content in Lion's Mane mushroom from different growth periods

Yields and compositions	Growth periods (days)		
	14	21	28
Yield (g dried LM/100g fresh LM)	14.35	16.175	13.5
Crude extract[ns] (g dried extract/100g dried LM)	11.73±1.13	14.58±0.58	13.38±1.89
Moisture (% dry weight basic)	9.46±0.29[a]	7.27±0.08[b]	6.79±0.09[c]
Total phenolic content[ns] (mg GAE/g dried LM)	0.85±0.15	0.90±0.19	0.87±0.23
Total flavonoid content[ns] (mg CE/g dried LM)	3.66±1.90	2.92±0.79	3.19±0.73

*Different superscripts in the same row mean significant difference at $p<0.05$
[ns]No significant differences at $p<0.05$

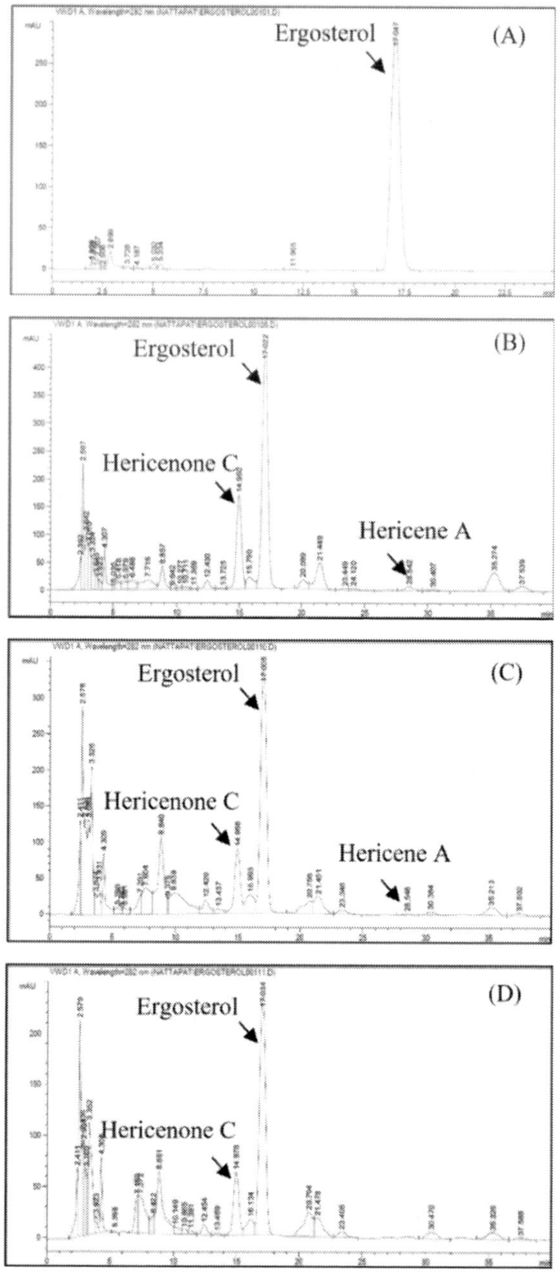

Figure 1. HPLC chromatogram of: ergosterol standard (A); and Lion's Mane mushroom from each growth periods, 14-days (B), 21-days (C) and 28-days (D).

3.3 Total phenolic content and total flavonoid content

Total phenolic content and total flavonoid content in the Lion's Mane mushroom from different growth periods were not significantly different ($p>0.05$) as presented in Table 3. The content of phenolic and flavonoid compounds tended to decrease along the growth period from 14-days to 28-days. The compounds may change their forms during each growth period [28]. The total phenolic content of the samples in this study was similar to the data demonstrated by Wong *et al.* 2009 [10]. However, the total phenolic content of the samples in our study was lower than wild Lion's Mane mushrooms [7]. Both total phenolic content and total flavonoid content from giant leucopax mushroom mycelia were reported to increase with increasing growth periods [23]. Different results for the effect of the growth periods on both total phenolic content and total flavonoid content might be due to the different mushroom species and cultivation environments in the mushroom farm.

Table 4. Bioactive compounds in Lion's Mane mushroom from different growth periods.

Bioactive compounds	Growth period*		
	14 days	21 days	28 days
Ergosterol (mg/g dried LM)	3.54±0.30[a]	2.62±0.31[b]	1.87±0.06[c]
Hericenone C (mg/g dried LM)	1.23±0.10[a]	0.73±0.09[b]	0.48±0.01[c]
Hericene A (mg/g dried LM)	0.09±0.01[a]	0.03±0.01[b]	-

*Different superscripts in the same row mean significant difference at $p<0.05$

3.4 Antioxidant activity

The inhibition capacity, IC_{50} measured by DPPH and ABTS radical scavenging activity of the 21-days sample was higher than those of 14-days and 28-days samples (Figure 2). The degree of antioxidant capacity (IC_{50}) of the samples in this study, 63.30 mg extract/mL were lower than $5x10^{-3}$ mg extract/mL of the Lion's Mane mushroom harvested from nature [7]. The correlation between DPPH and ABTS assays was found with total phenolic content as reported by Dudone *et al.* 2009 [29]. However, the antioxidant capacity in our study did not correlate well with the total flavonoid content in our study. Therefore, it is possible that there are other compounds not identified in this study, which can exhibit antioxidant activity such as β-glucans, caffeic acid, ergosterol derivatives, and hericenone derivatives in the extract. However, what exactly the compounds are should be investigated in further study. The antioxidant activity of Lion's Mane mushroom samples in this study is different from the giant leucopax mushroom mycelia. The growth periods of the giant leucopax mushroom mycelia correlated with the increase of antioxidant activity were demonstrated by Barros *et al.* 2007 [23].

4 Conclusions

Lion's Mane mushrooms were cultivated in the farms under controlled temperature and humidity. The research presented that the bioactive compounds of Lion's Mane mushrooms from different growth periods, including hericenone C, ergosterol and hericene A were highest at 14-days cultivation and decreased to lower content after that (21-days and 28-days). However, antioxidant activity measured by 2 different methods, total phenolic content and total flavonoid content were not significantly different between each growth period of the mushroom. The knowledge obtained from this study could be used for the cultivation as well as the criteria to select the proper harvesting time to achieve the mushroom with standard shape and health benefit compounds for further applications. The results and information obtained from the research would be beneficial to the production of high-quality mushrooms for consumption and to be used for value-added products. The farm management for maximum yield and high-quality products will also enhance sustainable development and natural bioresource management.

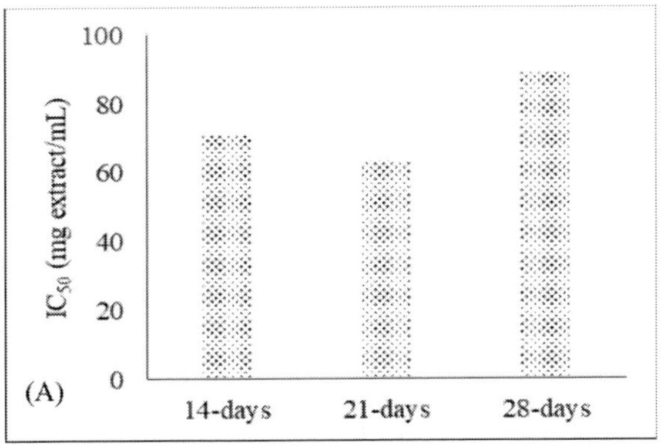

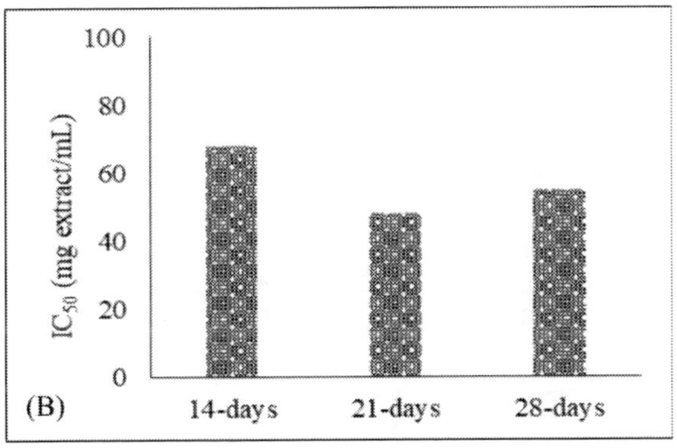

Figure 2. DPPH radical scavenging assay (A), and ABTS radical scavenging assay (B) of Lion's Mane mushroom from different growth periods.

Acknowledgements

This research was funded by King Mongkut's University of Technology North Bangkok and National Science and Technology Development Agency, Thailand (Contract no. 012/2563). This research was also partially supported by NRCT Senior Research Scholar Program (Contract No. 814-2020). Special thanks go to Mr. Wuttipong Raksawong, Managing Director, Fresh & Friendly Farm Co., Ltd., for his kind supports during the cultivation of the Lion's Mane mushrooms.

References

1. V. Sabaratnam, W. Kah-Hui, M. Naidu, P. Rosie David, Neuronal health - can culinary and medicinal mushrooms help, Journal of Traditional and Complementary Medicine, 3, 1 (2013): 62-68
2. N. Cohen, J. Cohen, M.D. Asatiani, V.K. Varshney, H.T. Yu, Y.C. Yang, Y.H. Li, J.L. Mau, S.P. Wasser, Chemical composition and nutritional and medicinal value of fruit bodies and submerged cultured mycelia of culinary-medicinal higher basidiomycetes mushrooms, International Journal of Medicinal Mushrooms, 16, 3 (2014). 273-291
3. S. Sharif. G. Mustafa, H. Munir, C.M. Weaver, Y. Jamil, M. Shahid, Proximate composition and micronutrient mineral profile of wild *Ganoderma iucidum* and four commercial exotic mushrooms by ICP-OES and LIBS, Journal of Food and Nutrition Research, 4, 11 (2016): 703-708
4. S. Dudhgaonkar, A. Thyagarajan, D. Sliva, Suppression of the inflammatory response by triterpenes isolated from the mushroom Ganoderma lucidum, International Journal of Immunopharmacology, 9, 11 (2009): 1272-1280
5. K. Hiwatashi, Y. Kosaka, N. Suzuki, K. Hata, T. Mukaiyama, K. Sakamoto, H. Shirakawa, M. Komai, Yamabushitake mushroom (*Hericium erinaceus*) improved lipid metabolism in mice fed a high-fat diet, Bioscience, Biotechnology, and Biochemistry, 74, 7 (2010): 1447-1451
6. X. Jiang, W. Meng, L. Li, Z. Meng, D. Wang, Adjuvant therapy with mushroom polysaccharides for diabetic complications, Frontiers in Pharmacology, 11, (2020): 168-180
7. E. Sharpe, A.P. Farragher-Gnadt, M. Igbanugo, T. Huber, J.C. Michelotti, A. Milenkowic, S. Ludlam, M. Walker, F. Bou-Abdallah, Comparison of antioxidant activity and extraction techniques for commercially and laboratory prepared extracts from six mushroom species, Journal of Agriculture and Food Research, 4 (2021): 100131-100139
8. M. Friedman, Chemistry, nutrition, and health-promoting properties of *Hericium erinaceus* (Lion's Mane) mushroom fruiting bodies and mycelia and their bioactive compounds, Journal of Agricultural and Food Chemistry, 63, 32 (2015): 7108-7123
9. B. Thongbai, S. Rapior, K.D. Hyde, K. Wittstein, M. Stadler, *Hericium erinaceus*, an amazing medicinal mushroom, Mycological Progress, 14, 10 (2015): 91-114
10. K.H. Wong, V. Sabaratnam, N. Abdullah, U.R. Kuppusamy, M. Naidu, Effects of cultivation techniques and processing on antimicrobial and antioxidant activities of *Hericium erinaceus* (Bull.:Fr.) Pers. Extracts, Food Technology and Biotechnology, 47, 1 (2009): 47-55
11. Y. Hou, X. Ding, W. Hou, Composition and antioxidant activity of water-soluble oligosaccharides from *Hericium erinaceus*, Molecular Medicine Reports, 11, 5 (2015): 3794-3799
12. E. Hiraki, S. Furuta, R. Kuwahara, N. Takemoto, T. Nagata, T. Akasaka, B. Shirouchi, M. Sato, K. Ohnuki, K. Shimizu, Anti-obesity activity of Yamabushitake (*Hericium erinaceus*) powder in ovariectomized mice, and its potentially active compounds, Journal of Natural Medicines, 71, 3 (2017): 482-491
13. M. Gasecka, M. Siwulski, Z. Magdziak, S. Budzynska, K. Stuper-Szablewska, P. Niedzielski, M. Mleczek, The effect of drying temperature on bioactive compounds and antioxidant activity of *Leccinum scabrum* (Bull.) Gray and *Hericium erinaceus* (Bull.) Pers, Journal of Food Science and Technology, 57, 2 (2020): 513-525
14. Y. Yu, Q. Hu, J. Liu, A. Su, H. Xu, X. Li, Q. Huang, J. Zhou, A.M. Mariga, W. Yang, Isolation, purification and identification of immunologically active peptides from *Hericium erinaceus*, Food and Chemical Toxicology, 151 (2021): 112111
15. P. Joradon, V. Rungsardthong, U. Ruktanonchai, K. Suttisintong, T. Iempridee, B. Thumthanaruk, S. Vatanyoopaisarn, N. Sumonsiri, D. Uttapap, Ergosterol content and antioxidant activity in Lion's Mane mushroom (*Hericium erinaceus*) and its induction to vitamin D_2 by UVC-irradiation, in Proceedings of the 8th International Conference on Agricultural and Biological Sciences - ABS 2022, Shenzhen (2022)
16. P. Joradon, V. Rungsardthong, U. Ruktanonchai, K. Suttisintong, T. Iempridee, B. Thumthanaruk, S. Vatanyoopaisarn, D. Uttapap, A comparative study of conventional and supercritical carbon dioxide extraction methods for the recovery of bioactive compound from Lion's Mane Mushroom (*Hericium erinaceus*), in Proceedings of International Conference on Research, Invention, and Innovation Congress - RI2C 2022, 4-5 August, Bangkok (2022)
17. K. Mori, Y. Obara, M. Hirota, Y. Azumi, S. Kinugasa, S. Inatomi, N. Nakahata, Nerve growth factor-Inducing activity of *Hericium erinaceus* in 1321N1 human astrocytoma cells, Biological and Pharmaceutical Bulletin, 31, 9 (2008): 1727-1732
18. H.C. Kuo, C.C. Lu, C.H. Shen, S.Y. Tung, M.C. Hsieh, K.C. Lee, L.Y. Lee, C.C. Chen, C.C. Teng, W.S. Huang, T.C. Chen, K.F. Lee, *Hericium erinaceus* mycelium and its isolated erinacine A protection from MPTP-induced neurotoxicity through the ER stress, triggering an apoptosis cascade, Journal of Translational Medicine, 14, 78 (2016)
19. C.O. Perera, V.J. Jasinghe, F.L. Ng, A.S. Mujumdar, The effect of moisture content on the conversion of ergosterol to vitamin D in Shiitake mushrooms, Drying Technology, 21, 6 (2003): 1091-1099
20. D. Huang, F. Cui, Y. Li, Z. Zhang, J. Zhao, X. Han, X. Xiao, J. Qian, Q. Wu, G. Guan, Nutritional

requirements for the mycelial biomass and exopolymer production by *Hericium erinaceus* CZ-2, Food Technology and Biotechnology, 45, 4 (2007): 389-395

21. A. Imtiaj, C. Jayasinghe, G.W. Lee, M.J. Shim, H.S. Rho, H.S. Lee, H. Hur, M.W. Lee, U.Y. Lee, T. S. Lee, vegetative growth of four strains of *Hericium erinaceus* collected from different habitats, Mycobiology, 36, 2 (2008): 88-92
22. Y. Nakano, H. Fujii, M. Kojima, Identification of blue-light photoresponse genes in oyster mushroom mycelia, Bioscience, Biotechnology, and Biochemistry, 74, 10 (2010): 2160-2165
23. L. Barros, P. Baptista, L. M. Estevinho, I. C.F.R. Ferreira, Bioactive properties of the medicinal mushroom *Leucopaxillus giganteus* mycelium obtained in the presence of different nitrogen sources, Food Chemistry, 105, 1 (2007): 179-186
24. AOAC (Association of Official Analytical Chemists), Official Methods of Analysis of AOAC International, 16thed. Washington, D.C.: AOAC International, (1999)
25. L.M. Cheung, P.C.K. Cheung, V.E.C. Ooi, Antioxidant activity and total phenolics of edible mushroom extracts, Food Chemistry, 81, 2 (2003): 249-255
26. K.M. Moon, B. Lee, W.K. Cho, B.S. Lee, C.Y. Kim, J.Y. Ma, Swertiajaponin as an anti-browning and antioxidant flavonoid, Food Chemistry, 252 (2018): 207-214
27. F. Hassan, Cultivation of the Monkey head mushroom (*Hericium erinaceus*) in Egypt. Journal of Applied Sciences Research, 3, 10 (2007): 1229-1233
28. D. Moore, G. D. Robson, A. P. J. Trinci, 21st Century Guidebook to Fungi (2 ed.). Cambridge: Cambridge University Press, (2020)
29. S. Dudonné, X. Vitrac, P. Coutière, M. Woillez, J. M. Mérillon, Comparative Study of Antioxidant Properties and Total Phenolic Content of 30 Plant Extracts of Industrial Interest Using DPPH, ABTS, FRAP, SOD, and ORAC Assays. Journal of Agricultural and Food Chemistry, 57, 5 (2009): 1768-1774

* Corresponding author: vilai.r@sci.kmutnb.ac.th

Investigation of low-cost media for *Bacillus thuringiensis* subspecies *israelensis*

Phuwana Mo-on[1], *Siriwan* Panprivech[1] and *Viyada* Kunathigan[2*]

[1]Department of Agro-Industry, Faculty of Biotechnology, Assumption University, Bangkok, Thailand, 10240
[2]Department of Food Biotechnology, Faculty of Biotechnology, Assumption University, Bangkok, Thailand, 10240

Abstract. *Bacillus thuringiensis* subspecies *israelensis* (*Bti*) has been used as a mosquito control agent in agricultural field and residences. The production of *Bti* by conventional medium was expensive. The objective of this study was to investigate low-cost media for *Bti*'s production with a cost analysis of the medium formulated. Soybean Casein Digest Medium (TSB) was used as a conventional medium to compare with Saberi medium at room temperature (approximately 21.5°C) on static and agitation (by obituary shaker at 100 rpm). Saberi medium was also modified to vary the percentage of nitrogen and carbon source to observe the growth rate and sporulation, compared to the Saberi medium. The growth of *Bti* in TSB with agitation for 24 hours yielded the highest growth rate. Sporulation of *Bti* in TSB with agitation at room temperature was 6 days which was faster than *Bti* in the Saberi medium. Modified Saberi nitrogen source and carbon source medium could not support the growth of *Bti* due to an imbalance in the percentage of nitrogen source and carbon source. Determination of sugar utilization showed sugar in the medium was utilized mostly in the exponential phase of *Bti*, from 3 hours to 6 hours. The price of TSB was 76.80 Thai baht per liter while Saberi medium's cost was 20.43 Thai baht per liter. In this study, the Saberi medium may not be able to replace TSB in terms of growth rate and period of sporulation. Further study will be needed to optimize the media that use agro-industrial wastes as ingredients following the concept of circular economy and sustainability.

Keyword. *Bti*, Bioinsecticide, Low-cost medium

1 Introduction

Mosquitoes impact human health and good living. They are the vector of transmitting diseases, such as Malaria and Dengue fever. Thailand is a country with tropical rainforest, Asian tiger mosquito (*Aedes albopictus*) is commonly found as a carrier of Chikungunya fever [1]. The solution to mosquito elimination may end up using insecticide, which may contain volatile chemicals that have adverse effects on human health and the environment. An example of a dangerous insecticide is DDT, dichloro-diphenyl-trichloroethane, which was banned in many countries. Other classes of chemical insecticides, such as organophosphates and carbamate were used instead of using DDT [2]. Environmental and health concerns have led to the alternative environmentally friendly bioinsecticide

Bacillus thuringiensis subspecies *israelensis* (*Bti*) has been used in many countries to control the population of mosquitoes [3]. Many studies demonstrated that *Bti* is environmentally safe and target-specific bioinsecticide. *Bti* targeted three main insect larvae: mosquitoes, black flies, and fungus gnats [3]. Protoxins of *Bti* is produced in the stationary phase when endospore formation takes place. After ingestion of toxins, protoxin in form of crystals, including Cry (crystal) and Cyt (cytolytic) toxins that are solubilized in the alkaline condition in the larva's gut, which is activated by midgut protease. Cry toxins bind to a protein receptor on the outer membrane of gut cells [4], while Cyt toxins bind to the phospholipids [5]. The function of the gut is disrupted, causing larval death from starvation [4]

The *Bti* production for mosquito control using laboratory media is expensive. Previous work by Saberi (2020) [6] optimized the culture media for the production of *B. thuringiensis* var. *tenebrionis*-BN1 using molasses from sugar beet and sugar cane as a carbon source while using corn steep liquor as a nitrogen source. The maximum CFU of 10^7 spore/ml was achieved with the amount of carbon source above 11% and nitrogen source less than 2% [6]. For *Bt*, the media that support the growth and sporulation consists of: glucose, 5 g/L; glycerol, 1 g/L; sodium acetate, 0 g/L; $(NH_4)_2SO_4$, 1 g/L; corn steep liquor, 2 mL/L; yeast extract, 1 g/L; hydrolyzed casein, 1 g/L; $CaCl_2$, 20 mg/L; $MgSO_4$, 50 mg/L; $MnSO_4$ 100 mg/L provided the growth 6.4 x 10^{-7} cell/mL and 99% of sporulation [10].

In this study, the use of sugar cane molasses and corn steep liquor which are agro-industrial waste were explored. These wastes are used in many industries, such as supplements of livestock's feed, to reduce the production cost [11]. The compositions of sugar cane molasses and corn steep liquor provide carbon and nitrogen source which can be used to promote bacterial growth. Hence, the use of both sugar cane molasses and corn steep liquor to formulate media for the production of *BTi* will enhance the circle of material usage following the concept of circular economy and sustainability.

* Corresponding author: viyadaKnt@au.edu

2 Materials and methods

2.1 Determination of kinetic growth of *Bacillus thuringiensis* subspecies *israelensis* in conventional medium

Soyabean Casein Digest medium TSB (HIMEDIA) was prepared for 300 mL per batch in triplicate. The *Bti* was subcultured in modified TSB [6] overnight as inoculum. Three percent (v/v) of *Bti* inoculum was added to the TSB and incubated at room temperature (approximately 21.5˚C) under shaking (100 rpm) and static conditions (Figure 1). The optical density was measured using UNICO 1200 spectrophotometer at wavelength 600 nm in triplicate to observed growth. The development of the endospore in *Bti* was observed by the standard endospore staining technique.

2.2 Preparation of Saberi medium and modified Saberi media

The production medium was prepared according to the formula from Saberi *et. al.* (2020) [6] and called as Saberi medium (Table 1). Corn steep liquor, NQ. QCA 3051/21, was sponsored by Friendship Corn Starch Company Limited. The solid content in the corn steep liquor was 38.05%. The protein content in the corn steep liquor was 48.46% analyzed by the Kjeldahl method. Sugar cane molasses was purchased from New Friend Farm Company Limited. Saberi medium was modified into 2 formulas: Modified Saberi Carbon Source (MSC) contains 8% (v/v) of sugarcane molasses and Modified Saberi Nitrogen Source (MSN) contains 3% (v/v) of corn steep liquor. Inoculum medium, Saberi medium and modified Saberi media were autoclaved for 10 minutes at 115˚C. The media were prepared in batches of 350 ml.

Table 1: Compositions of Saberi medium and modified Saberi media.

Media	Compositions of media				
	Sugar cane molasses (v/v)	Corn steep liquor (v/v)	$MgSO_4$ (w/v)	KH_2PO_2 (w/v)	K_2HPO_4 (w/v)
Saberi medium (SM) [6]	12.5% (5.33%w/v sugar)	1.5% (0.73%w/v protein)	0.3%	0.3%	0.4%
Modified Saberi carbon source (MSC)	8% (3.41%w/v sugar)	1.5% (0.73%w/v protein)	0.3%	0.3%	0.4%
Modified Saberi nitrogen source (MSN)	12.5% (5.33%w/v sugar)	3% (1.45%w/v protein)	0.3%	0.3%	0.4%

2.3 Culturing in Saberi medium and modified Saberi media

Bti was cultured overnight on the orbital shaker at 30˚C 100 rpm in the modified TSB [6] as inoculum culture. The 3% (v/v) of overnight *Bti* was inoculated in each 350 mL of the media (Figure 1). After the inoculum culture was added to the batch, 1 mL of the media was collected for growth measurement using OD at 600 nm and 9 mL from each treatment were collected from each time point (0, 1, 2, 3, 4, 5, 6 and 24 hour) for further analysis.

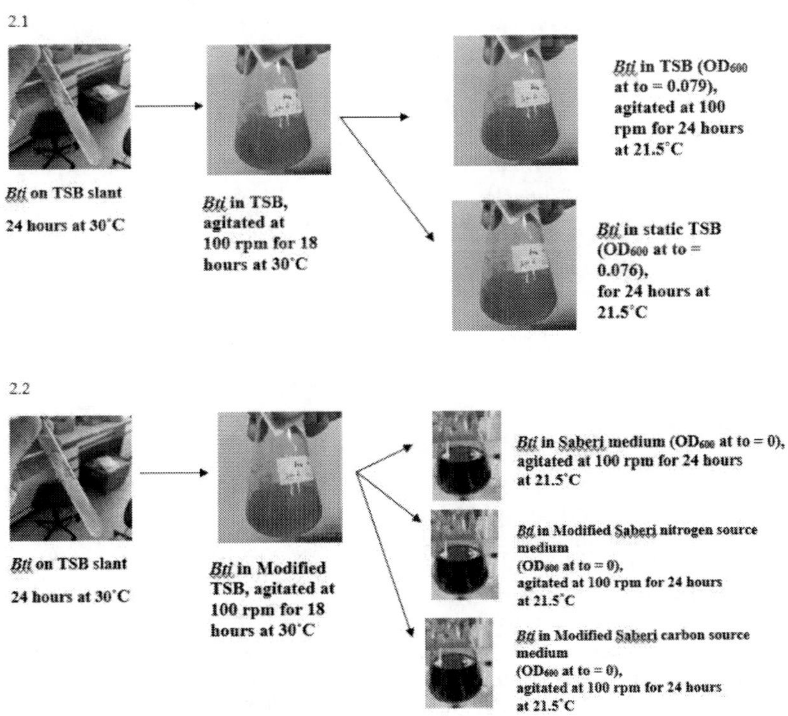

Figure 1 Schematic diagram of this experiment related to sections 2.1 and 2.2

 https://doi.org/10.1051/e3sconf/202235502017

2.4 Determination of growth kinetic of *Bacillus thuringiensis* subspecies *israelensis* in Saberi medium and modified Saberi media

For the growth of the bacteria, 1 mL of sample collected in 2.3 from each timepoint was centrifugated at 8000 rpm for 2 minutes. The supernatant was discarded; the pellet was washed with distilled water and vortexed until the mixtures became homogenized. The washing step was repeated before the mixtures were observed using a spectrometer at 600 nm. The growth curve was constructed and used to compare doubling time and specific growth rate using the following equation (eq 1-2).

Doubling time:
$$t_b = ln\ (2)/B \qquad (1)$$
Specific growth rate:
$$\mu = [2.303(log\ OD_2 - log\ OD_1)/(t_1 - t_2)] \qquad (2)$$

2.5 Determination of sugar utilization of *Bacillus thuringiensis* subspecies *israelensis* in Saberi medium and modified Saberi media

The sample was collected and centrifuged to remove solid particles and bacteria cells at 4,000 rpm for 5 minutes. The supernatant was kept at -20°C for the sugar analysis. For reducing sugar, samples were analyzed using the dinitrosalicylic colorimetric method (DNS) in duplicate. 1 mL of the sample was diluted in 100 mL of distilled water. 3 mL of DNS reagent were added to 3 mL of supernatant in a lightly capped test tube. The mixture was heated at 90°C for 15 minutes until the red brown color was developed. 1 mL of 40% (w/v) potassium sodium tartrate solution was added to stabilize the color. After cooling to room temperature in a cold-water bath, the absorbance was observed at 575 nm.

For non-reducing sugar, the Association of Official Agricultural Chemists (AOAC) Official method 923.09/2012 Invert Sugar in Sugar and Syrups was used for the preparation of reagents. AOAC Official Method 968.28/2012 [7]. Total Sugar in Molasses as Invert Sugar was used to analyze total sugar content as invert in the sample at each time point (0, 1, 2, 3, 4, 5, 6 and 24 hours). 8 mL of centrifugated sample was used for the total invert sugar analysis. Sugarcane molasses was also analyzed.

3 Results

The growth of *Bti* in TSB with agitation (100 rpm), at room temperature for 24 hours yielded the highest growth of *Bti* compared to other conditions and media (Figure 2). Growth of *Bti* in Saberi medium showed a higher growth rate compared to TSB without agitation at room temperature for 24 hours. When the percentage of nitrogen and carbon source from the Saberi medium was changed in the modified Saberi media, both MSC and MSN, the growth rate of *Bti* were lowered.

Reduction of reducing sugar was observed during the growth of *Bti* in TSB with and without agitation (Figure 3). The amount of reducing sugar in SM, MSC and MSN media however showed a constant level throughout the course of growth. Total sugar as inverted sugar was determined by AOAC Official Method 968.28/2012 in the SM, MSC and MSN during the growth of *Bti*. After 24 hours total sugar in SM and MSC reduced from 12.13% of sugar as inverted sugar to 5.68%. While in MSN, the total sugar was not reduced (Figure 4).

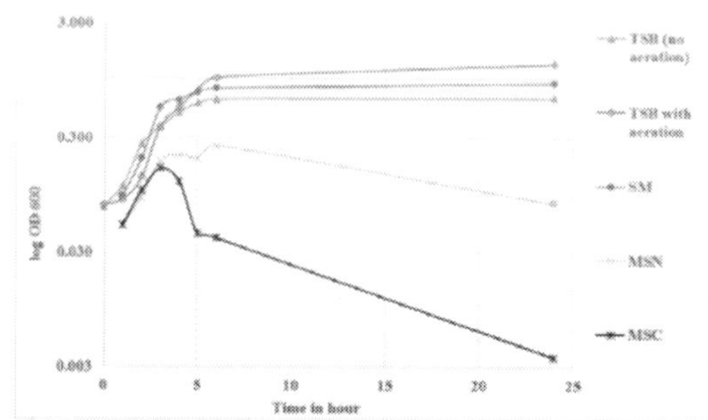

Figure 2. Growth of *B. thuringiensis* subspecies *israelensis* in TSB without agitation, TSB, Saberi medium (SM), modified Saberi carbon source (MSC) media, and modified Saberi nitrogen source (MSN) media with agitation (100 rpm) at room temperature (approximately 21.5°C), for 24 hours.

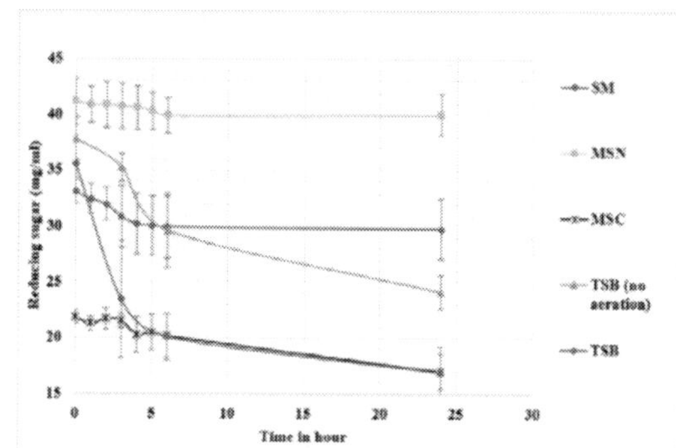

Figure 3. Reducing sugar during growth of *B. thuringiensis* subspecies *israelensis* in TSB without agitation, TSB, Saberi medium (SM), modified Saberi carbon source (MSC) media, and modified Saberi nitrogen source (MSN) media with agitation (100 rpm) at room temperature, for 24 hours.

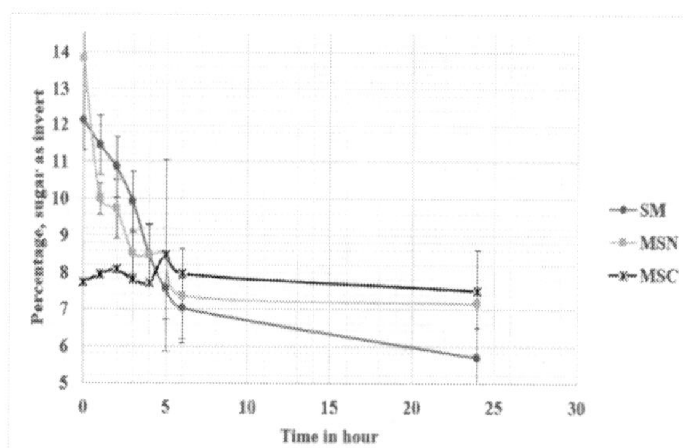

Figure 4. Non-reducing Sugar in Saberi medium (SM) and modified Saberi media (MSN, MSC) with agitation (100 rpm) at 21°C, room temperature, for 24 hours after additional *B. thuringiensis* subspecies *israelensis*.

4 Discussion

Growth of *Bacillus thuringiensis* subspecies *israelensis* was observed in the TSB medium with and without agitation. The growth rate and reducing sugar utilization were significantly faster when agitation was provided. This result confirmed the requirement of agitation as previously described [6]. The sporulation was also observed in the population of *Bti* that was grown in TSB and found that sporulation was faster in the TSB with agitation (6 days) (Table 2). As the sporulation was associated with the production of the *Bti* protoxin [4,5,6], the media that stimulate the sporulation in a shorter time would be more economically feasible. Hence, the agitation condition was further used to examine the growth and sporulation of *Bti* in the Saberi medium.

Table 2. Kinetic growth of *B. thuringiensis* subspecies *israelensis* in different types of media and conditions.

Types of media	Agitation (rpm)	Doubling time (td) (hour)	Specific growth rate (μ) (hour^{-1})	Sporulation (day)
TSB	-	3.04	0.419	16
TSB	100	1.46	0.440	8
Saberi medium	100	2.20	0.336	14
Modified Saberi nitrogen source	100	3.25	0.214	> 14
Modified Saberi carbon source	100	-	-	> 14

In previous work, different types of carbon sources and nitrogen sources were tested to formulate production mediums that have lower costs [6, 8]. These substrates include sugar beet and sugar cane molasses, sucrose, and potato. The sugar cane molasses and corn steep liquor were chosen to be used in this experiment due to the cost and availability of the raw materials in Thailand.

The Saberi media was chosen as described previously [6], and the amount of carbon source (sugar cane molasses) and nitrogen source (corn steep liquor) were varied to investigate the effect of these nutritional sources on the growth and sporulation of *Bti*.

Growth of *Bti* in TSB with agitation at room temperature for 24 hours shows the highest growth rate compared to all media tested. The doubling time of *Bti* on TSB with agitation (1.46 hours), was faster than the doubling time of *Bti* on TSB without agitation (3.04 hours) (Table 1). Growth of *Bti* in Saberi medium showed the doubling time of 2.20 hours, faster than *Bti* on TSB in static conditions. The Saberi medium consists of 12-14% (w/v) of sucrose, the bacteria can use this non-reducing sugar for growth. The sporulation of *Bti* that grew in the Saberi medium took place after 14 days. Modified Saberi carbon source (MSC) medium was formulated based on the utilization of total sugar demonstrated in the growth with Saberi medium. The total sugar as inverted sugar was reduced from 12.13% to 7.83% in the MSC medium. However, the MSC medium could not sustain the growth of *Bti* as the growth rate was reduced after 6 hours (Figure 2). Modified Saberi nitrogen source (MSN) medium consisting of 3% (v/v) of corn steep liquor was formulated and able to support the growth of *Bti* for a short period. The highest growth was reached within 6 hours. The increased percentage of nitrogen source in the modified Saberi nitrogen source medium could not sustained the growth of *Bti* which was corresponded to previous study in *B. thuringiensis* var. *tenebrionis*-BN1 [6]. In addition, the sporulation could not be observed in the *Bti* that grew in MSC or MSN medium within 14 days. The effect on the growth demonstrated in MSC and MSN medium may be caused by imbalanced carbon and nitrogen ratios. A carbon per nitrogen ratio of 7:1 is effective to support the growth of *Bti* and yields a high rate of Cry production about 150 g/L [9]. The development of the endospore on *Bti* and the doubling time is related. If the media are optimum for *Bti*, *Bti* will reach the stationary faster and have a shorter doubling time and shorter development of endospore.

DNS method was used to analyze reducing sugar to address sugar utilization of *Bti* in different media. The reducing sugar in TSB was utilized by the growth observed (Figure 2 and 3). *Bti* in TSB with agitation utilized reducing sugar from 35.61 mg/mL to 16.96 mg/mL while *Bti* in static TSB utilized from 37.80 mg/mL to 24.09 mg/mL with 24 hours. The reducing sugar observed from the growth of *Bti* in SM, MSC and MSN medium did not change over time, while the growth was observed hence indicating utilization of the other carbon source. Total sugar in SM, MSC and MSN was analyzed to monitor the utilization of non-reducing sugar (Figure 4). As *Bti* was able to utilize sucrose [8], the utilization of total sugar correlated with the growth of the *Bti*.

The cost of the medium was analyzed in which the TSB medium cost 76.80 Thai baht per liter, while the Saberi medium cost 20.42 Thai baht per liter. However, when considering the growth rate and the sporulation time, TSB despite the higher cost still has advantage of shorter operation time. Further optimization is needed to formulate a medium that is more feasible economically. Different carbon and nitrogen sources may be explored. In addition, other parameter like temperature, pH and agitation will also need to be optimized.

5 Conclusion

The Saberi medium (SM) formulated with 12.5% (v/v) sugar cane molasses and 1.5% corn steep liquor with supplementary was able to use for the production of *Bacillus thuringiensis* subspecies *israelensis* (*Bti*) when grown at room temperature (21.5°C) with agitation (100 rpm on rotary shaker) with 14 days to sporulate. Despite the cheaper cost of 20.42 Thai baht per liter of SM compared to 76.80 Thai baht per liter of TSB, further optimization is needed to formulate a more feasible media for the production of *Bti* as bioinsecticide. The utilization of agro-industrial wastes as the main ingredients of the production media would increase materials recycling ability that meets the concept of circular economy and sustainability.

References

1. N. Becker, An Ecological and Societal Approach to Biological Control, Biological control of mosquitoes; management of the Upper Rhine mosquito population as a model programme (Springer, Netherland, Dordrecht 2006): 227-245
2. C. F. Wurster, DDT Goes to Trial, Finally, in Washington, DC, DDT Wars (Oxford University Press, 2015).
3. L. Despres, C. Lagneau, R. Frutos, Using the bio-insecticide *Bacillus thuringiensis israelensis* in mosquito control, Pesticides in Modern World Pest Control and Pesticides Exposure and Toxicity Assessment (InTech, Rijeka, Croatia, 2011)
4. V. Vachon, R. Laprade, J. L. Schwartz, Current models of the mode of action of *Bacillus thuringiensis israelensis* insecticidal crystal proteins: a critical review. Journal of Invertebrate Pathology 111, 1, (2012): 1-12
5. E. Ben-Dov, *Bacillus thuringiensis* subsp. *israelensis* and Its Dipteran-Specific Toxins. Toxins, 6, 4, (2014): 1222-1243
6. F. Saberi, R. Marzban, M. Ardjmand, F. P. Shariati, O. Tavakoli, Optimization of culture media to enhance the ability of local *Bacillus thuringiensis* var. *tenebrionis*. Journal of the Saudi Society of Agricultural Sciences, 19, 7, (2020): 468-475
7. AOAC 968.28-1969, Invert sugar in sugars and syrups. Lane-eynon, Chapter 44 (2021): p.9-20
8. S. Poopathi and K. A. Kumar. Novel Fermentation Media for Production of *Bacillus thuringiensis* subsp. *israelensis*. Journal of Economic Entomology. 96, 4, (2004): 1039 -1044
9. R. R. Farrera, F. Pérez-Guevara, M. De la Torre, Carbon:nitrogen ratio interacts with initial concentration of total solids on insecticidal crystal protein and spore production in Bacillus thuringiensis HD-73. Applied Microbiology and Biotechnology, 49, 6, (1998): 758–765
10. S.M. Hossein, Nutritional Requirements Of Bacillus thuringiensis During Different Phases Of Growth, Sporulation And Germination Evaluated By Plackett-Burman Method. Iranian Journal of Chemistry & Chemical Engineering-international English Edition, 31, (2012): 131-136.
11. A. Azizi-Shotorkhoft, A. Sharifi, D. Mirmohammadi, H. Baluch-Gharaei,, J. Rezaei,. Effects of feeding different levels of corn steep liquor on the performance of fattening lambs. Journal of Animal Physiology and Animal Nutrition, 100, 1, (2016): 109–117.

* Corresponding author: viyadaKnt@au.edu

https://doi.org/10.1051/e3sconf/202235502018

Optimization of Microwave-Assisted Extraction of Palm Kernel Cake Protein

Thanida Tungchaisin[1,2], *Sudathip* Sae-Tan[3], and *Kittipong* Rattanaporn[1,2]

[1]Fermentation Technology Research Center (FTRC), Department of Biotechnology, Faculty of Agro-industry, Kasetsart University, Bangkok, Thailand
[2]Department of Biotechnology, Faculty of Agro-industry, Kasetsart University, Bangkok, Thailand
[3]Department of Food Science and Technology, Faculty of Agro-industry, Kasetsart University, Bangkok, Thailand

Abstract. Palm kernel cake (PKC) is an abundant by-product of the palm oil industry. It is used as an ingredient in feed due to the high amount of protein and fiber content. In order to increase the value of PKC, the PKC protein can be extracted and may be able to be used as an alternative protein for plant-based food. This study aims to optimize the PKC protein extraction using the microwave-assisted extraction (MAE) method with a response surface methodology (RSM). MAE is a green extraction method due to less chemicals needed, less time and less energy consumption when compared to the traditional thermal extraction method. The experiment was designed by the Box-Behnken method with 3 factors; microwave power (A), extraction time (B) and solid-liquid ratio (C). The optimum condition was at the microwave power of 700.16 W, extraction time of 543.08 s and the solid-liquid ratio of 1:7.73 g PKC/ ml water resulting in a theoretical yield of protein extraction of 32.46%.

Keyword. Palm kernel cake, Microwave-assisted extraction, Protein, Response surface methodology

1 Introduction

Protein is an essential nutrition, which can be obtained from animals, insects, plants, and microorganisms. Nowadays, protein consumption is continuously increasing while protein sources are limited. In order to solve this problem, the by-products or wastes from food and agro-industry become alternative protein sources for human foods and feeds. Utilization of by-product as the protein source is not only add additional value to the by-product but also relieves the environmental problems.

Thailand is one of the top five palm oil-producing countries which can annually produce 16.40 million tons and the production rate is expected to be higher [1]. One thousand kilograms of oil palm produces 250 kilograms of oil which includes 200 kilograms of crude palm oil and 50 kilograms of crude palm kernel oil. Therefore, the by-products and waste from palm oil process are approximately up to 750 kilograms. Palm kernel is the pulp within the oil palm seed which is composed of lipid up to 49%. The residue after the palm kernel oil extraction process is called palm kernel cake (PKC). The PKC consists of carbohydrate 50.3%, fiber 16.7%, protein 14.8%, lipid 7.9%, moisture 6.4% and ash 3.9% [2].

Normally, it is used as a feed supplement for ruminants due to the high amount of protein and fiber [3-5]. Thus, the PKC is an interesting protein source. Moreover, PKC contains essential amino acids around 6.98%, such as arginine, leucine, valine, phenylalanine and threonine [6]. Therefore, PKC protein is an interesting alternative source for future food. Currently, there are several research works investigating the optimization of PKC protein extraction. One of the most popular methods of protein extraction is the chemical extraction method, which uses solvent, such as water, alkali, acid, and organic solvent. The alkaline extraction method is widely used to extract protein from plant material due to its high protein yield and easy operation. However, there are some problems including brown color from Maillard reaction, lack of some functional properties and protein denaturation caused by a high concentration of alkaline [7].

The microwave-assisted extraction (MAE) is an environmental-friendly extraction technique, which is the combination of the chemical extraction method and the physical extraction method. Microwave generates heat by two simultaneous mechanisms: ionic conduction and dipole rotation, thus the whole sample is heated up at the same time. The rotation of dipolar molecules can break the weak hydrogen bonds. Moreover, microwaves can reduce the viscosity and induce the solvent to penetrate the matrix and release the target product [8]. There are many advantages of MAE, such as short extraction time, small amount of solvent needed, and low cost of energy. The applications of MAE are typically used for the extraction of some natural products, such as essential oil [9-11], flavonoid [12-14], phenolic compounds [15], alkaloids [16-17], saponins [18-20], polysaccharide [21-22] and protein, etc. Many studies have also reported the applications of MAE technique for protein extraction from plant feedstocks, such as rice bran [23-24], sesame bran [25], coffee silverskin [26] and watermelon seeds [27]. Some of the main factors affecting the efficiency of MAE are sample characteristics, temperatures, microwave powers, extraction times, solid-liquid ratios, and solvents [23,28]. According to the previous studies the MAE was applied to extract rice bran protein by optimizing the microwave power, extraction time and solid-liquid ratio. The results showed that the protein yield obtained from the optimal condition (1000 W of microwave power, 90 s of extraction time and 0.89 g of rice bran/ 10 mL of water) was 1.45-fold higher than the yield of alkaline extraction [23].

To obtain the maximum yield of protein extraction, statistical process optimization is required to reduce the number of experiments and to predict the response using the mathematical model. The experiments can be designed by many methods such as full factorial design, Taguchi and response surface methodology. Full factorial design is used for optimizing the effects of each factor which suits a low number factor experiment. This method is one of the most precise methods because of the number of experimental runs. On the other hand, the more factors or levels are studied, the more runs are generated which consume a lot of time and labor [29]. Taguchi method or robust design method can estimate the most effective factor with a minimum amount of experiment which can save time and resources. It can be divided into 4 phases including planning, conducting, analysis and validation. However, the main advantage of this method is the bias of the studies which require testing all relationships of all factors, because it does not test all combinations [30-31]. Response surface methodology (RSM) is a well-known optimization method. It can identify the factors and the combinations, which have an effect on the response. The linear regression model shown in equation 1 is generated based on the experiment data to predict the response.

$$S=a_0+\sum_{i=1}^{k} a_i x_i+\sum_{i=1}^{k} a_{ii} x_i^2+\sum_{i,j=1,\, j\neq i}^{k} a_{ij} x_i x_j \quad (1)$$

where S is a response, a_0 is an average of responses; a_i, a_{ii} and a_{ij} are response coefficients [32].

Comparing to full factorial design and Taguchi method, RSM is more compromise due to the number of experimental runs which is not too much like the full factorial design or too less like the Taguchi method. Furthermore, RSM can study the effect of a combination of variables on the response.

Overall, the aim of this study was to extract the protein from PKC using microwave-assisted extraction technique which is an eco-friendly method. The optimal condition was obtained by a response surface methodology based on Box-Behnken design. Three variables including microwave powers, extraction times and liquid-solid ratios were studied.

2 Material and method

2.1. Sample preparation

Defatted PKC from a palm oil factory in Southern Thailand was ground and sieved in order to get the PKC size between 250-500 μm (Figure 1). The protein content of PKC was analyzed by the Kjeldahl method and the moisture content was obtained using a moisture analyzer (Halogen Moisture Analyzer HE53, Mettler-Toledo Ltd., Ohio, United States) before extraction.

a. b.

Figure 1. Defatted PKC before (a.) and after (b.) grinding and sieving

2.2. Experimental design and statistical analysis

The response surface methodology (RSM) based on the Box-Behnken design (BBD) was used to optimize the PKC protein extraction using MAE method due to the BBD requiring a small number of experimental runs and suitable for the limitation of microwave power generated by the machine. A three-level-three-factor BBD generated a response surface model using the Design-Expert software (V.7.0.0, Stat- Ease, Inc., Minneapolis, USA). The independent variables for this experimental design are microwave power (A), extraction time (B) and liquid-solid ratio (C) which are some of the main extraction factors. The range of values of the variables used in the experimental design from MAE was presented in Table 1 which was chosen according to the preliminary experiment. The levels of each parameter coded -1, 0 and +1 were the lowest, the middle and the highest values, respectively.

The experimental design consisted of 17 experimental runs which were 5 replicates of the center point shown in Table 2. The protein extraction yield (%) was selected as the response (Y).

Table 1. Values range of the variables used in the PKC protein extraction by MAE method.

Parameters	Codes	Levels		
		-1	0	+1
Microwave power (W)	A	100	450	800
Extraction time (s)	B	30	315	600
Liquid-solid ratio (ml/g)	C	3:1	6.5:1	10:1

Table 2. Box-Behnken design (BBD) and PKC protein extraction yield.

Run	Factors			Y: Yield (%)
	A: Microwave power (W)	B: Extraction time (s)	C: Liquid-solid ratio (ml/g)	
1	100	30	6.5:1	21.08
2	800	30	6.5:1	23.59
3	100	600	6.5:1	28.08
4	800	600	6.5:1	30.07
5	100	315	3:1	23.87
6	800	315	3:1	23.98
7	100	315	10:1	23.61
8	800	315	10:1	30.86
9	450	30	3:1	24.40
10	450	600	3:1	25.31
11	450	30	10:1	21.52
12	450	600	10:1	31.80
13	450	315	6.5:1	27.35
14	450	315	6.5:1	31.12
15	450	315	6.5:1	31.36
16	450	315	6.5:1	29.28
17	450	315	6.5:1	32.29

2.3. Protein extraction using MAE method

The PKC protein extraction was carried out using a home-used microwave oven (Samsung ME711K, Thai Samsung Electronics Co., Ltd., Thailand). Five grams of PKC were mixed with the distilled water with the liquid-solid ratio of each run in a 50 ml-centrifuge tube, and put the tubes in a microwave oven. The PKC protein was extracted following the condition in Table 2. Afterward, the distilled water was added to the slurry to get the final volume of water at 50 ml. Then, the supernatant was separated by centrifugation (Centrifuge 5804R, Eppendorf Co., Ltd., Hamburg, Germany) at 10,000 rpm at a temperature of 4 °C for 10 min and the soluble protein in the supernatant was determined. The experiment was replicated for each run.

2.4. Protein determination

The soluble protein content in the supernatant was determined by the modified Folin-Lowry method [33] for 96-well plate. The difference concentration of albumin (BSA) (Bio Basic Canada Inc., Ontario, Canada), including 0, 1, 5, 25, 125, 250, 500, 1000, 1500 and 2000 µg/ml, was prepared and used for the standard. Forty µl of standard and diluted samples were added to each well of 96-well plate. The Lowry reagent was made from the combination of reagent A (2% sodium carbonate and 0.4% sodium hydroxide), 1% copper sulfate and 2% potassium sodium tartrate with the ratios of 100:1:1, and added 200 µl of the reagent into the samples. The mixtures were shaken for 30 s and incubated at room temperature for 10 min. Then, 20 µl of 1X Folin-Ciocalteu reagent was added immediately to each well. The mixtures were shaken for 30 s and incubated at room temperature for 30 min. The absorbance was detected at 750 nm. In order to calculate the protein concentration, the standard curve was plotted between the BSA concentrations and wavelengths, and the equation was used to calculate. All experiments were triplicated. The extraction yield (%) was calculated by equation 2:

$$\text{Yield (\%)} = \frac{\text{Protein concentration} \times \text{Extract volume}}{\text{mg of PKC} \times \text{Raw protein content}} \times 100 \quad (2)$$

3 Results and discussion

Microwave-assisted extraction (MAE) was performed to extract the PKC protein. The effect of factors, including microwave power (A), extraction time (B) and liquid-solid ratio (C) on the protein extraction yield (%) were studied. Box-Behnken design (BBD) of response surface methodology (RSM) was us as a statistical tool for optimization. The crude protein of PKC before extraction determined by the Kjeldahl method was 18.49% ± 0.05 (w/w). The experimental conditions were generated including 17 experimental runs presented in Table 2. The lowest extraction yield was obtained in the first run with 21.08% (w/w) and the highest yield was 32.29% (w/w) from the run number 17.

The statistical analysis of the model was performed in the form of analysis of variance (ANOVA) with backward elimination regression shown in Table 3. The p-value of this model was 0.0011 which implied the model was significant.

The coefficient of determination or R-squared (R2) value of 0.9216 represented a strongly fit of the model with the experimental data. Moreover, the lack of fit was not significant at the p-value of 0.9342, which could clearly confirm that the model was acceptable.

Table 3. ANOVA for the RSM quadratic model

Source	SS	df	MS	F value	p-value Prob > F
Model	216.18	8	27.02	11.75	0.0011
A	17.61	1	17.61	7.66	0.0244
B	76.04	1	76.04	33.06	0.0004
C	13.09	1	13.09	5.69	0.0441
AC	12.75	1	12.75	5.54	0.0464
BC	21.98	1	21.98	9.55	0.0149
A^2	23.78	1	23.78	10.34	0.0123
B^2	20.35	1	20.35	8.85	0.0178
C^2	22.73	1	22.73	9.88	0.0137
Residual	18.40	8	2.30		
Lack of Fit	2.88	4	0.72	0.19	0.9342
Pure Error	15.52	4	3.88		
Core Total	234.59	16			

In this case microwave power (A), extraction time (B), liquid-solid ratio (C), the interaction of microwave power and liquid-solid ratio (AC), the interaction of extraction time and liquid-solid ratio (BC) and the quadratic term of all variables (A2, B2 and C2) had an effect on the response value (p<0.05). The interaction of microwave power and extraction time (AB) was removed because it was the insignificant variable (p>0.10). The statistic model represented the relation of response variable (extraction yield) and the significant model terms, which could be expressed by the quadratic equation 3:

$$Y = 17.03441 + 0.012227A + 0.012594B + 1.43535C + 0.00145735AC + 0.00234987BC - 0.0000194006A^2 - 0.0000270646B^2 - 0.18968C^2 \quad (3)$$

where A, B, C and Y coded for microwave power, extraction time, liquid-solid ratio and extraction yield, respectively.

The three-dimensional response surface plots explained the interactive effects of the independent variables on the amount of protein extraction yields. Two variables were chosen to observe when the other variable was fixed at the middle level. These plots should indicate the optimal value of variables, which resulted in the highest response. Figure 2 and 3 were the three-dimensional response surface plot and contour plot showing the effect of variables on the response.

Figure 2 showed the effect of microwave power and liquid-solid ratio on extraction yield, when extraction time was fixed at 315 s. Increasing microwave power to around 600-700 W could increase the yield. It might be the effect of the microwave-induced dipole rotation and the penetration of water to the matrix which could help release the intracellular product. Moreover, the extraction yields slightly decreased when microwave power was higher than the optimal value. It was possibly caused by the denaturation of protein from the heat generated by microwaves [23].

a.

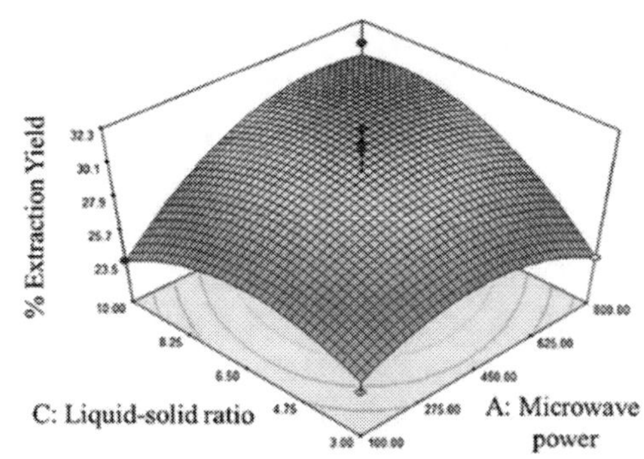

b.

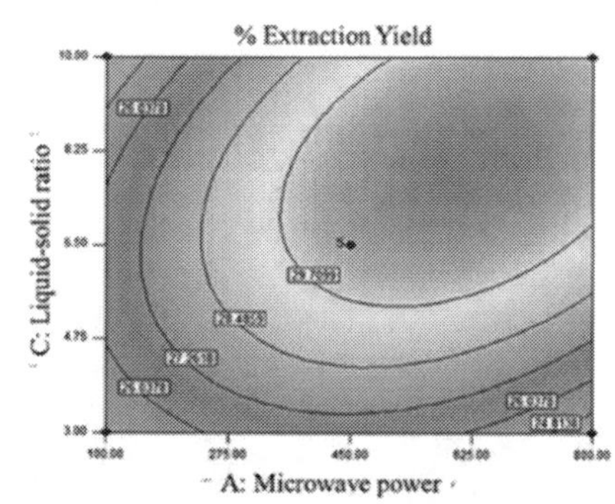

Figure 2. The three-dimensional response surface plot (a.) and contour plot (b.) show the effect of microwave power and liquid-solid ratio on extraction yield.

The effect of extraction time and liquid-solid ratio on extraction yield, when microwave power was fixed at 450 W was presented in Figure 3. The more extraction time was increased, the more extraction yield was obtained. The time increased from 30 to 600 s. showed the increasing extraction yield. It might be due to the duration of diffusion.

In both figures 2 and 3, the yield was increased when the liquid-solid ratio increased to the optimum, and the yield slightly decreased when the ratio was higher than the optimum. It might be due to increasing water not only containing more extract but also water could absorb more microwave power and dispersing microwave energy [19].

a.

b.

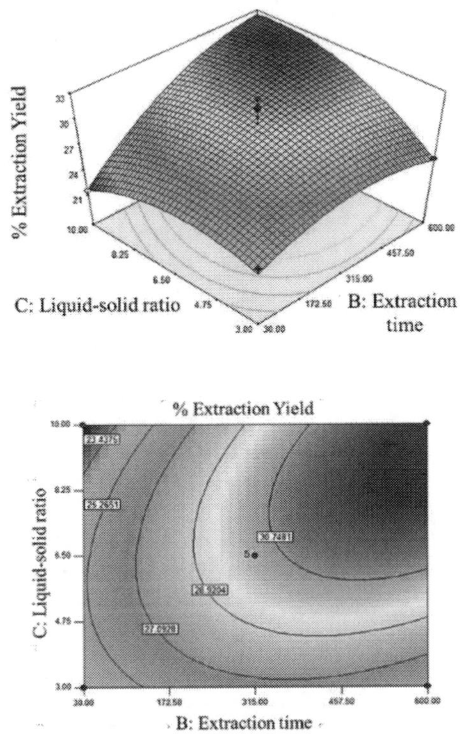

Figure 3. The three-dimensional response surface plot (a.) and contour plot (b.) showing the effect of extraction time and liquid-solid ratio on extraction yield.

The optimized model was validated by performing several experiments. The predicted and experimental values of optimum condition of PKC protein extraction using MAE method were presented in Table 4. The optimum condition was predicted with the maximum desirability of 32.4554%. The model validation was performed at microwave power of 700.00 W, extraction time of 543 s and liquid-solid ratio of 7.73:1. The protein extraction yield at 31.03% (w/w) was obtained from this condition, which missed from the prediction about 4.38%.

The experimental results were in a good correlation between predicted and actual response values. Thus, the response surface methodology could be applied to optimize the condition of palm kernel cake protein extraction using the microwave-assisted method.

Table 4. The optimal predicted and experimental values of variables.

Variables	Predicted values	Experimental values
Microwave power (W)	700.16	700.00
Extraction time (s)	543.08	543.00
Liquid-solid ratio (ml/g)	7.73:1	7.73:1
Extraction yield (%)	32.4554	31.03

From the previous studies, PKC protein was extracted by the chemical extraction method using saline and alkaline. The optimum condition for saline treatment was at pH 9, 0.2 M NaCl and liquid ratio of 60:1, and the optimum condition for alkaline treatment was at 0.03M NaOH, 35°C and liquid ratio of 30:1 gave the protein recovery yield of 78.59% and 73.07%, respectively [34]. In another study, PKC protein was extracted using the trypsin-assisted extraction method. The optimum condition was 50 °C, pH 9.5, PKC 1.1 g/ 100 ml and trypsin concentration 1.36 g/ 100 g which resulted in protein yield of 61.99 g/ 100 g [35].

According to previous studies, the high protein yield was got from the high pH value (9-9.5) condition. Moreover, they have to consume a lot of time and chemicals (NaCl, NaOH and phosphate buffer) which could cause a high production cost.

4 Conclusion

Response surface methodology (RSM) was used to optimize the condition of palm kernel cake (PKC) protein extraction using the microwave-assisted extraction (MAE) method. The PKC protein could be extracted by using water as a solvent with the MAE method because the microwave enhanced the penetration of water to the matrix to release the intracellular protein. The optimum condition was obtained from quadratic modeling. The extraction yield was 31.03% at a microwave power of 700 W, extraction time of 543 s and liquid-solid ratio of 7.73:1. However, further study of the functional properties of PKC protein will be investigated for functional food and feed applications.

5 Acknowledgment

The Authors would like to thank the Department of Biotechnology, Faculty of Agro-industry, Kasetsart University.

References

[1] Office of Agricultural Economics, *Agricultural statistics of Thailand 2020* (Office of Agricultural Economics, Bangkok, 2021)

[2] N.A. Ibrahim, Characteristics of malaysian palm kernel and its products, Journal of Oil Palm Research, 25(2) (2013):245-252.

[3] R. Oliveira, M. Faria, R. Silva, L. Bezerra, G. Carvalho, A. Pinheiro, J. Simionato, A. Leão, Fatty acid profile of milk and cheese from dairy cows supplemented a diet with palm kernel cake, Molecules, 20(8) (2015):15434-15448.

[4] A.O.K. Adesehinwa, Utilization of palm kernel cake as a replacement for maize in diets of growing pigs: effects on performance, serum metabolites, nutrient digestibility and cost of feed conversion, Bulgarian Journal of Agricultural Science, 13(5) (2007):593-600.

[5] S.W.A. Rhule, Growth rate and carcass characteristics of pigs fed on diets containing palm kernel cake, Animal feed science and technology, 61(1-4) (1996): 167-172.

[6] A.M. Marini, M.Y. Ayub, B.A. Salam, H. Hadijah, E.A.E. Azahan, S.A. Tarmizi, Protein quality of *Aspergillus niger*-fermented palm kernel cake, J Trop Agric and Food Science, 36(2) (2008):227-237.

[7] G. Franca-Oliveira, T. Fornari, B. Hernández-Ledesma, A review on the extraction and processing of natural source-derived proteins through eco-innovative approaches, Processes, 9(9) (2021):1626.

[8] A. Delazar, L. Nahar, S. Hamedeyazdan, S.D. Sarker, Microwave-assisted extraction in natural products isolation, Methods Mol Biol, 864 (2012): 89-115.

[9] K. Liu, M.L. Lota, J. Casanova, F. Tomi, The essential oil of *Bupleurum fruticosum* L. from Corsica: a comprehensive study, Chem Biodivers, 6(12) (2009):2244-2254.

[10] M.L. Presti, S. Ragusa, A. Trozzi, P. Dugo, F. Visinoni, A. Fazio, D. Giovanni, L. Mondello, A comparison between different techniques for the isolation of rosemary essential oil, Journal of separation science, 28(3):273-280.

[11] N. Tigrine-Kordjani, B.Y. Meklati, F. Chemat, Contribution of microwave accelerated distillation in the extraction of the essential oil of Zygophyllum album L. Phytochemical Analysis, 22(1) (2011):1–9.

[12] L. Chen, L. Ding, A. Yu, R. Yang, X. Wang, J. Li, H.Y. Jin, H.Q. Zhang, Continuous determination of total flavonoids in Platycladus orientalis (L.) Franco by dynamic microwave-assisted extraction coupled with on-line derivatization and ultraviolet-visible detection, Analytica chimica acta, 596(1) (2007):164–170.

[13] M. Gao, B.Z. Song, C.Z. Liu, Dynamic microwave-assisted extraction of flavonoids from Saussurea medusa Maxim cultured cells, Biochemical Engineering Journal, 32(2) (2006):79–83.

[14] Z.E. Huma, M.A. Vian, J.F. Maingonnat, F. Chemat, Clean recovery of antioxidant flavonoids from onions: optimizing solvent free microwave extraction method, Journal of Chromatography A (2009):7700–7707.

[15] A. Liazid, M. Palma, J. Brigui, C.G. Barroso, Investigation on phenolic compounds stability during microwave-assisted extraction, Journal of Chromatography A, 1140(1-2) (2007):29–34.

[16] C.C. Teo, S.N. Tan, J.W. Yong, C.S. Hew, E.S. Ong, Evaluation of the extraction efficiency of thermally labile bioactive compounds in Gastrodia elata Blume by pressurized hot water extraction and microwave-assisted extraction, Journal of Chromatography A, 1182(1) (2008):34–40.

[17] H. Li, B. Chen, Z. Zhang, S. Yao, Focused microwave-assisted solvent extraction and HPLC determination of effective constituents in Eucommia ulmodies Oliv. (E. ulmodies), Talanta, 63(3) (2004):659–665.

[18] Y. Chen, M.Y. Xie, X.F. Gong, Microwave-assisted extraction used for the isolation of total triterpenoid saponins from Ganoderma atrum, Journal of Food Engineering, 81(1) (2007):162–170.

[19] J. Li, Y.G. Zu, Y.J. Fu, Y.C. Yang, S.M. Li, Z.N. Li, M. Wink, Optimization of microwave-assisted extraction of triterpene saponins from defatted residue of yellow horn (Xanthoceras sorbifolia Bunge.) kernel and evaluation of its antioxidant activity, Innovative Food Science & Emerging Technologies, 11(4) (2010):637–643.

[20] J.H. Kwon, J.M.R. Belanger, J.R.J. Pare, V.A. Yaylayan, Application of the microwave-assisted process (MAPTM) to the fast extraction of ginseng saponins, Food Research International, 36(5) (2003):491–498.

[22] Y. Wang, F. Leng, X. Liu, W. Zhang, M. Yang, Optimization of Microwave-Assisted Extraction of Water-Soluble Polysaccharides from Piteguo Fruit by Response Surface Methodology, Food Science and Technology Research, 20(4) (2014):755-764.

[23] D. Wang, B.B. Zhang, X.X. Qu, F. Gao, M.Y. Yuan, Microwave-assisted extraction of polysaccharides from Yupingfeng powder and their antioxidant activity, Pharmacognosy Magazine, 11(43) (2015):546-554.

[24] S. Phongthai, S.T. Lim, S. Rawdkuen, Optimization of microwave-assisted extraction of rice bran protein and its hydrolysates properties, Journal of cereal science 70 (2016):146-154

[25] S. Bedin, K. Zanella, N. Bragagnolo, O. P. Taranto Implication of microwaves on the extraction process of rice bran protein, Brazilian Journal of Chemical Engineering, 36, 4 (2019):1653-1665.

[26] A. Görgüç, C. Bircan, F.M. Yılmaz, Sesame bran as an unexploited by-product: Effect of enzyme and ultrasound assisted extraction on the recovery of protein and antioxidant compounds, Food chemistry, 283 (2019):637–645.

[27] L. Wen, C. Álvarez, Z. Zhang, M. M. Poojary, M. N. Lund, D.W. Sun, B. K. Tiwari, Optimisation and characterisation of protein extraction from coffee silverskin assisted by ultrasound or microwave techniques, Biomass Conversion and Biorefinery, 11(5) (2021):1575-1585.

[28] M. Behere, S. S. Patil, V. K. Rathod, Rapid extraction of watermelon seed proteins using microwave and its
functional properties, Preparative Biochemistry & Biotechnology, 51(3) (2021):252-259.

[29] H. Dominguez, M.J.G. Muñoz, Water Extraction of Bioactive Compounds: From Plants to Drug Development, Elsevier, Amsterdam, 2017.

[30] L. Kumar, M.S. Reddy, Managuli, R.S. Managuli, G. Pai, Full factorial design for optimization, development and validation of HPLC method to determine valsartan in nanoparticles, Saudi Pharmaceutical Journal, 23(5) (2015):549-555.

[31] Y.H. Tan, M.O. Abdullah, C. Nolasco-Hipolito, N.S.A. Zauzi, Application of RSM and Taguchi methods for optimizing the transesterification of waste cooking oil catalyzed by solid ostrich and chicken-eggshell derived CaO, Renewable Energy, 114 (2017):437-447.

[32] A.E. Tontowi, L. Ramdani, R. Erdizon, D.K. Baroroh, Optimization of 3d-printer process parameters for improving quality of polylactic acid printed part. International Journal of Engineering and Technology, 9(2) (2017):589-600.

[33] S.J.S. Chelladurai, K. Murugan, A.P. Ray, M.Upadhyaya, V. Narasimharaj, S. Gnanasekaran, Optimization of process parameters using response surface methodology: A review, Materials Today: Proceedings, 37 (2020).

[34] O. H. Lowry, N. J. Rosebrough, A. L. Farr, R. J. Randall, Protein measurement with the Folin phenol reagent, J Biol Chem, 193 (1951):265-275.

[35] B. Arifin, A. Bono, Y.Y. Farm, A.L.L Ling, S.Y. Fui, Protein extraction from palm kernel meal, Journal of Applied Sciences, 9(17) (2009):2996-3004.

[36] K.L. Chee, H.K. Ling, M.K. Ayob, Optimization of trypsin-assisted extraction, physico-chemical characterization, nutritional qualities and functionalities of palm kernel cake protein, LWT-Food Science and Technology, 46(2) (2012):419-427.

Paddy straw: an economical substrate for oyster mushroom (*Pleurotus ostreatus*) cultivation

Anne Sahithi Somavarapu Thomas[1], *Sona* Arivu[2], *Vinodhini* Shanmugam[2], *Marttin Paulraj* Gundupalli[3], *Suksun* Amornraksa[3,*]

[1] School of Biosciences and Technology (SBST), Vellore Institute of Technology (VIT) Vellore, Tamilnadu, India
[2]Department of Biotechnology, D.K.M College for Women, Vellore, Tamilnadu, India
[3]Biorefinery and Process Automation Engineering Center, Department of Chemical and Process Engineering, The Sirindhorn International Thai-German Graduate School of Engineering, King Mongkut's University of Technology North Bangkok, Bangkok, Thailand

Abstract. Cultivation of the *Pleurotus ostreatus*, oyster mushroom on paddy straw without supplements was investigated to follow circular economy concept to convert agricultural waste to value added products. Substrate nutrients, mushroom yield, and biological efficiency were determined. Three different extracts were used in this study (methanol, ethyl acetate, and hexane). Antioxidant and scavenging activity was determined using DPPH and H_2O_2. To find the essential compounds present in the mushrooms, GC-MS was analyzed. It was found that mushroom growth on paddy substrate was less than five days with excess mushroom yield. The biological efficiency was found between 54.5-130.9%, with the moisture of 93%. It was found that C, P, N, and K were integrated into mushrooms with these elements than in the utilized substrate. In DPPH results, the minimum concentration was 37.07 µg/ml, and the maximum was 67.2 µg/ml. IC50 value of 42.6 µg/ml were 50% for inhibition concentration. In H_2O_2, the minimum concentration was found to be 72.57 µg/ml, and the maximum was 98.02 µg/ml. This concentration indicates that the IC50 value of 84.07 µg/ml can be used in the biological process or component by 50% for inhibition concentration. The compounds include Oxirane, 2-Methyl-3-(1-Methylethyl)-, O-Methylisourea Hydrogen Sulfate, Diethyl Phthalate, 1,1,3,3-Tetrapropoxy- were found commonly in all three extracts. Hence, analysis of mushroom extracts is needed to determine the mechanisms of action of the various components for antimicrobial activity and inhibitory activity. Therefore, paddy straw could be used as an effective and economical substrate for oyster mushroom cultivation.

Keyword. *Pleurotus ostreatus*, Paddy straw, *DPPH,* Mushroom, Cultivation

1 Introduction

Oyster mushrooms are generally valued for their taste and nutritional and medicinal properties, which have been extensively studied [1]. It contains a vital nutritional source of protein, vitamins, carbohydrates, iron and calcium. In addition, these mushrooms obtain nutrients from organic materials, such as straw, deadwood, dung, manure, and decaying organic matter. Growing these mushrooms with lignocellulosic agricultural waste results in value-added food for humans. It has been cultivated widely in different parts of the world and is known as fruiting bodies of fungi. Therefore, it can grow over a wide range of temperatures using a variety of lignocelluloses [2]. In India, the cultivation of mushrooms began in the 19th century. According to the Food and Agriculture Organization (FAO) of United Nations statistics, China is the world's top leader in mushroom production, as reported in 2018 [3]. Consequently, they are widely used in the bioremediation of pollutants and decomposition of lignocellulosic residues under the action of various enzymes.

Mushrooms require inorganic compounds, carbon and nitrogen as a food source and primary nutrients, such as cellulose, hemicellulose and lignin. Paddy or rice straw as a substrate for mushroom cultivation is produced as an agricultural by-product [4]. Currently, rice fields are disposed of by burning in the open air, leading to serious environmental problems. If paddy straw can nourish the growth of oyster mushrooms, this could convert the non-edible waste into profitable and edible biomass. These fungi can be collected in the wild during the last the wettest part of the rainy season, where they meet and grows on deeply decaying organic matter. However, not all mushrooms growing in the wild are suitable for human consumption. In addition, this biomass could serve as an inexpensive substrate for mushroom growers. However, it is reported that the use of rice straw in oyster mushroom cultivation is not widespread in China due to its low biological competence and yield [3]. However, it also has medicinal benefits for people who have cancer and diabetes. In particular, *Pleurotus* species constitutes high potassium and sodium, edible food for

* Corresponding author: suksun.a@tggs.kmutnb.ac.th

patients suffering from high blood pressure and heart disease [1]. In addition, mushroom cultivation practice could improve the straw quality by reducing the lignocellulosic biomass composition. Thus, this current study focuses on using paddy straw as a basal substrate for the cultivation of oyster mushrooms.

2 Materials and Methods

2.1. Substrate preparation

Experimental design involves four steps in mushroom cultivation production: substrate preparation, inoculation, growing, and harvesting. Substrate preparation includes spawning production, in which wheat grains were soaked in water for 7-8 h and boiled in water for 30 min in the ratio of [1:2 w/v]. Later, it was cooled, and calcium carbonate was mixed in the ratio of [50:1 w/w]. This mixture was autoclaved at 121°C for 15 min. Thereafter, it was incubated at 21°C for 15- 20 days. Lignocellulosic biomass, i.e. paddy straw, was used as the substrate [5]. The growth of mycelium was seen in the production of mushrooms. The substrate was pre-treated with hot water at 70°C for a few hours. Thereby, the substrate was soaked with benzoyl to avoid contamination of *Trichoderma*. Paddy straw was obtained from Velapadi village, Vellore. It was stored for three months after harvesting **Fig. 1**. The straw was chopped into 2.5cm for every cultivation using a hand cutter.

(a)

(b)

Fig. 1 (a) Substrate preparation **(b)** Paddy straw

2.2. Inoculation

In order to avoid contamination, 15 g formalin was mixed with 1 kg batch culture.

2.3 Cultivation of mushroom

For the cultivation of mushrooms, bedding preparation was done. Initially, 200-250 g of the substrate was taken in a polythene bag, 1st layer of the substrate was about 12 cm. 2nd layer of spawns was scattered for about 12-20 days for incubation. During the incubation period, the temperature and pH were maintained. An optimized temperature of about 20-25°C was maintained. Casein soil was used for the bedding of mushrooms [6]. The casing mixture was prepared using soil, coco-peat and sand in the ration of [3:2:1]. Earlier, this casing mixture was sterilized for 30 min at 121°C. Temperature and moisture were maintained by spraying the water between 21-23°C.

2.4 Harvesting

The mushrooms were harvested using an environmental chamber (temperature, ventilation, and humidity). Relative humidity and temperature were controlled at 70-80% and 24 ± 1°C, respectively. These conditions were determined using the preliminary test carried out using different conditions. The inoculated bags were kept in the environmental chamber to develop the fruit bodies. After harvesting, the mushrooms were gently twisted, and the end parts were cut off and later weighed. Also, the substrate in the bag was weighed. The samples of both substrate and mushrooms were analyzed for the contents of biological efficiency.

$$BE,\% = \frac{W2}{W1} \times 100 \quad (1)$$

Where, BE (Biological efficiency, %) W1- Total dry weight of the compost, W2- Total weight of fresh mushroom.

2.5 Analysis of nutrient elements of substrate and mushroom

To understand the concentration of nutritional elements such as nitrogen (N), carbon (C), phosphorus (P), and potassium (K) in the substrate were analyzed using an inductively coupled plasma atomic emission spectrometer (ICP-AES) [7].

2.6 *DPPH* radical scavenging activity

Antioxidant activity of *P. ostreatus* was determined using the purple DPPH radical solution. 1 mg/ml of extract was dissolved in methanol, ethyl acetate, and hexane using different concentrations (25, 50, 75, and 100 µg/ml). For control, DPPH and methanol were used without the extract mixture [8]. The mixture was vigorously mixed and stored in the dark for 30 min. The scavenging activity was measured at 517 nm using a UV-VIS spectrophotometer. For positive control, ascorbic acid was used.

2.7 Hydrogen peroxide scavenging activity assay

Hydrogen peroxide (H_2O_2) solution was prepared using phosphate buffer with the pH of 7.4 methods proposed by [9]. Different concentrations of extract (25, 50, 75, and 100 µg/ml) in 1 ml of phosphate buffer with addition of H_2O_2 solution 0.6 ml. The mixture was stored at room temperature for 10 min. Using a spectrophotometer, the absorbance of the mixture was determined at 230nm, and ascorbic acid was used as control. The scavenging activity was calculated using the following equation.

Scavenging effect (%)/% Inhibition=
A0 -A1 /A0 × 100

Where A_0= the absorbance of control. A_1 absorbance of standard.

2.8 GC-MS analysis

5g of dry powder sample was extracted with different solvents systems based on polar (methanol and ethyl acetate) and non-polar (hexane). The samples were kept in a shaker overnight was repeated in triplicates. Later, the extract was filtered and condensed under vacuum at 60°C using a rotary evaporator and stored at 4°C. The sample was filtered using a 0.2 µm nylon membrane and injected into Gas Chromatrography-Mass Spectrophotometry (GC-MS). The analysis was carried out on a Perkin-Elmer workstation, with model Clarus 600 GC coupled to a mass spectrometer (Perkin Elmer Technologies, Inc., Wilmington, DE) [10]. Elite-5MS (30 m x 0.25 mm width film depth of 250 µm capillary tube was used with the following conditions. Oven initial temperature of 55°C for 3 min, ramp program for 6°C/min up to 310 °C, further 3 min isothermal hold. Helium (He) gas was used with a flow rate of 10:1 ratio. 2 µl of the sample was injected into the injector was maintained at 250°C. Results were analyzed using mass spectra from the National Institute of Standards and Technology (NIST-LIB 0.5) library. The individual phytochemicals present in the crude extract were separated by GC column. The MS spectrum displays the molecular weight of individual molecules accurately.

3 Results and discussion

In this work, the results discussed the mushroom yield, biological efficiency, nutrient elements of the substrate, antioxidant activity, and extracted essential compounds. The cost for the production was calculated in each step (spawn preparation, bedding supplements and packing). Mushroom bedding and fruiting are shown in **Fig. 2.** The mycelium was spread all over the mushroom bedding during incubation (10-20 days).

(a) (b)

Fig. 2 (a) Bedding preparation of mushroom **(b)** Incubation and fruiting

For harvesting the mushrooms, casein soil was used to induce the mushroom growth. This mixture was prepared as above mentioned. Instead of coco peat, farmyard manure can also be used for mushroom production due to its unique water holding capacity. After the incubation period, the beds were cut into two halves, and 3-4 cm thickness of casein soil was spread over the substrate and temperature was maintained under 21± 1°C as shown in **Fig. 3.** Similarly, the budding of mushrooms and matured mushrooms are shown in **Fig. 4.** The total cost for the production of mushrooms is listed in **Table 1.**

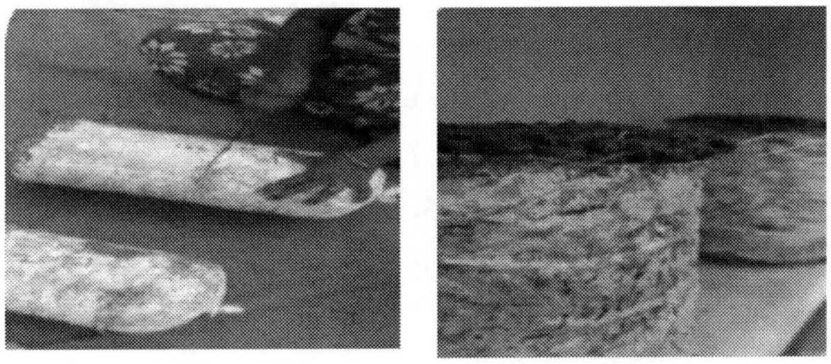

(a) (b)

Fig. 3 (a) Mushroom bedding was cut into 2 halves **(b)** beds with casein soil

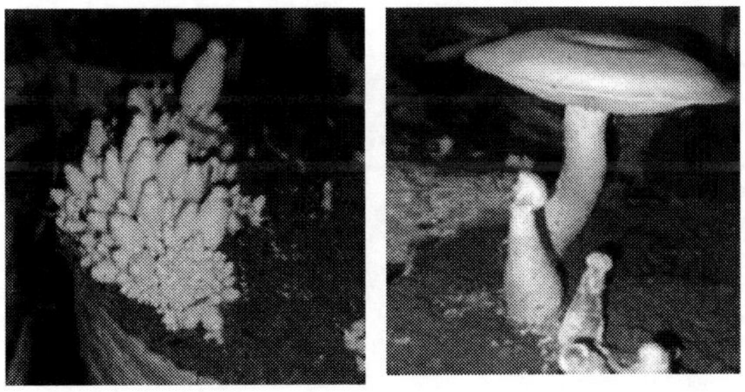

(a) (b)

Fig. 4 (a) Budding of Mushroom **(b)** Matured species *P. ostreatus*

Table 1 Production cost of Mushrooms

S.No	Items	Cost
1	Spawn preparation	105
2	Preparation of 80 beds	270
3	Supplements	30

4	Packaging bag	30
	Total	**435/-**

In addition, to increase mushroom marketing, storage, packaging, and canning are essential sources for consumption. It has been reported that development and economic mushroom consumption are gradually generating. Hence, this method of cultivation and production could be user-friendly for farmers and consumers.

Substrate degradation and nutrient analysis determined the mushroom's yield and biological efficiency. The dry matter loss was determined by comparing mushroom yield to its biological efficiency. The fungus partially assimilated the loss of dry matter and partially lost to the atmosphere as carbon dioxide due to the respiration of the fungus. The biological efficiency was found between 54.5-130.9%, with the moisture of 93%. The nutrient elemental analysis results were determined for the chopped substrate and mushroom. The initial utilized weight was also measured. It was found that C, P, N, and K were integrated into mushrooms with these elements than in the utilized substrate. In addition, the C and N loss may be due to mushroom respiration and volatilization during the N mineralization process [11]. P and K contents were found to be well balanced among initial, utilized, and mushrooms

Table 2. Scavenging activity was determined using DPPH and H_2O_2 assay. In DPPH free radical scavenging, the mechanism for screening the antioxidant activity of the extracts (methanol, ethyl acetate, and hexane) was done, and it was found that the violet color of the DPPH solution was reduced to yellow color. The minimum concentration was 37.07 µg/ml, and the maximum was 67.2 µg/ml **Fig. 5**. IC50 was 42.6 µg/ml, where this concentration indicates can be used in the biological process or component by 50% for inhibition concentration.

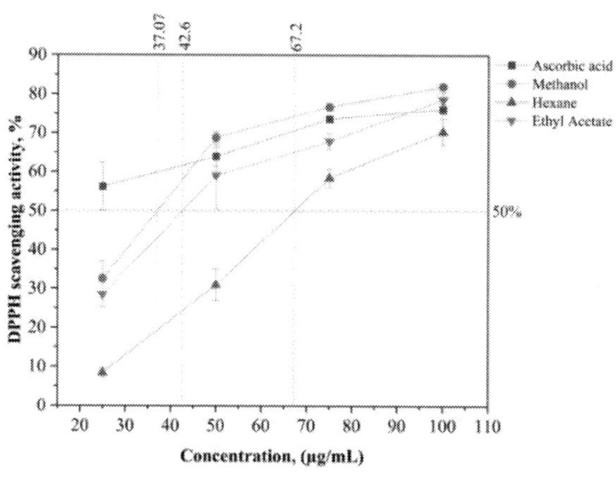

Fig. 5 DPPH radical scavenging assay

Similarly, H_2O_2 scavenging activity was determined to know the reactive oxygen metabolic, which is the key regulator of oxidative stress-related states. It was found that the quantitative determination of hydrogen peroxide in extracts (methanol, ethyl acetate and hexane) were disappeared at the wavelength of 230nm. The minimum concentration was found to be 72.57 µg/ml, and the maximum was 98.02 µg/ml **Fig. 6**. IC50 was 84.07 µg/ml, indicating that this concentration can be used in the biological process or component by 50% for inhibition concentration. This method is convenient, highly accurate, and suitable for quickly quantifying the ability of standard and natural antioxidants present in plant extracts to absorb H_2O_2.

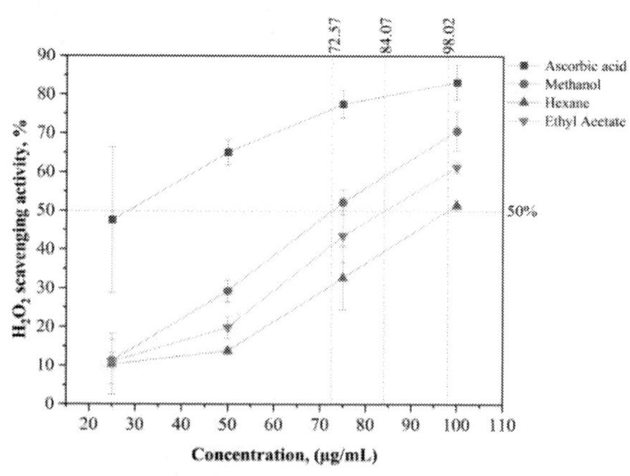

Fig. 6 H_2O_2 scavenging activity

Table 2 Nutrients elements of both substrate and mushroom

Paddy straw	Materials	Dry weight (g)	Nutrient Content C	P	N	K
Chopped	Initial substrate	100	34.8	0.10	0.80	1.15
	Utilized substrate	70.3	32.9	0.05	0.63	1.25
	mushroom	6.8	40.5	0.95	4.32	3.57

GC-MS chromatograms of the crude extracts were determined by ruling the similar compounds present in all three extracts (methanol, ethyl acetate, and hexane) using the presence of a peak at retention time [12]. On comparison of the mass spectra of the compound with the NIST library, the compound was matched. (NIST

 https://doi.org/10.1051/e3sconf/202235502019

library; Molecular weight (m/z). The common compounds present in extracts were given in Error! Reference source not found.. The compounds include Oxirane, 2-Methyl-3-(1-Methylethyl)-, O-Methylisourea Hydrogen Sulfate, Diethyl Phthalate, 1 Methyl-2-Phenoxyethylamine, 1,3 Bis(Hydroxymethyl) Urea, 2-Propanone, 1,1,3,3-Tetrapropoxy- were found commonly in all three extracts and methanol extract showed the accurate results than other extracts **Fig. 7**.

Table 3 GC-MS analysis of mushroom crude extracts

S.No	Retention Time (min)	Name of Component	Molecular weight	Area%	Molecular Formula
1.	3.31	OXIRANE, 2-METHYL-3-(1-METHYLETHYL)-	100	2.84	$C_6H_{12}o$
2.	3.92	O-METHYLISOUREA HYDROGEN SULFATE	74	0.70	$C_2H_6ON_2$
3.	4.60	DIETHYL PHTHALATE	222	1.5	$C_{12}H_{14}O_4$
4.	3.18	1-METHYL-2-PHENOXYETHYLAMINE	151	3.65	$C_9H_{13}ON$
5.	13.296	1,3-BIS(HYDROXYMETHYL)UREA	120	1.17	$C_3H_8O_3N_2$
6.	17.104	CYCLOBUTANONE, 2-METHYL-4-HYDROXY-	100	2.59	$C_5H_{12}N_2$
7.	19.30	2-PROPANONE, 1,1,3,3-TETRAPROPOXY-	290	7.05	$C_{15}H_{30}O_5$
8.	27.153	9-1-OCTADECADIENOIC ACID, METHYL ESTER	280	2.84	$C_{18}H_{32}O_2$
9.	30.769	HEPTANOIC ACID, 2-METHYL-2-BUTYL ESTER	200	6.43	$C_{12}H_{24}O_2$

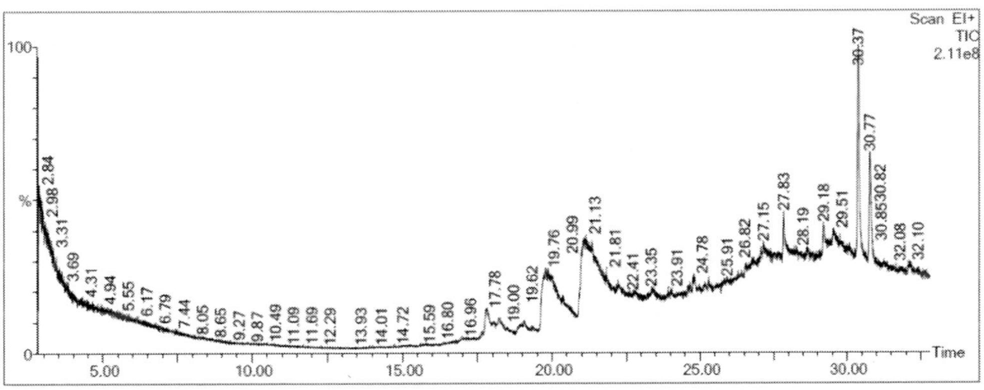

Fig. 7 Chromatogram profile methanol extract

4 Conclusion

The major finding of this study is the paddy substrate showed a better yield of mushroom production in a cheap and faster way. The mushrooms grew faster on the substrate with less than five days of growth cycles. There are several types of edible mushrooms that, however, have not been researched for their potential use in foods. These new fungal species may have different organic processes, sensory organs, and environmental effects than those exhibited by the species used so far. Therefore, there is a need to analyze these different species. In addition, any analysis is needed to determine the mechanisms of action of the various components of the fungus, for example, their antimicrobial activity, inhibitory activity, structure, textural characteristics and taste characteristics. The benefits of mushrooms are relatively economical because mushrooms are grown on

a variety of cheap agricultural or forest waste, such as rice straw, corncobs, and sawdust. Anti-fungal inoculants can be manufactured in the factory using today's simple techniques that usually do not produce fungi. When looking for economic and environmentally friendly strategies for environmental restoration, the use of mushrooms can be an excellent approach and solution.

Acknowledgements

The authors would like to thank VIT Vellore and D.K.M College for supporting this work.

References

[1] J. Raman, K. Y. Jang, Y. L. Oh, et al., "Cultivation and Nutritional Value of Prominent Pleurotus Spp.: An Overview," *Mycobiology*, **vol**. 49, no. 1, pp. 1–14, (2021).

[2] M. A. Kabel, E. Jurak, M. R. Mäkelä, and R. P. de Vries, "Occurrence and function of enzymes for lignocellulose degradation in commercial Agaricus bisporus cultivation," *Applied Microbiology and Biotechnology*, **vol**. 101, no. 11, pp. 4363–4369, (2017).

[3] V. V. Pandey, A. Kumari, M. Kumar, J. Saxena, C. Kainthola, and A. Pandey, "Mushroom cultivation: Substantial key to food security," *Journal of Applied and Natural Science*, **vol**. 10, no. 4, pp. 1325–1331, (2018).

[4] L. V. Thuc, R. G. Corales, J. T. Sajor, et al., "Rice-Straw Mushroom Production," *Sustainable Rice Straw Management*, pp. 93–109, (2020).

[5] Y. S. Cheng, P. Mutrakulcharoen, S. Chuetor, et al., "Recent situation and progress in biorefining process of lignocellulosic biomass: Toward green economy," *Applied Science and Engineering Progress*, **vol**. 13, no. 4, p. (2020).

[6] P. Soimaloon, P. Tinchan, and L. Horng-Liang, "Effect of Extraction Conditions on Color, pH, Flavor Profile and Ribonucleotide Contents of Limnophila aromatica (Lam.) Merr. Extracts," *KMUTNB International Journal of Applied Science and Technology*, p. (2018).

[7] N. Chaichana, "Nutrition Composition and Analysis of Medicinal Herbal Potential of Horsfieldia glabra Warb. Seeds," *KMUTNB International Journal of Applied Science and Technology*, pp. 61–69, (2015).

[8] H. C. Tran, H. A. T. Le, T. T. Le, and V. M. Phan, "Effects of enzyme types and extraction conditions on protein recovery and antioxidant properties of hydrolysed proteins derived from defatted lemna minor," *Applied Science and Engineering Progress*, **vol**. 14, no. 3, pp. 360–369, (2021).

[9] S. Nabavi, R. Fox, C. D. Proulx, J. Y. Lin, R. Y. Tsien, and R. Malinow, "Engineering a memory with LTD and LTP," *Nature*, **vol**. 511, no. 7509, pp. 348–352, (2014).

[10] E. Jayex Panakkal, M. Sriariyanun, J. Ratanapoompinyo, et al., "Influence of Sulfuric Acid Pretreatment and Inhibitor of Sugarcane Bagasse on the Production of Fermentable Sugar and Ethanol," *Applied Science and Engineering Progress*, **vol**. 15, no. 1, p. (2022).

[11] A. S. S. Thomas, W. Pongprayoon, K. Cheenkachorn, and M. Sriariyanun, "Plant-Microbe Interactions - Insights and Views for Applications in Sustainable Agriculture," *Applied Science and Engineering Progress*, **vol**. 15, no. 1, p. (2022).

[12] J. Korkitpoonpol, N. Suwannaprom, S. Keeratipibul, et al., "Flavor Profile in Fresh-squeezed Juice of Four Thai Lime Cultivars: Identification of Compounds that Influence Fruit Selection by Master Chefs," *Applied Science and Engineering Progress*, **vol**. 13, no. 2, pp. 146–157, (2020).

AUTHOR INDEX

EDP Sciences – Web of Conferences
17, Avenue du Hoggar
Parc d'Activité de Courtabouf
BP 112
F-91944 Les Ulis Cedex A
France

ISBN 978-1-7138-5912-3

CURRAN ASSOCIATES INC.
proceedings
.com